DOUBLE AWARD
SCIENCE for GCSE

GRAHAM HILL

Hodder & Stoughton

A MEMBER OF THE HODDER HEADLINE GROUP

Preface

Science for GCSE: Double Award will help students as they study Key Stage 4 of the National Curriculum and prepare for examinations in Double Science.

In a single volume, I have covered the syllabuses of all the examination boards thoroughly whilst maintaining a concise and informative presentation page by page.

Each of the thirty chapters covers a major theme and ends with a page of GCSE questions, most of them taken from recent examinations.

My own name may appear on the cover of this book, but I must thank two other people who have contributed very considerably, creatively and highly professionally to its success – Emma Broomby, the project editor from Hodder and Stoughton Educational and Elizabeth, my wife. Without their input, efficiency and encouragement, the book would not have been possible.

Graham Hill
January 1998

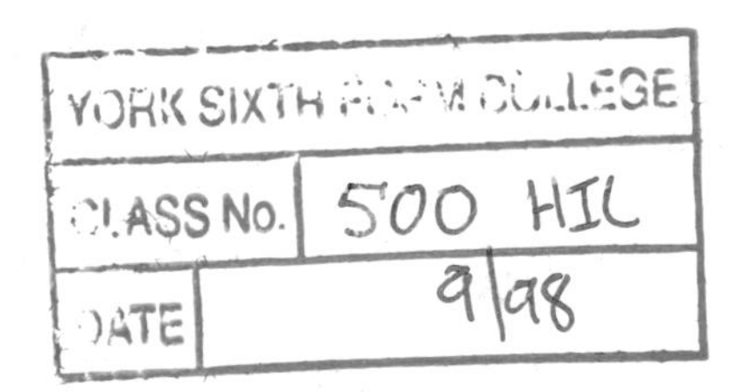

Orders: please contact Bookpoint Ltd, 39 Milton Park, Abingdon, Oxon OX14 4TD. Telephone: (44) 01235 400414, Fax (44) 01235 400454. Lines are open from 9.00 – 6.00, Monday to Saturday, with a 24 hour message answering service. Email address: orders@bookpoint.co.uk

British Library Cataloguing in Publication Data
A catalogue record for this title is available from The British Library

ISBN 0 340 67936 0

First Published 1998
Impression number 10 9 8 7 6 5 4 3 2 1
Year 2004 2003 2002 2001 2000 1999 1998

Typeset by Wearset, Boldon, Tyne and Wear
Printed in Italy by Printer Trento for Hodder & Stoughton Educational, a division of Hodder Headline Plc, 338 Euston Road, London NW1 3BH by Scotprint Ltd, Musselburgh

Contents

A female brimstone butterfly rests well camouflaged amongst hazel leaves. Other examples of how animals and plants have adapted to their environment are described in Chapters 11 and 12, whilst photosynthesis in green plants like hazels is considered in Chapter 5. The whole of this section is about **Biology** – the study of living things and life processes.

SECTION 1

Life Processes and Living Things

CHAPTER

Cells and life

1.1 The variety of life

Walk through the park, along the banks of a stream or in your garden. Notice the number and variety of living things – birds, mammals, insects, fish, grasses, flowers and trees.

A park, a stream and a garden are just three examples of the many different places or **habitats** where living things can be found. Habitats can be very large, like a city or a forest, much smaller like a garden or a stream and even tiny like a puddle or a stone.

The conditions in one habitat (the **environment**) may be very different to those in another. Because of this, the living things are different too. Living things are often **adapted** to the habitat in which they live. Some **adaptations** are very obvious, like fish to rivers and birds to trees. Other adaptations are less obvious, like polar bears to the Arctic and camels to the desert.

A variety of living things

A hummingbird collecting pollen from a flower

Wildebeest and zebra in the Serengeti National Park, Africa

Feeding geese in the park

1.2 Life processes common to animals and plants

Most of the time it is easy to decide whether a thing is living or not. It is obvious that animals and plants are living and that stones and water are not living.

Living things are called **organisms** and all organisms carry out seven important processes to stay alive. These are sometimes called **life processes**.

1 Movement

Most animals can move quite quickly from one place to another. Plants can also move but much more slowly. For example, plants move as they grow and they move in response to light from the Sun. However, plants cannot take up root and move to a new place.

This chick hatching from an egg carries on the next generation

Wildebeest on the move

2 Reproduction

No species of plant or animal can survive unless its members produce young (offspring). **Sexual** reproduction involves the union between a male and a female of the same species. **Asexual** reproduction occurs when an organism reproduces on its own like a strawberry plant which produces runners from which new plants grow.

3 Sensitivity

If someone claps their hands in front of your face, you are bound to blink. The rapid movement of their hands and the sound of the clap provide a **stimulus** to which your **response** is a blink.

Human beings respond to movement, sound, light, touch, heat and chemicals. Organisms vary in their sensitivity to stimuli. Plants are usually less sensitive and respond more slowly than animals.

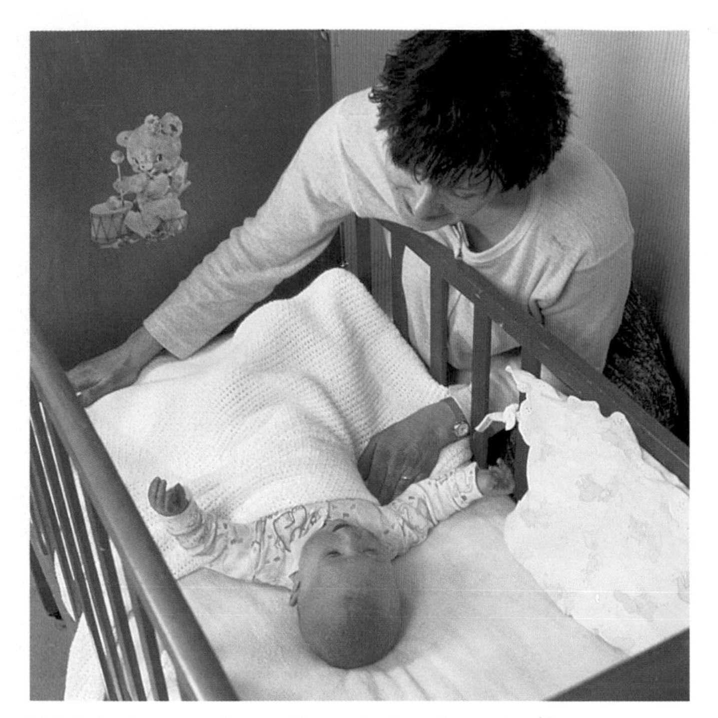

This baby is responding to his mother's voice and smile

4 *Growth*

Growth is a very obvious characteristic of living things. Young animals increase in size until they become adults. Most plants continue to grow all their life.

Buds and then flowers emerge as this honeysuckle grows

5 *Respiration*

Organisms need energy to grow and move. They obtain this energy by 'burning food' in a process called **respiration**. When respiration occurs, the food is broken down usually in a process which requires oxygen. As the food is broken down, energy is released.

Respiration provides Yvonne Murray with energy to win the 2000 m

6 *Nutrition*

Organisms must feed to obtain the substances they need for growth, for energy and for the replacement of worn-out parts. This process of taking in necessary substances for survival is called **nutrition**.

Plants and animals feed in very different ways. Animals obtain complex substances by eating other animals and plants. They break down these substances obtaining energy and producing simpler chemicals for growth. Green plants are able to make their own food from carbon dioxide and water which are simple chemicals. In order to do this, the plants need energy which comes from sunlight. This process of making food is called **photosynthesis**.

A monkey eating an orange

An elephant urinating

7 *Excretion*

As organisms feed, respire and grow, they produce waste products. These waste products must be removed and this process is called **excretion**. For example, animals get rid of excess water and nitrogen compounds by urinating. Carbon dioxide is excreted during breathing.

1.3 Animal cells and plant cells

All living things are composed of cells. Cells are the basic units of life. They are the building blocks for organisms in the same way that bricks are the building blocks for houses.

Your body contains about one hundred million cells. Each cell is about one thousandth of a centimetre wide ($\frac{1}{1000}$ cm). You cannot see them with the naked eye, but they can be easily viewed under a microscope.

Figure 1.1 shows a typical animal cell from the human liver side by side with a typical plant cell from a leaf.

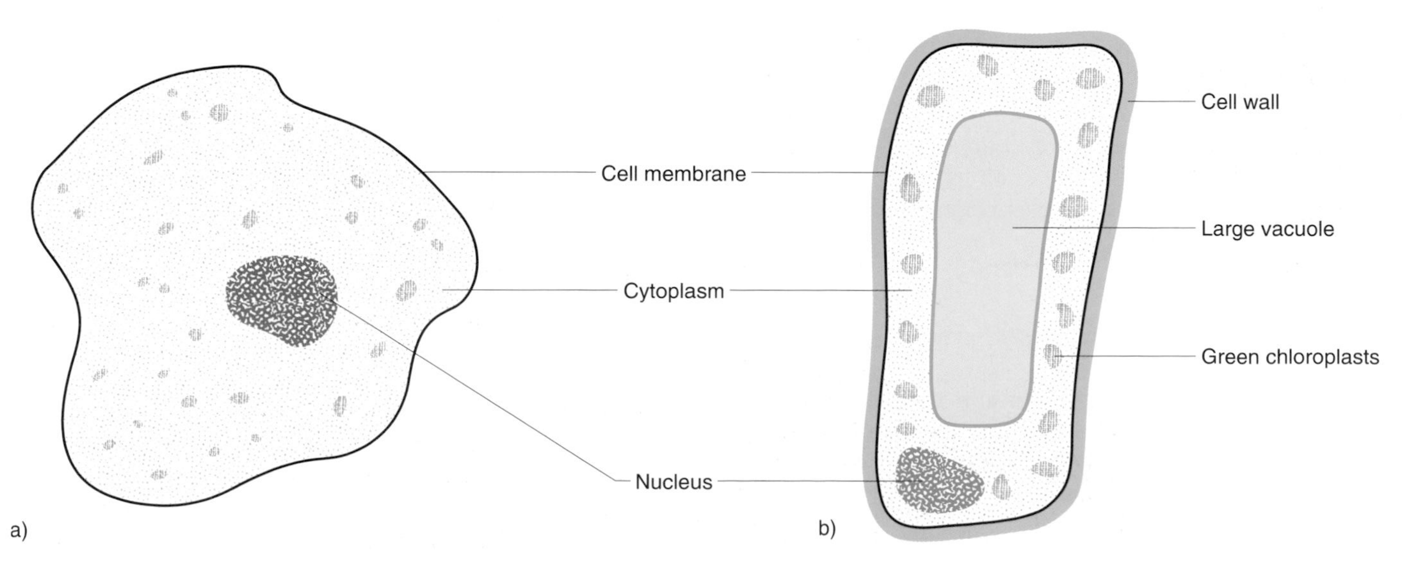

Figure 1.1 A typical animal cell and a typical plant cell

Notice that animal cells and plant cells have three important similarities in structure.

- A **nucleus** which contains long thin structures called **chromosomes**. The chromosomes are composed of nucleic acids (DNA) which control all the chemical reactions inside the cell. Chromosomes also play an important part when a cell divides.
- **Cytoplasm** in which most of the chemical reactions of living things take place. The cytoplasm is a jelly-like watery liquid. It contains a large number of the smaller parts of the cell called **organelles**. Organelles look like dots under an ordinary microscope. They include **mitochondria** in which energy-producing reactions occur, and small granules of stored food such as starch in plants and glycogen in animals.
- A **cell membrane** which acts as a thin boundary layer and keeps the cell together. The cell membrane also controls the movement of substances such as water and other small molecules into and out of the cell.

In addition to these three structures which are common to all cells, plant cells have three other important features.

- A **cell wall** outside the cell membrane made of cellulose. The cell wall is much thicker than the cell membrane, but it is still porous and flexible. The porous nature of the cell wall allows water and dissolved substances to pass through easily. The main function of the cell wall is to support and protect the cell.
- **Chloroplasts**, which are small organelles containing the green pigment chlorophyll. Chlorophyll has a key role in photosynthesis. It absorbs sunlight and uses it to synthesize food for the plant.
- A **vacuole** in the centre of the cell containing a watery liquid called **cell sap**. The vacuole takes up a large volume of the cell and is separated from the cytoplasm by a thin membrane. The vacuole has two main functions. First, it acts as a storage place for dissolved plant foods such as sugars and salts. Secondly, it creates a pressure on the cell wall and keeps the cell wall rigid.

1.4 From cells to organisms

We have already noted three major differences between animal cells and plant cells. There are also other differences within these two groups as different cells are adapted to their special functions.

Muscle cells, for example, need to contract and then relax rather like a rubber band (Figure 1.2a).

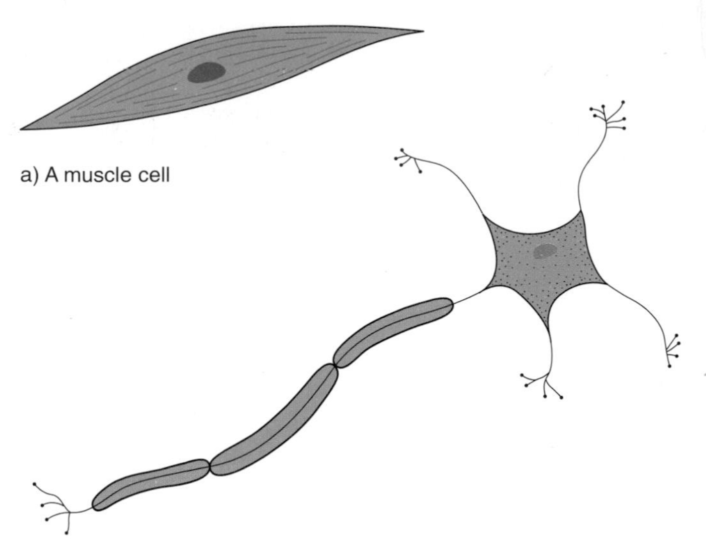

Figure 1.2 Muscle cells and nerve cells are specially adapted to their functions in our bodies

This is why they are long and thin. Nerve cells are much rounder with long thin fibres leading from them (Figure 1.2b). This is because they have to carry messages from one part of the body to another.

When lots of cells of a particular type are grouped together, they form a **tissue**. For example, in our bodies we have muscle tissue, skin tissue and nerve tissue.

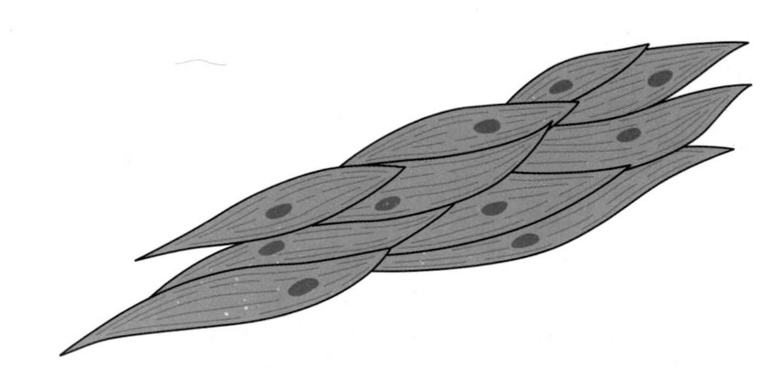

Figure 1.3 Muscle cells group together to form muscle tissue

In the same way, groups of tissues are co-ordinated to form the **organs** in our bodies. An organ is a complex part of a living organism which has a particular job. For example, the eye enables us to see, the heart is a pump involved in circulating the blood and the lungs are the organs of respiration. In the body, several organs often work together to form an **organ system**. So the heart, arteries, veins and capillaries form the *circulatory system* and the brain, spinal chord and the nerves form the *nervous system*.

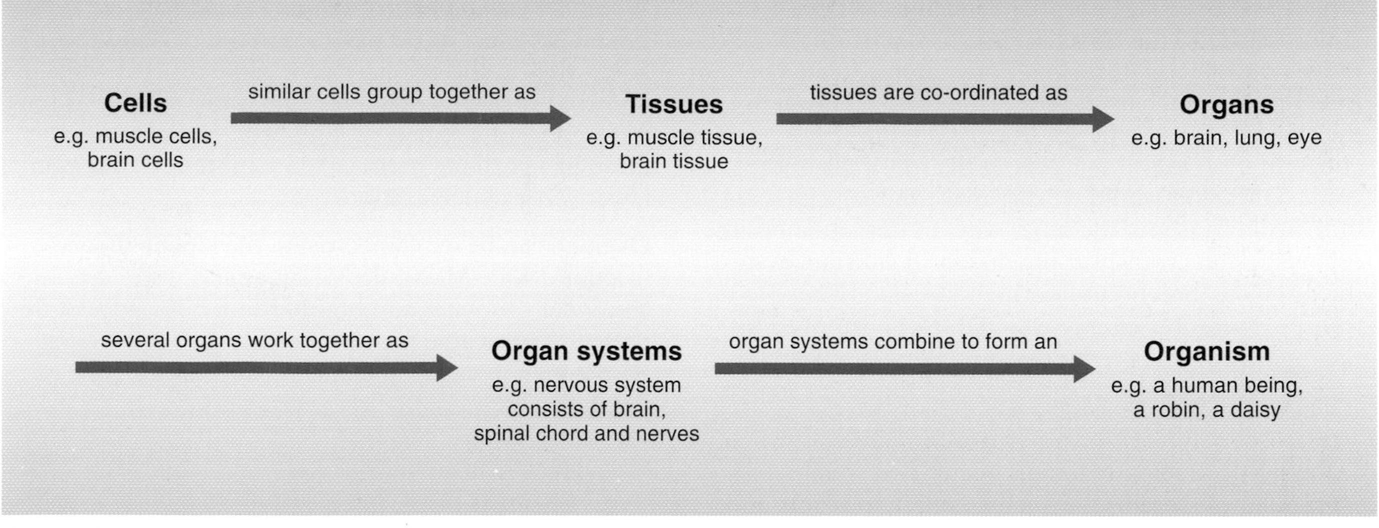

Figure 1.4 From cells to organisms

Organ systems, like cells, are specially adapted for their roles in life processes. So, for example, the digestive system consists of a long muscular tube (the gut) along which food is moved by muscles. As food moves through the gut, it is broken down (digested), absorbed into the blood system and then the waste is excreted. The length of the tube ensures that the food has time to be properly digested and the muscular nature keeps the food moving.

1.5 Movement of substances into and out of cells

Water is the major component of all animal and plant cells. This water in the cytoplasm and in the cell sap contains many dissolved substances. Water and other substances enter and leave cells through the cell membrane. In fact, the cell membrane *controls* the movement of particles of different substances into and out of the cell. The cell membrane will let certain particles pass through it, but it will block the passage of others. Because of this, it is described as **partially permeable**.

Section 14.5 describes how the particles move in different states of matter.

- *In a solid*, the particles vibrate about fixed positions.
- *In a liquid*, the particles roll around each other like tennis balls in a bag.
- *In a gas*, the particles move around very rapidly, colliding with each other and the walls of their container.

Particles enter and leave cells by three processes:

1 **diffusion**, 2 **active transport** and 3 **osmosis**.

1 Diffusion

This occurs because of the free (passive) movement of particles from a region of high concentration to a region of low concentration.

The process happens simply because of the random motion of the particles and the difference in concentration. The difference in concentration creates a **concentration gradient** from high to low and the rate of diffusion increases if:

- the concentration gradient increases;
- the temperature rises.

Examples of diffusion in living things include:

- small molecules from digested food diffusing from the gut into blood capillaries;
- oxygen diffusing from the air sacs (alveoli) in the lungs into blood capillaries;
- carbon dioxide diffusing from blood capillaries into the air sacs (alveoli) in the lungs.

Substances with small molecules, such as oxygen, carbon dioxide and water will usually diffuse easily through a cell membrane and down a concentration gradient. Substances with larger particles, such as sugar molecules, diffuse much more slowly through membranes because of their partially permeable character.

2 Active transport

This is the process whereby an organism moves dissolved substances from a region where they are in low concentration to a region of higher concentration.

This, of course, is against the concentration gradient and opposite to the direction in which particles would move naturally by diffusion. For example, plants obtain mineral salts, such as nitrates, from the soil through their roots. These salts are usually at a lower concentration in the soil than in the root cells of the plant. Diffusion would tend to make the salts pass *out* of the roots and not the other way. In this situation, the organism must transport the dissolved substances in the opposite direction to normal diffusion. This transport requires energy which must be supplied by the organism. It is therefore called active transport. It uses the energy from respiration and the cells capable of active transport usually have more mitochondria for aerobic respiration than other cells (section 3.5). Table 1.1 compares diffusion with active transport.

Table 1.1

Diffusion	Active transport
Non-selective (random) movement of particles	**Selective** movement of particles
Particles move **down** a concentration gradient	Particles move **against** a concentration gradient
No energy required	Energy required

3 Osmosis

Osmosis can be regarded as a special case of diffusion. Look at the experiment shown in Figure 1.5. Visking tubing consists of a thin membrane made from cellulose. This membrane has tiny, invisible holes which allow water particles to pass through, but not the larger sugar particles. The visking tubing is an example of a partially permeable membrane, like a cell membrane.

When the visking bag containing sugar solution is placed in water, water passes into the bag and the liquid rises in the capillary tube. The process which has occurred is called **osmosis**. The water molecules have diffused from a region of high water concentration through a partially permeable membrane to a region of lower water concentration.

Osmosis can be explained using the kinetic theory (section 14.4). Molecules of sugar and water are continually moving and bombarding both sides of the partially permeable membrane (Figure 1.6). Sometimes, a water molecule passes through one of the tiny holes.

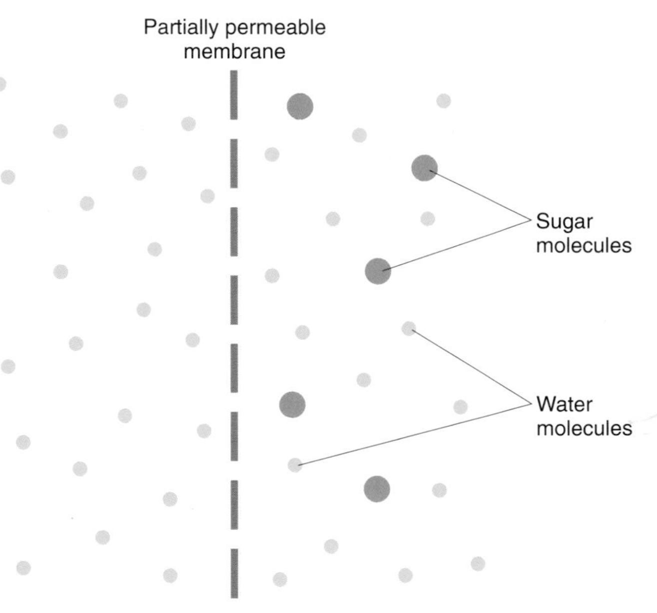

Figure 1.6 How does osmosis occur?

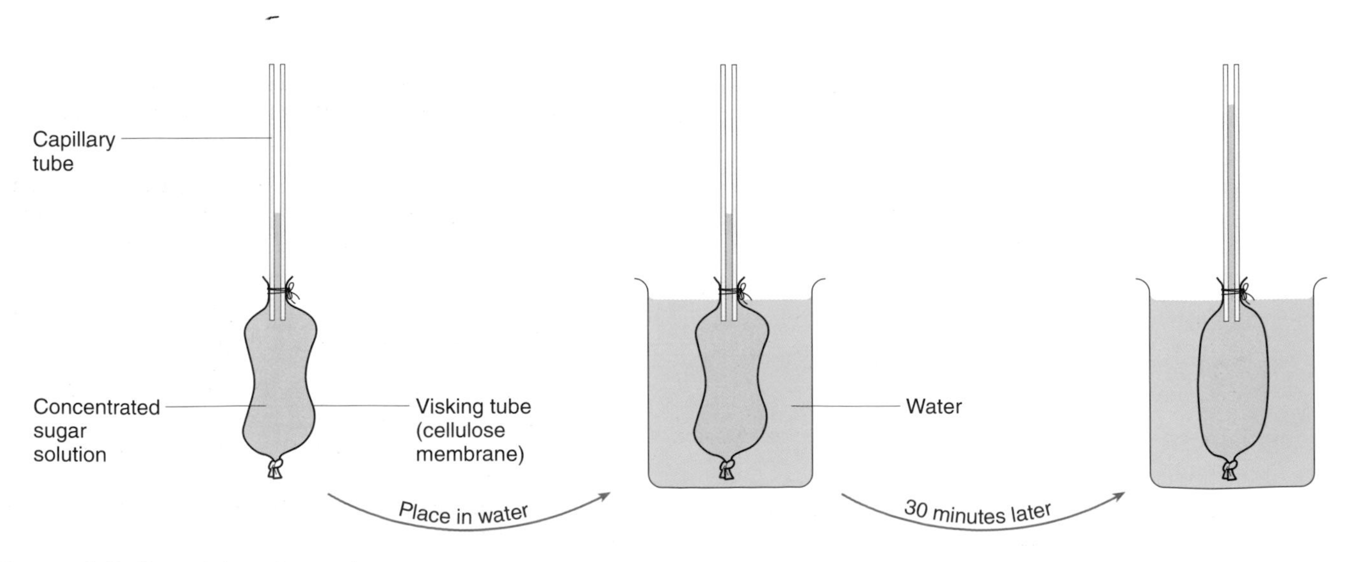

Figure 1.5 Osmosis in action

If there was pure water on both sides of the membrane, equal numbers of water molecules would flow in both directions and there would be no overall change.

However, in the sugar solution, large sugar molecules hinder the movement of water molecules and block their passage through the holes. So, more water molecules flow from the pure water to the sugar solution than the other way. The overall effect is that water flows through the membrane and into the sugar solution.

Osmosis in living things

Osmosis is very important in living organisms because the membranes of cells are partially permeable. Suppose some cells, such as red blood cells are placed in salt solutions of different concentrations (Figure 1.7).

Cell cytoplasm contains a solution of various substances (salts, sugars, proteins, etc.) in water. When the red blood cells are placed in water or a very dilute salt solution, water flows through the cell membrane and into the cell. The cell swells and may even burst (Figure 1.7a). The opposite process happens in concentrated salt solution (Figure 1.7b) and the cell shrinks.

From this you will appreciate how important it is to have the right concentration of solutes in the liquid around cells.

In animals, a number of mechanisms make sure that body fluids are kept at the right concentration. The kidneys play an important part in this process which is called **osmo-regulation**. A solution containing 0.6% salt in water (usually called **saline**) is just the right concentration to prevent osmosis in humans. Saline is used widely in hospital treatments and especially after major surgical operations.

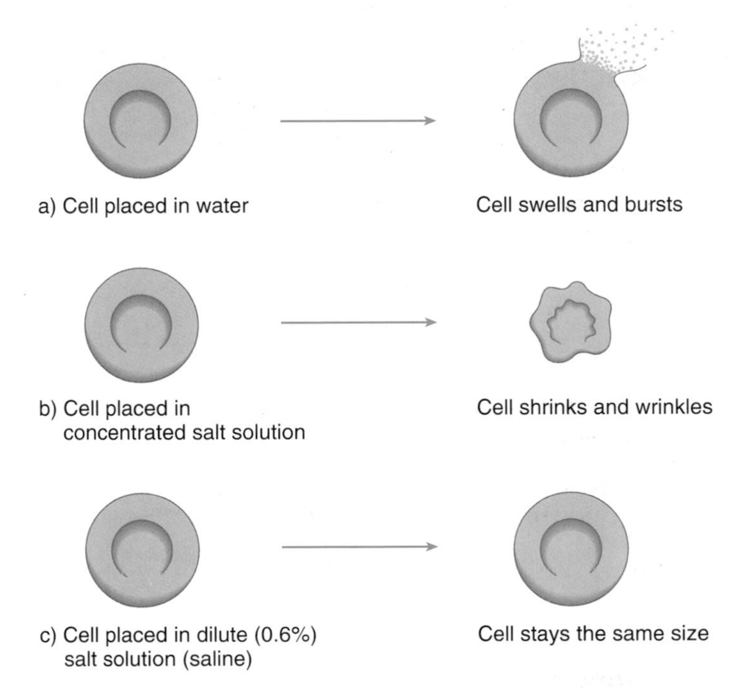

Figure 1.7 The effect of different concentrations of salt solution on red blood cells

QUESTIONS

1 Living things carry out seven important life processes to stay alive. Certain things that are not alive also carry out some of these life processes. Cars, for example, move and excrete.

a) Explain why it is reasonable to think that cars excrete.

b) Name two other processes that cars carry out which are like life processes.

c) Explain why the two processes you chose in part b) are like life processes.

d) How would you convince someone that cars are **not** living things?

e) Are viruses living things? Say 'yes' or 'no' and explain your answer.

2 Animals, like plants, are mainly made of water. Throughout their lives they must maintain water balance.

a) As humans, we obtain some of the water we need from chemical reactions in our body tissues. How do we obtain most of the water we need from day to day?

b) Humans lose water from their bodies in four distinct processes. Name three of these processes.

c) Camels are specially adapted for life in the desert where rainfall is light and patchy.
i) How are camels adapted for the dry desert climate?
ii) State two other adaptations that camels have for survival.

3 a) The diagram below shows a typical plant cell.

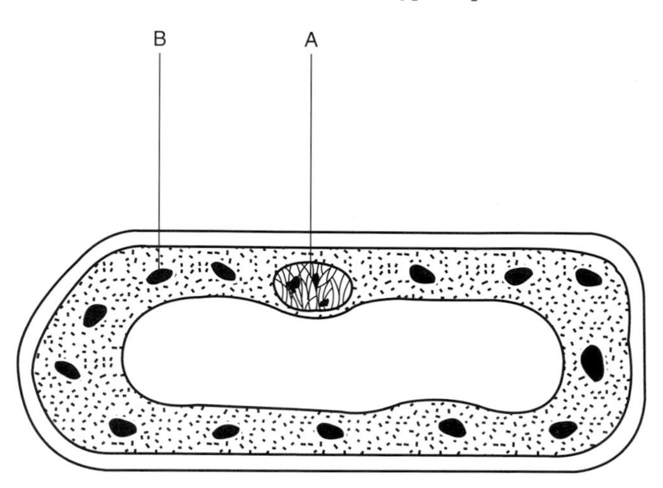

i) Name the part labelled A.
ii) What is the function of the part labelled B?

b) Two experiments were set up as shown to investigate osmosis.

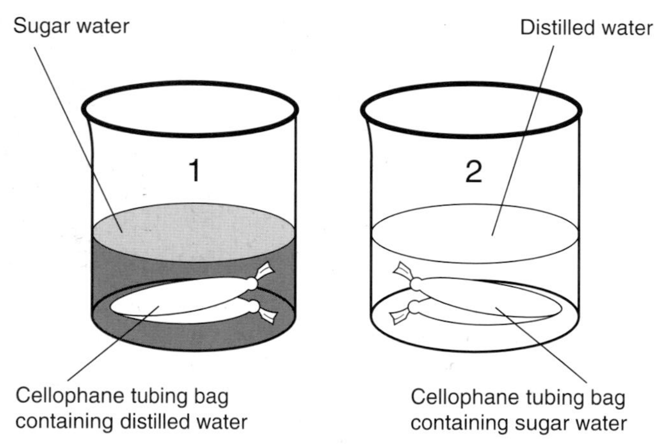

In which experiment did the cellophane tubing bag lose weight? Explain why. **SEG**

4 a) Explain fully what is meant by the term **osmosis**.

b) Two students investigated how the mass of slices of potato changed when in different concentrations of sugar solution. The slices were weighed, placed in different concentrations of sugar for 30 minutes, then weighed again without drying. Suggest **three** ways in which this investigation could be improved.

c) In the experiment the students calculated the percentage mass gain or loss. Why was this better than simply using the actual mass gain or loss?

d) Explain why there was no change in mass of boiled potato slices when these were used for the same investigation as described.

e) Water enters the roots of a plant through the many root hair cells.

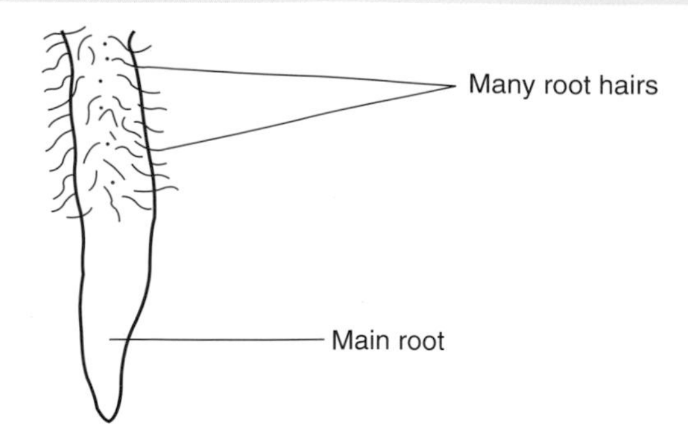

Explain why these root hair cells are more use to the plant than just a single root. **WJEC**

5 Plant cells have two extra structures which are **not** found in animal cells – a cell wall and chloroplasts.

a) What is the main function of the cell wall in plants?

b) What problems would animals have if their cells had a tough cell wall?

c) What is the main function of choroplasts in plant cells?

d) Plant cells are usually arranged more regularly than animal cells. What is the advantage of this regular arrangement in plants?

e) It is often difficult to see the boundaries of animal cells under a microscope. Why is this?

6 Here is a list of the different ways in which living organisms can get food.
A – by making their own food,
B – by eating animals (carnivores),
C – by eating plants (herbivores),
D – by eating plants and animals (omnivores),
E – by feeding off living organisms (parasites),
F – by feeding off dead or decaying organisms (saprophytes).

Use one of the letters A to F above to describe the way each of the following get their food.
a) blackbirds b) daisies c) deer d) mistletoe
e) oak trees f) sharks g) toadstools

7 Explain the following observations.

a) Lettuce leaves go limp and floppy a few hours after being torn up.

b) Limp lettuce can be made crisper by placing in fresh, cold water.

c) Lettuce leaves become very limp very quickly if placed in salty water in order to get rid of insects and slugs.

CHAPTER

Nutrition and digestion

2.1 Our diet – nutrition

Our diet is composed of the food that we eat. We must eat to survive and the process of taking in food for survival is called **nutrition**. To be healthy, we must eat a **balanced diet**. Our diet must include seven important constituents:

1 **Carbohydrates**, such as starch, which are present in bread, potatoes and rice. Carbohydrates are needed to provide **energy**.

2 **Fats**, which are present in foods such as milk, cheese, butter and red meats. Fats are needed to provide **energy** and for making **cell membranes**.

3 **Proteins** which are present in foods such as fish, meat, eggs, peas and beans. Proteins provide the chemicals which we need for **growth** and for **replacing cells**.

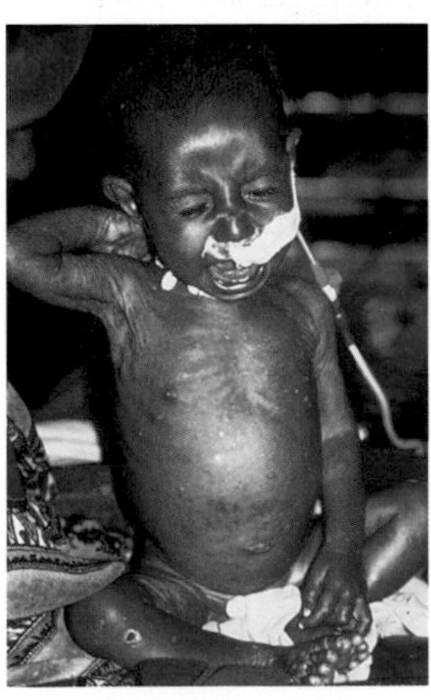

Insufficient protein in the diet causes poor general health. Children are very seriously affected. This child has skin sores, a bloated abdomen and is weak and unhappy. The illness is called Kwashiorkor

A chip buttie is not a nutritious meal as it is composed mostly of carbohydrate and fat. The meal of a salmon steak with vegetables contains many different food types to provide a healthy diet

In addition to these three key energy and body-building foods, we also need:

4 **Vitamins**, such as vitamin C in oranges.

5 **Minerals**, such as salt (sodium chloride).

Vitamins and minerals are sometimes described as **maintenance foods**. They are needed in small amounts to control our body processes (metabolism) and to keep our bodies running smoothly.

6 **Water** is needed to **maintain cell processes** and to keep the **concentration** of the substances in cells at a **steady level**. We can survive for weeks without food, but only a few days without water. We take in most water by drinking, but there are significant amounts of water in solid foods, particularly fruit and vegetables. We should drink at least two pints (about 1000 cm^3) of water every day.

7 **Fibre** is important in our diet as **roughage**. Fibre is present in fruit, vegetables and brown bread. It helps the **movement of food** through the gut and prevents constipation.

2.2 Food testing

It is important to test for the different constituents in foods.

Testing for fat

Figure 2.1 shows how you should carry out the test for fat and the result from a positive test.

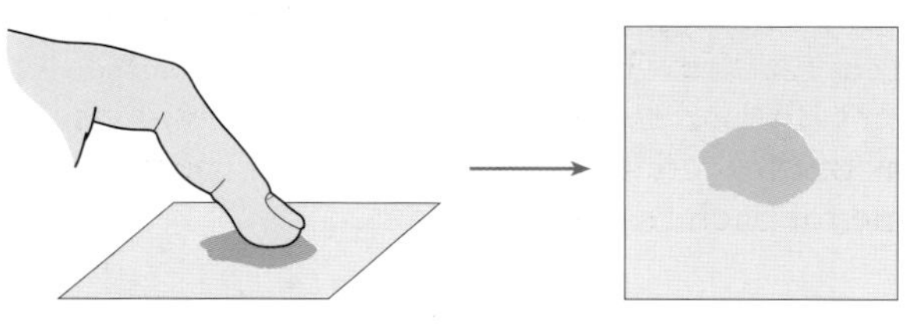

Figure 2.1 Testing for fat

Tests for starch, sugars and proteins

Before testing for starch, sugars or proteins, it is necessary to prepare the sample of food. This is shown below in Figure 2.2.

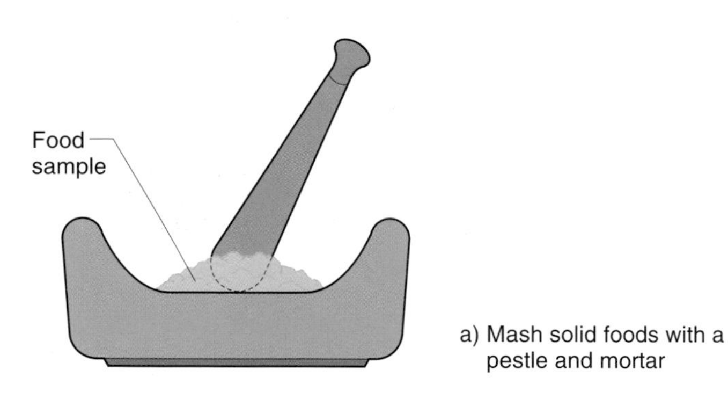

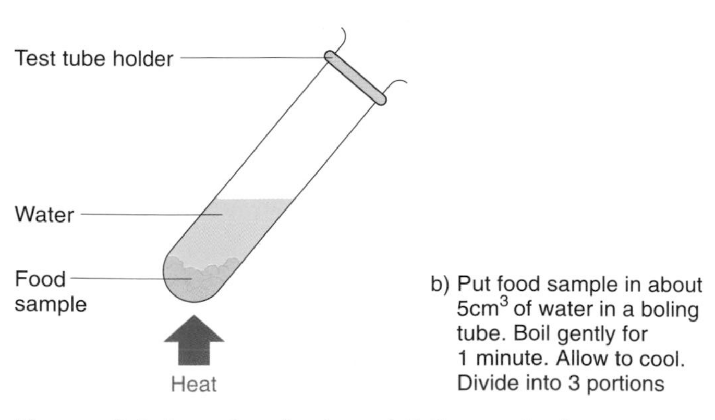

Figure 2.2 Preparing a food sample before testing for starch, sugar or protein

The specific tests for starch, sugars and proteins are shown in Figures 2.3, 2.4 and 2.5.

Testing for starch with iodine solution

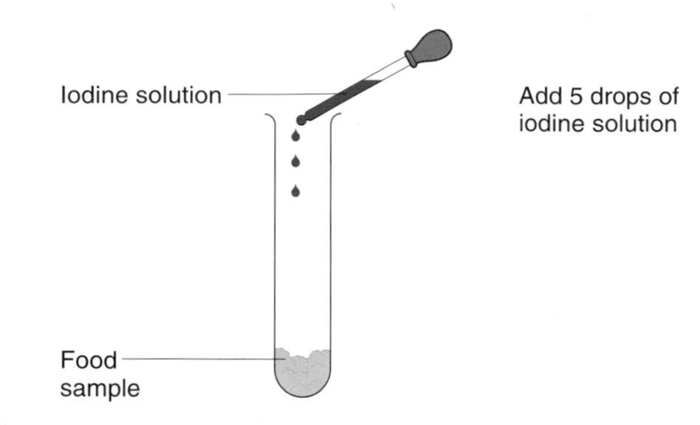

Figure 2.3 Testing for starch

A blue-black colour shows that starch is present in the food sample.

Testing for simple sugars with Benedict's solution

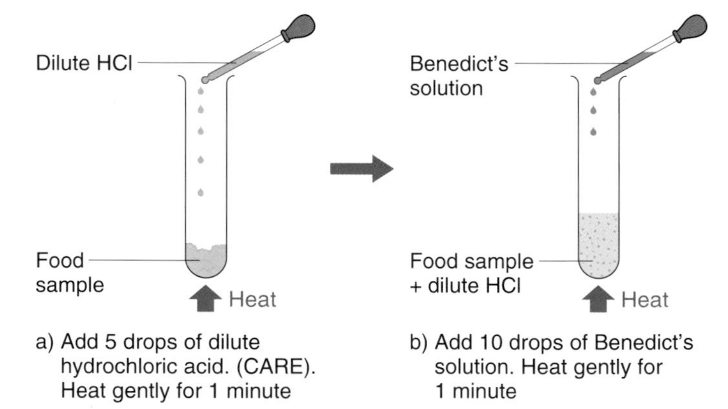

Figure 2.4 Testing for simple sugars

A green or orange-red precipitate shows that a simple sugar (e.g. glucose or sucrose) is present in the food sample.

Testing for protein (the Biuret test)

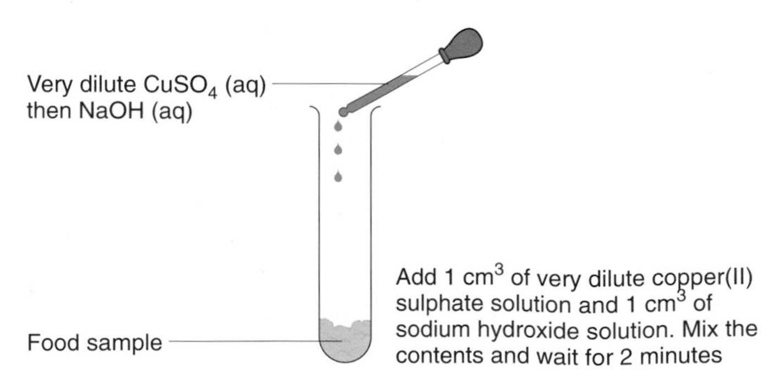

Figure 2.5 Testing for proteins

A purple or violet colour shows that a protein is present in the food sample.

2.3 Digestion

Food is needed by every part of your body. But, how do you think that the food you eat that is then taken into the digestive system reaches the different parts of your body? The answer to this question can be explained from the experiment illustrated in Figure 2.6.

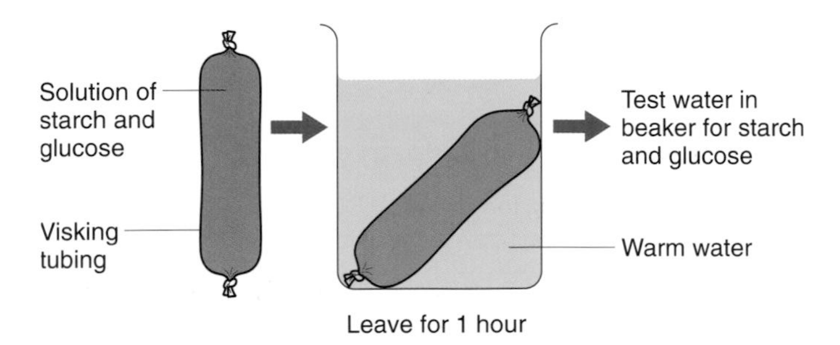

Figure 2.6 An experiment to illustrate digestion

The visking tubing containing a solution of starch and glucose (a simple sugar) is left in pure, warm water for one hour. At the end of this time, the water in the beaker is tested for starch with iodine and for glucose using Benedict's solution. Glucose is found to be present in the water, but no starch.

Starch contains very large molecules. It is formed when hundreds of smaller glucose molecules join together. The large starch molecules cannot pass through the tiny holes in the visking tubing, but the glucose molecules can. Because of this, all the starch remains inside the visking tubing but glucose can pass through the tubing into the warm water.

The digestive system is like a long visking tube which runs from the mouth to the anus. It is usually called the **gut** or the **alimentary canal** (Figure 2.7).

The main purpose of digestion is to break down the large molecules of carbohydrates, proteins and fats in our food into smaller particles. These smaller particles (like glucose) are soluble in water and can pass through the walls of the gut and into the bloodstream just as the glucose molecules pass through the visking tubing. Once in the bloodstream, these smaller molecules can be carried to other parts of the body where they are needed. Undigested food, which cannot be broken down, passes along the gut, finally passing out through the anus as faeces.

Digestive enzymes

The main processes in digestion are chemical reactions in which carbohydrates, proteins and fats, which have large, insoluble molecules are broken down into smaller, soluble molecules. This breakdown of large molecules into smaller molecules is speeded up (catalysed) by **digestive enzymes**. The enzymes are produced by glands which open into the gut.

Enzymes are biological catalysts (section 18.8). Their molecules are themselves proteins which are affected by changes in temperature and pH. Almost all the chemical reactions in our bodies are catalysed by enzymes. In fact, every cell in our bodies contains dozens of different enzymes as there is a specific enzyme for each reaction.

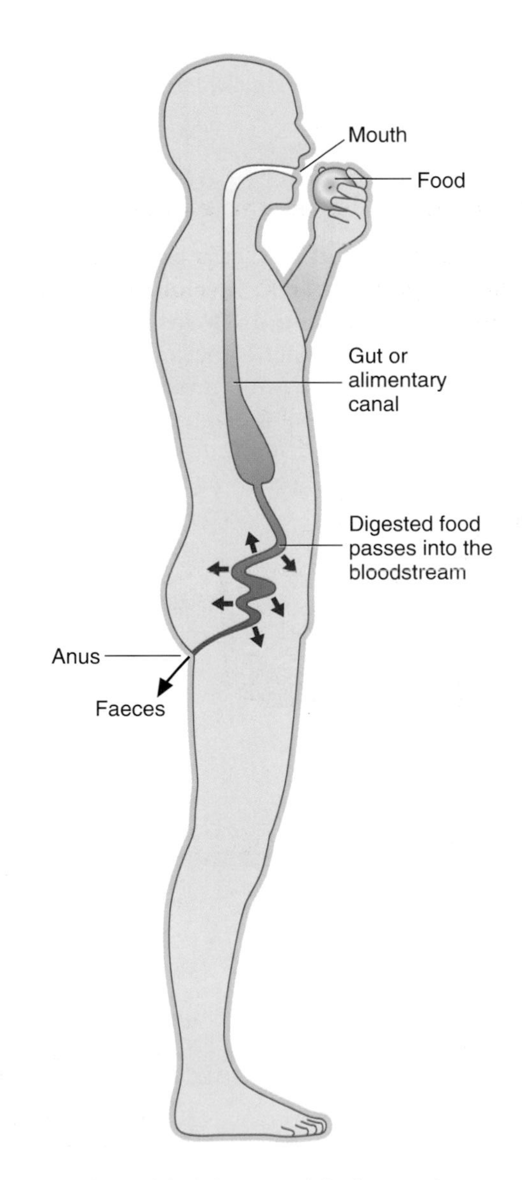

Figure 2.7 A simplified diagram of the human digestive system

The substance on which an enzyme works is called its **substrate**. Enzymes are usually named by adding the ending **-ase** to the name of their substrate. So, for example, **carbohydrases** are enzymes which catalyse the breakdown of carbohydrates like starch, into glucose.

Amylase, a carbohydrase enzyme in saliva begins the breakdown of starch in food. The starch molecules (composed of glucose molecules joined together) are broken down by amylase to maltose. Maltose molecules are composed of two glucose molecules.

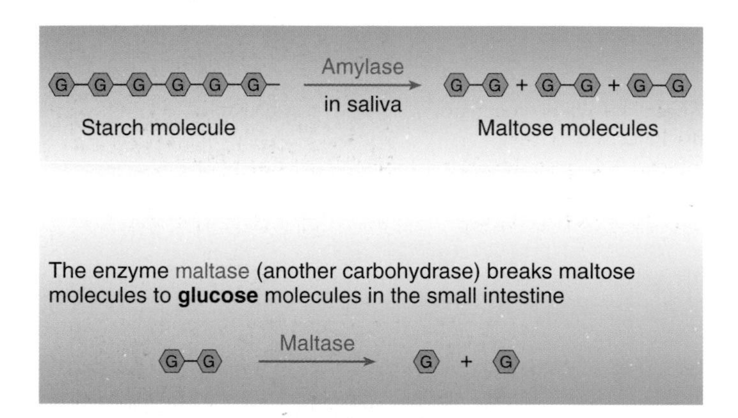

Proteases are enzymes which catalyse the breakdown of proteins into **amino acids**. **Pepsin**, a protease enzyme, in gastric (stomach) juices begins the breakdown of proteins. Protein molecules are composed of similar small molecules, called amino acids, joined together. Proteins are broken down by pepsin to shorter chains of amino acids called **peptides**.

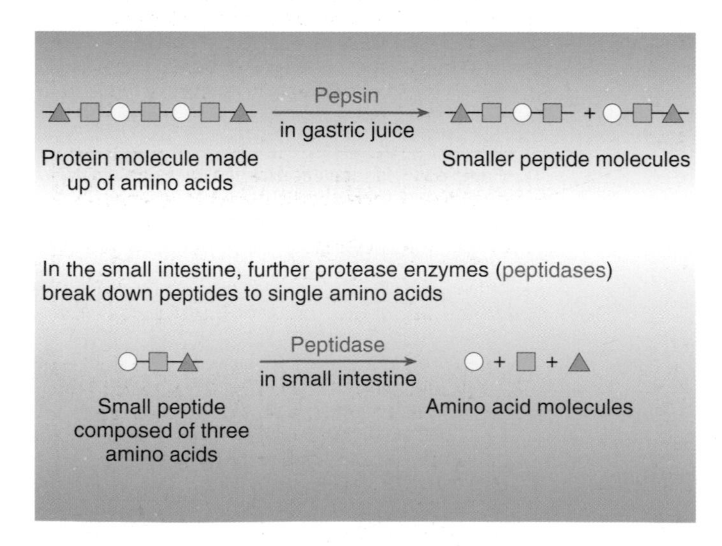

Lipases are enzymes in the small intestine which catalyse the breakdown of fats (lipids) into **fatty acids** and **glycerol**.

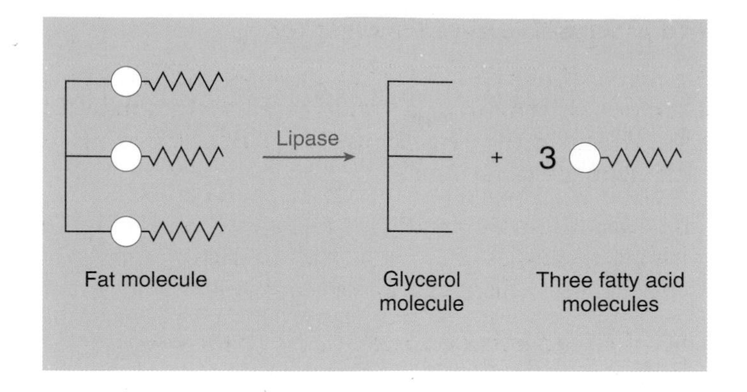

2.4 The human digestive system

Although the alimentary canal can be described in simple terms as a long tube, it has different parts. Each of these parts has a different role in the process of digesting food. Figure 2.8 shows the structure of the human digestive system as it is arranged in the body.

Organs which are important in digestion, but not actually part of the alimentary canal (the salivary glands, the liver and the pancreas) are labelled on the right hand side of the diagram.

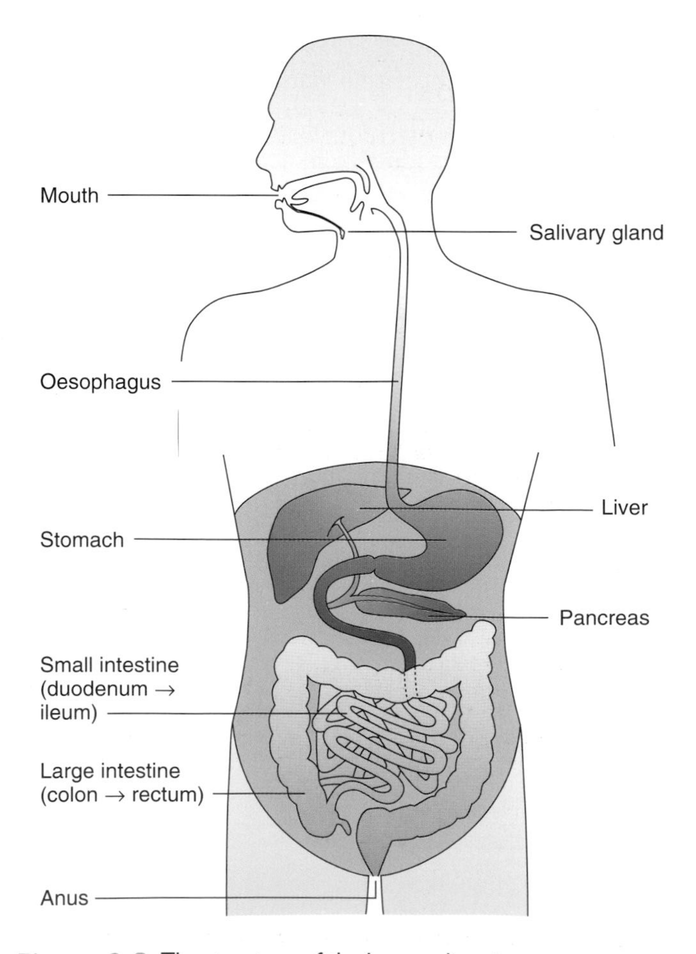

Figure 2.8 The structure of the human digestive system

The processes involved in digestion

Figure 2.9 shows the alimentary canal as if it has been taken out of the body and spread out from mouth to anus. Alongside the diagram there is a description of what happens in each part.

The walls of the alimentary canal are thick and muscular. Muscles contract behind the food, pushing it along the gut. This is called **peristalsis**.

Once the food has been broken down into small, soluble molecules like glucose and amino acids, it is absorbed into the blood through the lining of the small intestine. The inside surface of the small intestine has hundreds of small, finger-like projections called **villi**. These increase the surface area and enhance the absorption of small molecules. Just below the surface of the villi, there are tiny blood capillaries. The soluble food molecules diffuse into these capillaries and are then carried by the blood to other parts of the body.

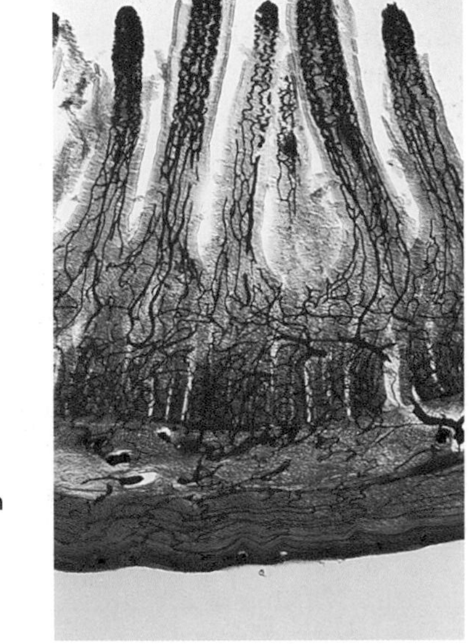

Finger-like villi project from the inner wall of the intestines. Digested food is absorbed through the villi into the blood capillaries

Mouth – food is chewed by teeth, breaking it into smaller pieces. This increases the surface area so that enzymes can react with the food more easily.

Salivary glands produce saliva (spit). This contains:
- *mucus* which covers the food and makes it easier to swallow;
- *amylase*, a carbohydrate enzyme which breaks down starch into maltose.

To lungs

Oesophagus (gullet) has muscular walls which push the food along. The movement is called *peristalsis*.

Stomach – the stomach wall is thick and muscular. Contractions of these muscles churn up the food. Glands in the stomach wall (gastric glands) produce *gastric juice*. This contains *pepsin*, an enzyme which breaks down proteins to peptides. The gastric glands also produce *hydrochloric acid* which kills most bacteria taken in with food and adjusts the pH for the most effective enzyme action.

Bile duct – the liver produces *bile* which is stored in the *gall bladder* before being released into the small intestine after a meal. Bile neutralizes the acid from the stomach and provides the best conditions for enzymes in the small intestine. Bile also emulsifies fats (breaks up fat into smaller droplets). This increases the surface area of fats for lipase enzymes to act upon.

Pancreas produces *pancreatic juice* containing enzymes. Pancreatic juice enters the small intestine and causes the breakdown of:
- starch to sugars
- proteins to peptides
- fats to fatty acids and glycerol.

Small intestine – glands in the small intestine produce enzymes which break down:
- sugar to glucose
- peptides to amino acids
- fats to fatty acids and glycerol.

Small molecules of digested food are absorbed through the walls of the small intestine into the blood stream.

Large intestine – by this stage, digestible foods have been broken down and absorbed into the blood. Indigestible food (*roughage*) and water remain. Water is absorbed into the blood. Semi-solid matter collects in the rectum as *faeces* which pass out through the *anus* at regular intervals.

Figure 2.9 The processes involved in digestion along the alimentary canal

1 Use the table below to answer the questions which follow.

Dairy produce	Amount per 100 g				
	Protein (g)	Fat (g)	Energy content (kJ)	Calcium (mg)	Iron (mg)
Cow's milk, whole	3.4	3.7	276	120	0.1
Cow's milk, skimmed	3.5	0.2	146	124	0.1
Butter	0.4	85.1	3318	15	0.2
Margarine	0.2	85.3	3326	4	0.3
Cheese	25.4	34.5	1778	810	0.6

a) i) Which product gives the greatest energy per 100 g?
ii) Which class of food is the **main** source of energy in margarine?

b) Explain why cheese contains more minerals than milk.

c) i) Which of the foods in the table would be **best** for someone with heart disease?
ii) Give a reason for your answer. **WJEC**

2 a) A student studied the time taken for saliva to digest starch. The diagram below shows an investigation that he set up. Draw and label a suitable control tube.

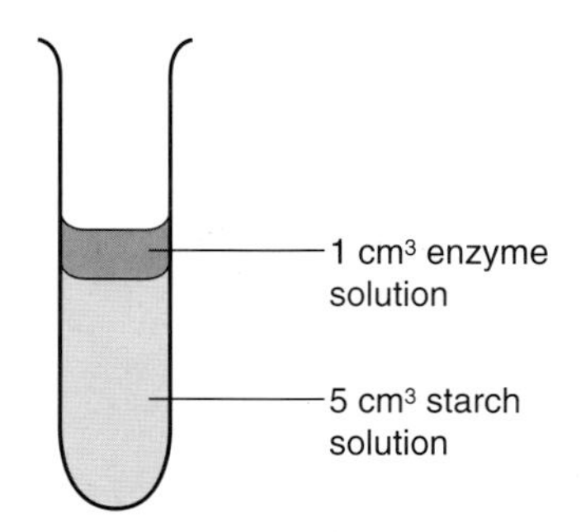

b) Graph 1 at the top of the page shows the results obtained by changing the temperature, and keeping the pH the same.
Graph 2 shows the results obtained by changing the pH (acidity or alkalinity) and keeping the temperature constant.

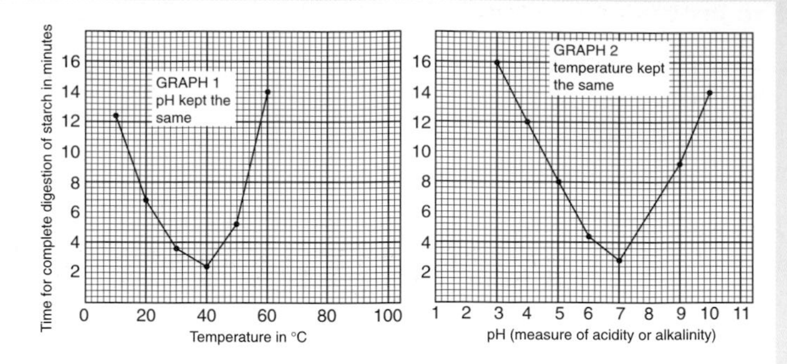

i) What is the best temperature for the reaction?
ii) What would you expect to happen to the enzyme if it were heated to 100 °C for 5 minutes?
iii) What is the best pH for the reaction?
iv) Suggest why this enzyme would not work in the stomach of a mammal. **WJEC**

3 a) The diagram is a height/weight chart. Four pupils measured their heights and weights. They each marked them on the chart with an X.

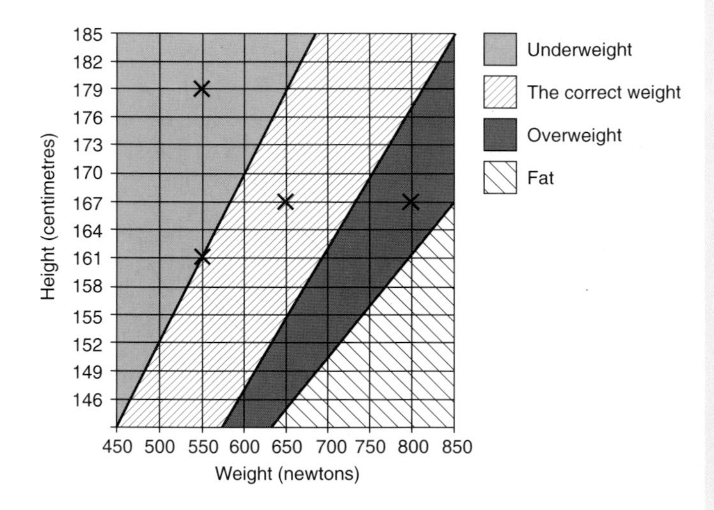

Peter is one of the pupils who marked his result on the chart. He is overweight. Decide which X belongs to Peter and write down his:
i) height,
ii) weight.

b) The table below shows the food content of some of Peter's favourite foods.
Peter wants to lose weight.
i) Which food in the table should he not eat?
ii) Explain your answer. **Edexcel**

Table for question 3

Food	Energy (kJ)	Protein (g)	Fat (g)	Carbohydrate (g)	Vitamin C (mg)
Chicken	599	26.5	4.0	0.0	0
Chips	1065	3.8	10.9	37.3	10
Baked beans	270	5.1	0.5	10.3	0
Apple	196	0.3	0.0	11.9	5

(All values are for 100 g of food.)

CHAPTER

Respiration and breathing

3.1 What is respiration?

Respiration is the term used to summarise three very important processes in the body. These are:

- **Breathing**, in which a regular pattern of chest movements ensures that air (oxygen) is taken into the lungs and carbon dioxide is breathed out (section 3.2);
- **The exchange of gases** which takes place in the capillaries surrounding the thousands of air sacs in the lungs (section 3.3);
- **Cellular respiration** in which chemical reactions in the cells throughout the body result in the release of energy (section 3.5).

Remember that all living things respire in order to obtain the energy they need. In plants, the intake of oxygen and release of carbon dioxide takes place through the stomata (section 5.4). This chapter focuses on respiration in animals.

3.2 How does breathing take place?

All animals breathe in and out, whether they are awake or asleep. We breathe automatically in order to inhale oxygen from the air and exhale carbon dioxide. If this process stops, we die.

Inhalation

During inhalation the ribs move up and the diaphragm becomes flattened so that the chest cavity (thorax) increases in volume (Figure 3.1). The external intercostal muscles (upper chest muscles) have contracted to raise the ribs. At the same time, the diaphragm muscle sheet contracts causing it to move downwards. The increased volume of the chest cavity results in a lower air pressure inside than that outside the body, so air moves in, inflating the lungs.

Exhalation

In order to exhale, the internal intercostal muscles contract lowering the ribs and the diaphragm relaxes, resuming its dome shape. The combined effect reduces the space and the elastic lungs deflate, squeezing air out.

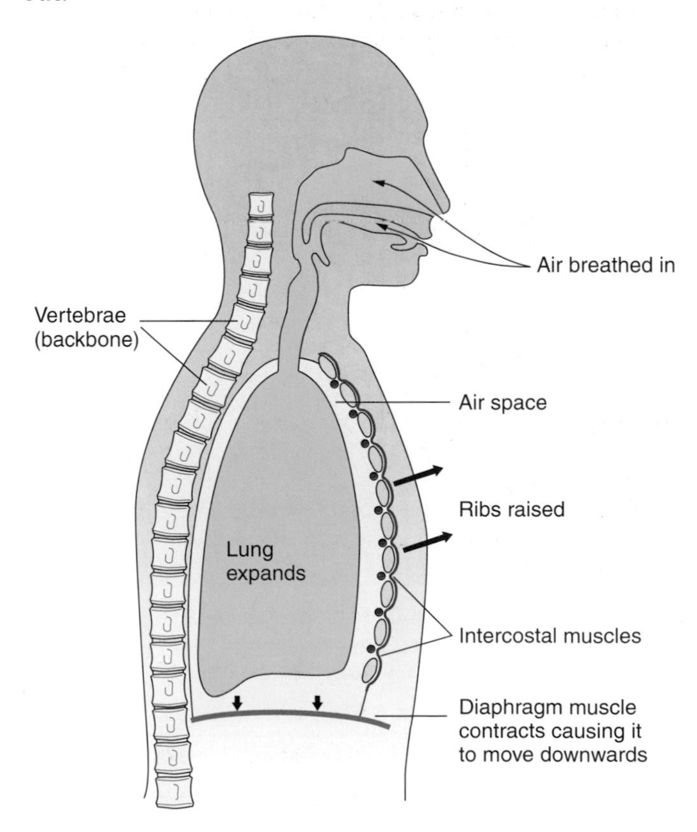

Figure 3.1 Movement of the ribs and diaphragm during inhalation

3.3 Gas exchange in the lungs

Air is composed mostly of nitrogen (78% by volume) which is not used by our bodies. What we do need is oxygen and this makes up 21% by volume in clean, dry air. Oxygen enables the chemical process of cellular respiration to produce the energy which we need to stay alive.

When we are involved in energetic pursuits, such as climbing, cycling and other sports, our bodies respond in two ways ensuring that we breathe more often and more deeply. This allows us to take in more oxygen and expend more energy.

- A special part of the brain contains cells that are very sensitive to the concentration of carbon dioxide in the blood. High concentrations of carbon dioxide in the blood act as a stimulus on these cells which in turn cause our breathing to increase automatically.
- During energetic activity, lactic acid will accumulate in our muscles if there is an insufficient supply of oxygen (section 3.6). The lactic acid gets into the bloodstream and triggers brain cells, similar to the carbon dioxide sensitive cells, which make the lungs work harder.

These runners in the Borrowdale Fells Race are breathing faster than normal so that extra oxygen, combined with food (glucose) can provide extra energy for the climb

The process of breathing is designed to allow oxygen to reach the body cells so that aerobic cellular respiration can take place (section 3.5).

Inside the cells of animals, oxygen combines with food such as glucose in the diet to produce energy, carbon dioxide and water.

$$\text{glucose} + \text{oxygen} \rightarrow \text{carbon dioxide} + \text{water} + \text{energy}$$
$$C_6H_{12}O_6 + 6O_2 \rightarrow 6CO_2 + 6H_2O + \text{energy}$$

Table 3.1 shows the temperature and composition of gases in inhaled and exhaled air.

Table 3.1 The temperature and composition of gases in dry, inhaled and exhaled air. What do these numbers tell you about the chemical processes during respiration?

	Dry, inhaled air	Exhaled air
% oxygen	21	14
% carbon dioxide	0	5
% water vapour	0	2
temperature (°C)	20	25

Figure 3.2 shows the structure of the human breathing system in more detail.

When we inhale, air is taken in through the nose or mouth. The **nasal cavity** is divided up by bony partitions with membranes covering a large surface area and producing a shiny **mucus**. This mucus is warm and moist, filtering dust and germs from the air before it reaches the lungs. The nasal membranes also have tiny hairs, called **cilia**, which sweep the mucus towards the throat to be swallowed.

From the nasal cavity, the air passes down the **trachea** and the two bronchi (singular, **bronchus**) into the **lungs**. From each bronchus, a branched system of elastic air passages, called **bronchioles**, become smaller and smaller and lead eventually to thousands of tiny air sacs, called **alveoli**. These are collapsible and inflate and deflate as we inhale and exhale.

A network of **blood capillaries** covers each alveoli and it is here that the exchange of oxygen and carbon dioxide occurs.

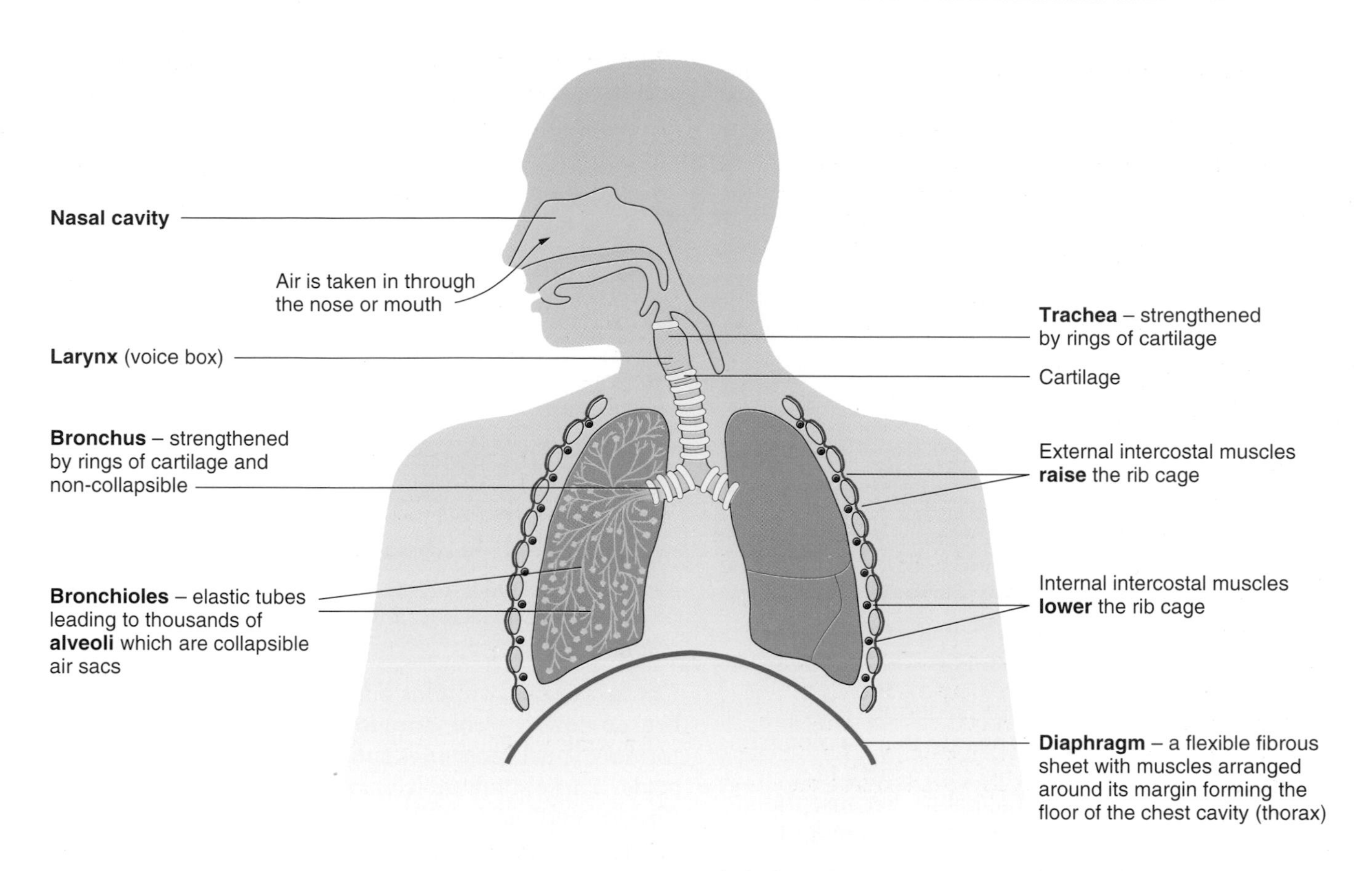

Figure 3.2 The structure of the human breathing system

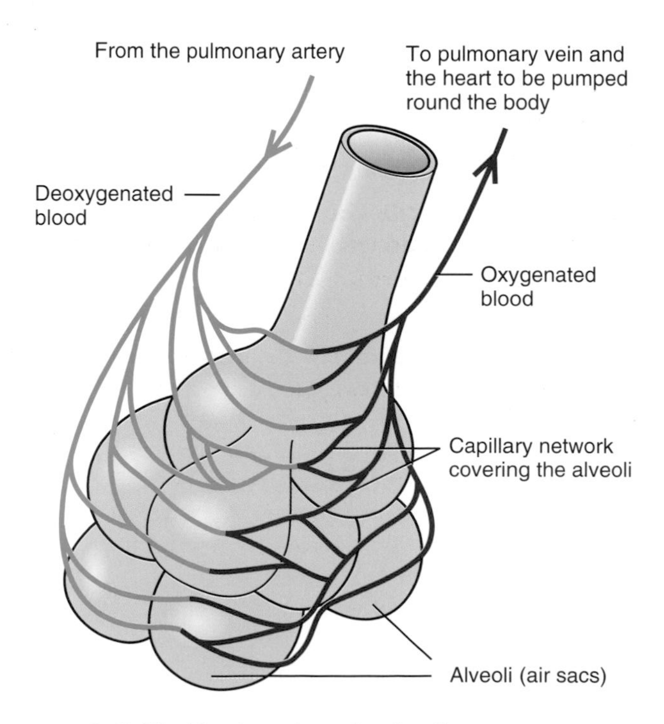

Figure 3.3 The blood supply to the alveoli

The alveoli provide an enormous moist surface area where gas exchange can take place. Their thin walls, one cell thick and clothed in blood capillaries, allow oxygen to diffuse out of the lungs and into the bloodstream. This oxygen is picked up by the pigment **haemoglobin** in the red blood cells and then circulates as **oxygenated blood**. **Deoxygenated blood** in the capillaries, returning from the rest of the body, comes to the alveoli. Carbon dioxide diffuses across their thin walls into the lungs and is then excreted as we exhale.

3.4 Smoking and health

Most people now recognise that smoking is a dangerous habit and that smokers cannot really be classed as healthy individuals. Every cigarette packet and advert must carry a health warning.

Tobacco smoke contains carbon monoxide, nicotine, tar and other toxic substances. The carbon monoxide reacts readily with haemoglobin in the blood during gas exchange in the lungs. The carbon monoxide forms a strong bond with the haemoglobin preventing the oxygen from bonding and so oxygen intake is seriously affected.

A smoker is not a healthy person

General health suffers and energy is bound to be reduced. This is illustrated by the fact that women who smoke during their pregnancy may give birth to underweight and sometimes frail babies because the foetus has been deprived of the full complement of oxygen during its development in the uterus.

Nicotine in tobacco is **physically addictive**. It is also **carcinogenic**. A carcinogen is an inducing agent linked with cancerous growths. Normal cells are affected by carcinogens which cause them to divide and grow as an abnormal mass or tumour. What may start as a tumour of malignant cells seated in the lungs can spread to other parts of the body and eventually cause death.

The tar in cigarette smoke paralyses the ciliated cells in the mucous membranes of the nasal cavity. This means that harmful bacteria and viruses are not removed and these may cause infections. Even a common cold can develop into bronchitis or pneumonia.

Emphysema is another serious and disabling condition caused by smoking and related to bouts of bronchitis. After a long period of time, tar causes the alveoli to break down, reducing the surface area available for oxygen absorption. Sufferers of emphysema cannot do the simplest task without becoming out of breath.

Over the years, smoking causes irreparable damage to the heart and arteries. In particular, carbon monoxide in smoke causes fatty deposits to occur in the linings of the arteries. This is called **arteriosclerosis**. These fatty deposits reduce the blood circulation. Poor circulation in the feet or lower limbs can lead to gangrene, a condition in which the body cells begin to die due to an insufficient supply of oxygen. The only solution is amputation of the diseased part.

If the arteriosclerosis leads to a blockage in the coronary arteries, this will deprive the heart muscles of oxygen and a heart attack may occur.

3.5 Cellular respiration

Respiration involves the overall process of breathing, gas exchange in the lungs and cellular respiration.

> Cellular respiration covers the chemical reactions which occur in cells and which result in the release of energy from foods.

A supply of energy is needed every minute of the day and night by every living cell. This energy for the life processes in plants and animals is provided by respiration.

Carbohydrates, proteins and fats in our food are first broken down by enzymes in the digestive system forming small molecules such as glucose and amino acids. These small molecules react with oxygen forming carbon dioxide, water and energy.

Aerobic respiration

In the presence of oxygen (aerobic conditions), small molecules react with the oxygen to produce carbon dioxide and water and energy is released.

glucose + oxygen → carbon dioxide + water + energy
(food)

$$C_6H_{12}O_6 + 6O_2 \rightarrow 6CO_2 + 6H_2O + 2900\ \text{kJ}$$

The rate of oxygen consumption during respiration can be used as an indication of metabolic rate.

Notice the large amount of energy in the equation above. This energy is used in five important ways in living cells.

- We all need energy to run, skip and jump. Even the slightest muscular action like the blinking of an eye needs **mechanical energy**.
- Most muscles are under the direction of nerve impulses which are like tiny electric currents. So, nerve cells need **electrical energy**.
- Some of the energy from food must be released as heat, **thermal energy**, to keep us warm and maintain our body temperature.
- When plants absorb ions from the soil and when animal cells absorb ions, the process of active transport is involved (section 1.5). This movement of ions across cell membranes and against a concentration gradient requires **energy**.

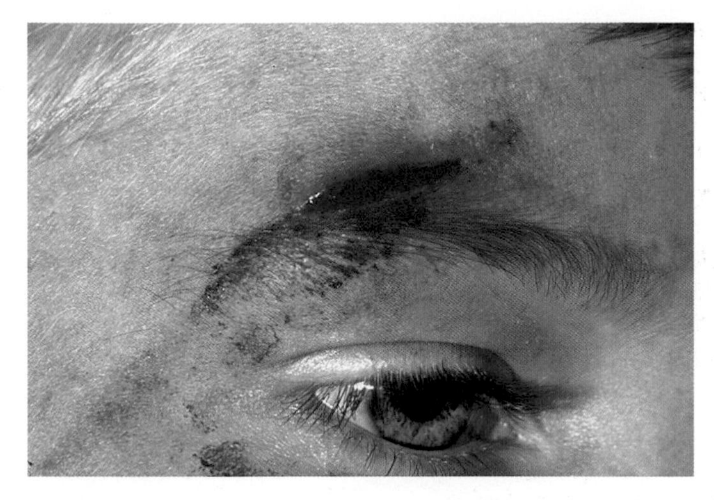

Damaged tissues, like this cut above the eyebrow, are repaired using the chemical energy produced during cellular respiration

- As animals and plants grow or repair tissues that have been damaged, **chemical energy** is needed for the synthesis of important chemicals – proteins, nucleic acids and cellulose. This chemical energy comes as a result of respiration.

During respiration, we use up foods and oxygen producing carbon dioxide, water and energy. A similar process occurs during combustion when fuels burn. The foods and fuel react with oxygen, so these reactions are called *oxidation reactions*. The foods and fuels are said to be **oxidised**.

food or fuel + oxygen → carbon dioxide + water + energy

As a result of the similarities between foods and fuels, foods are sometimes called 'body fuels' or 'biological fuels'. Although aerobic cellular respiration and burning are similar processes, there are important differences. These similarities and differences are summarised in Table 3.2.

Table 3.2 Similarities and differences between cellular respiration and burning

Similarities	Differences
• reactants contain carbon and hydrogen • reactants combine with oxygen • carbon dioxide is produced • water is produced • energy is released	• cellular respiration is relatively slow but burning is usually a rapid reaction with flames at a high temperature • cellular respiration involves a large number of separate enzyme-catalysed reactions but burning is a more direct process without enzymes

3.6 Anaerobic respiration

Respiration normally occurs in the presence of air (oxygen) and this is called aerobic respiration. However, respiration can also occur in the absence of oxygen and this is known as **anaerobic respiration**.

Anaerobic respiration in fermentation and bread making

A major application of anaerobic respiration is in **fermentation**. Fermentation is used to make beer and wines by allowing yeast to metabolise sugar.

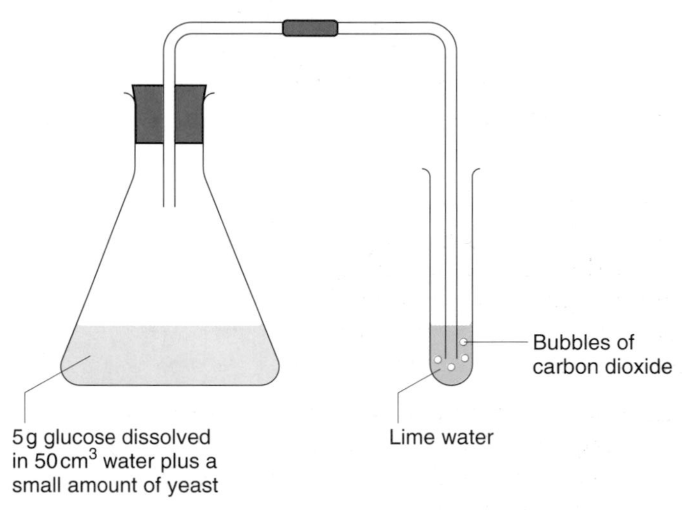

Figure 3.4 In a closed container, yeast respires aerobically to start with. It respires anaerobically when all the oxygen is used up

Yeast contains single-celled organisms which respire aerobically if oxygen is available. When yeast is mixed with a sugar or glucose solution, it soon starts to respire (Figure 3.4). The yeast uses sugar and oxygen dissolved in the water to produce carbon dioxide, water and energy by aerobic respiration.

$$\text{glucose} + \text{oxygen} \xrightarrow{\text{yeast}} \text{carbon dioxide} + \text{water} + \text{energy}$$

$$C_6H_{12}O_6 + 6O_2 \xrightarrow{\text{yeast}} 6CO_2 + 6H_2O + 2900\text{ kJ}$$

When all the oxygen has been used up, the yeast continues to respire anaerobically. Under anaerobic conditions, the yeast produces carbon dioxide and ethanol (alcohol) rather than carbon dioxide and water.

$$\text{glucose} \xrightarrow{\text{yeast}} \text{carbon dioxide} + \text{ethanol} + \text{energy}$$

$$C_6H_{12}O_6 \xrightarrow{\text{yeast}} 2CO_2 + 2C_2H_5OH + 84\text{ kJ}$$

Although yeast can survive during anaerobic respiration, it does not grow and multiply as it would during aerobic metabolism. Anaerobic respiration releases much less energy than aerobic respiration – only 84 kJ compared to 2900 kJ. In anaerobic conditions, most of the energy in the glucose remains 'locked in' the ethanol. This impairs the yeast's growth.

Brewing and wine making rely on the enzymes in yeast for fermentation to take place.

Bread making relies on the carbon dioxide that the yeast produces in order for the dough to rise.

Brewing
In brewing, a sweetish liquid, called wort is first produced by dissolving the sugars in barley in warm water. The wort is then fermented using yeast which converts the sugars to alcohol (ethanol) and carbon dioxide. Different types of beer are obtained by varying the strain of yeast used, the length of the fermentation time and the additives such as hops.

Beer is produced by fermenting wort in large vats. Why is there so much froth on the liquid?

Wine making
In this case, the juice from grapes is fermented using the natural yeasts that occur on grapes. Different varieties of grapes produce wines of differing flavours. The climate, the soil in which the vines grow and differences in the actual processes used make for hundreds of varieties of wine. But, essentially fermentation is the key.

Bread making
Here, yeast is used because of its quick production of carbon dioxide. Bakers add yeast to warm water which is then mixed into flour to make a dough. The yeast respires using sugars in the flour. This produces bubbles of carbon dioxide which are trapped in the dough. This causes the dough to 'rise'. The baker puts this risen dough into a very hot oven. The heat kills the yeast, preventing any further rising, any alcohol evaporates and the bread is baked.

It is the respiration of yeast that makes bread light and palatable

Anaerobic respiration in muscle cells

During a sprint race, muscles have to work very hard for a short period of time. Under these conditions, we cannot breathe fast enough to get sufficient oxygen to our muscles. If insufficient oxygen is reaching our muscles, aerobic respiration slows down and will eventually stop. Then anaerobic respiration takes over. The product of anaerobic respiration in the muscles is **lactic acid**.

$$\text{glucose} \rightarrow \text{lactic acid} + \text{energy}$$
$$C_6H_{12}O_6 \rightarrow 2C_3H_6O_3 + 120\text{kJ}$$

Because the breakdown of glucose is incomplete, much less energy is released during anaerobic respiration so we have to work much harder to get the same energy as we would under aerobic conditions. This is why we feel so fatigued. The process also results in an 'oxygen debt' in our muscles which must be paid back in order to oxidize the lactic acid to carbon dioxide and water.

Any build up of lactic acid in our muscles causes aches and cramps. When we rest, the blood brings oxygen to the muscles which can then respire aerobically. This uses up the lactic acid and the pain is relieved.

1 Bronchitis is an infection of the bronchial tubes which lead to the lungs.

a)

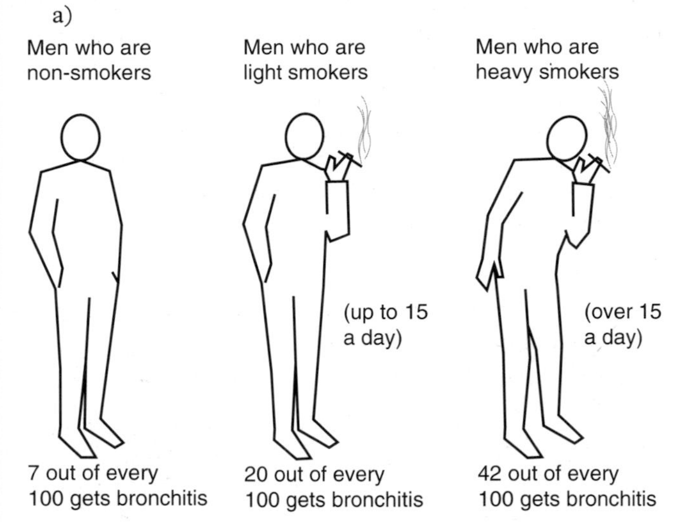

What general pattern is shown by the information in the diagram above?

b) The diagrams below show the effects of cigarette smoke on small hairs called cilia in the bronchial tubes.

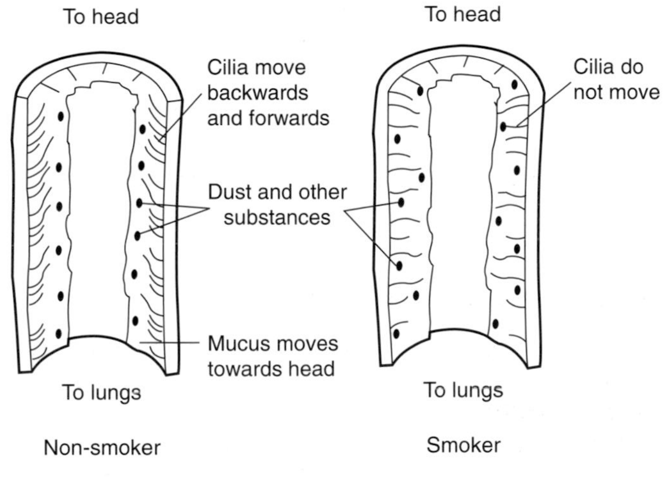

i) What effect does cigarette smoke have on the movement of mucus?
ii) Suggest why smokers tend to cough more than non-smokers. **MEG**

2 Anaerobic respiration is a form of respiration which takes place in the absence of oxygen.

a) Copy and complete the word equation which represents:
i) anaerobic respiration in yeast;

glucose → ________ + carbon dioxide + energy transferred

ii) anaerobic respiration in human muscle cells.

glucose → ________ + energy transferred

b) Explain **one** problem caused when muscle cells respire anaerobically. **SEG**

3 The doctor below is examining the thorax of a patient who smokes 20 cigarettes every day.

Give **three** diseases of the thorax that are more common amongst smokers than non-smokers. **NEAB**

4 The figure below shows an experiment to investigate whether germinating peas produce carbon dioxide.

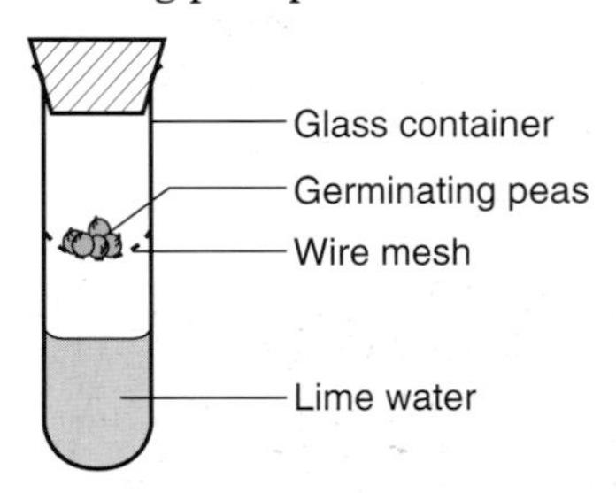

a) What change will you see in the glass container as well as the germination of the peas?

b) What control experiment should also be carried out?

c) State one other change that would occur in the glass container and design an experiment to measure this change.

d) Will germination of the peas in the tube continue as normal? Explain your answer.

5 a) Why is it better to:
i) breathe through your nose rather than your mouth?
ii) breathe as deeply as possible,
iii) blow you nose when necessary?

b) i) State two essential features of the surface of alveoli which makes respiration as efficient as it can be.
ii) Why must the alveoli have these features?

6 a) Mr Jones and Mr White both make their own wine. Mr Jones leaves his to ferment on a sunny window ledge. Mr White puts his in a dark shed.
i) Whose wine would you expect to ferment first?
ii) Explain your answer.

b) Mr Jones makes jam. When he opened one of the jars of jam several months later, the jam looked frothy and smelt of alcohol.
i) What do you think had happened?
ii) Why had it happened?
iii) What must Mr Jones do to prevent this happening when he next makes jam?

Blood and circulation

The human circulatory system
The structure of the heart
The composition and function of blood
The effects of exercise, diet, smoking and stress on the circulatory system

4.1 The human circulatory system

Organisms must be able to take in nutrients (food) and oxygen and get rid of waste products if they are to stay alive. In animals, this taking in of nutrients and oxygen and getting rid of waste products is helped by the circulatory system.

The structure of the human circulatory system is shown in Figure 4.1. The circulatory system transports substances around the body in the blood. The blood carries oxygen and food to all parts of the body and carries away waste products. The main organs in our bodies through which the blood **circulates** are shown in Figure 4.1.

- The **heart** pumps blood around the body.
- In the **lungs**, the blood picks up oxygen and gets rid of carbon dioxide.
- The blood absorbs digested food (nutrients) through the walls of the **small intestine**.
- Digested food is processed and stored in the **liver** which is well supplied with blood vessels.
- Blood flows to the **kidneys** where waste products are removed from it.
- The blood supplies **the rest of the body** (brain, muscles, skin etc.) with food and oxygen. The blood carries away the waste products.

Notice in Figure 4.1 that there are two blood circuits from the heart. One circuit goes to the lungs and back to the heart. This is called the **pulmonary circuit**. The other circuit goes to the other organs and the rest of the body and then back to the heart. This is called the **systemic (body) circuit**. Because there are two circuits from the heart, it is described as a **double circulatory system**. The blood passes through the heart twice for each circulation through the body.

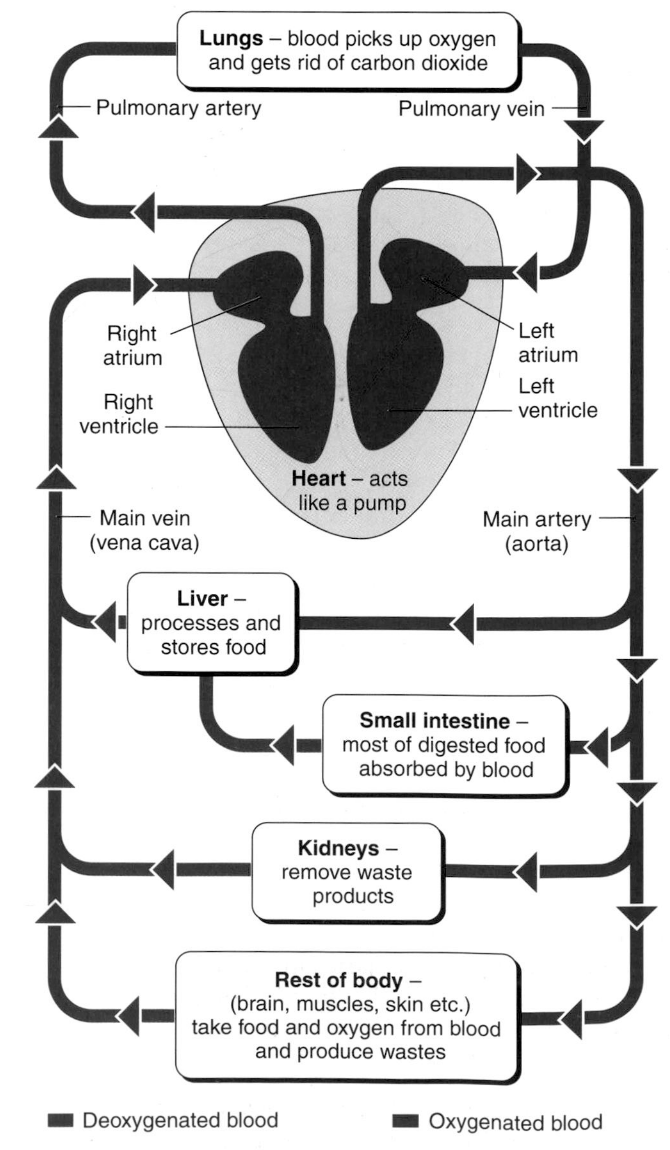

Figure 4.1 The human circulatory system

The heart pumps blood round the body through a system of tubes called **blood vessels**. There are three different types of blood vessel. Each of these has a structure related to its function.

1 **Arteries**, like the pulmonary artery (going to the lungs) and the aorta (the main artery serving the rest of the body) carry blood *away* from the heart at high pressure. They have thick muscular walls with elastic fibres to withstand the pumping pressure from the heart.

2 **Veins**, like the pulmonary vein (from the lungs) and the vena cava (the main vein) carry blood *back* to the heart at lower pressure. Veins have thinner, less muscular walls than arteries. They tend to be closer to the surface of the body than arteries and they contain **valves** to prevent the backward flow of blood.

3 **Capillaries** are very narrow tubes with walls that are only one cell thick. Capillaries divide from the arteries and then rejoin to form veins. There are thousands of capillaries running through every organ in your body and every cell is close to a capillary.

As the blood flows through the capillaries of most organs, dissolved foods and oxygen diffuse through the thin capillary walls to the nearby cells. At the same time, waste materials diffuse out of the cells and into the capillaries.

As you might expect, the reverse of this process can also occur in the lungs. Oxygen diffuses into the capillaries surrounding the alveoli. So blood which is leaving the lungs along the pulmonary vein is rich in oxygen (Figure 4.1). It is described as **oxygenated blood**. This oxygenated blood enters the heart to be pumped via the main artery (the aorta) to the small intestine, the liver, the kidneys and to the rest of the body. Its oxygen is used up by the various organ systems. The blood becomes **deoxygenated** and returns to the heart via the main vein (the vena cava).

4.2 The structure of the heart

In order to provide for the double circulatory system, the heart acts as two synchronized pumps. The pump on the right of your heart drives blood via the pulmonary artery to capillaries in the lungs where oxygen is absorbed into the blood. The pump on the left of your heart drives oxygenated blood via the aorta to the rest of the body.

All body processes need this oxygenated blood. Oxygen is used up and the deoxygenated blood is returned to the heart and the cycle begins again.

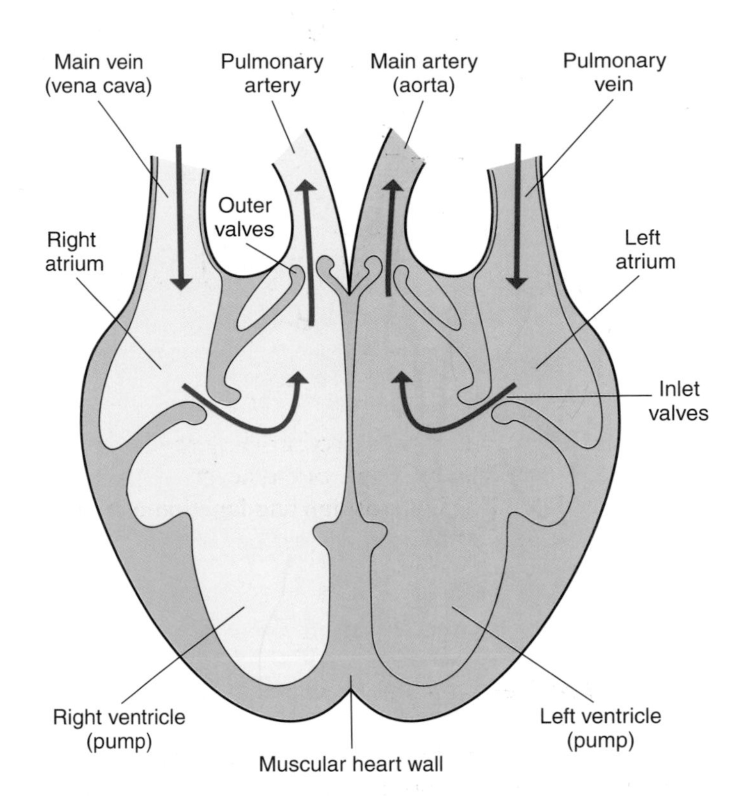

Figure 4.2 The structure of the heart

Notice that each pump in the heart has two chambers – an entry chamber called an **atrium** (plural, atria) and an exit, pumping chamber called a **ventricle**.

Blood returning in the veins to the heart collects in the atria until the pressure causes the one-way inlet valves between each atrium and ventricle to open. Blood moves from the atria to the ventricles and the one-way inlet valves prevent the blood from flowing the wrong way. Muscles in the walls of the ventricles now contract very strongly. The pressure increases and blood is forced into the arteries. Every time the ventricles contract, the pumping effect produces a regular heartbeat with a dull throb in the chest and a surge of blood into your main arteries. You can feel this as a **pulse** in the artery at your wrist. One-way outlet valves at the base of the aorta and the pulmonary artery prevent blood from flowing back.

In order to maintain this lifelong pumping action, the walls of the heart are made of very strong muscle fibres. In particular, the walls of the ventricles need to be powerful and thick in order to pump blood out of the heart to the rest of the body.

4.3 The composition and function of blood

Blood consists of four constituents – plasma, red blood cells, white blood cells and platelets. The functions of these constituents are summarised in Figure 4.3 on the following page.

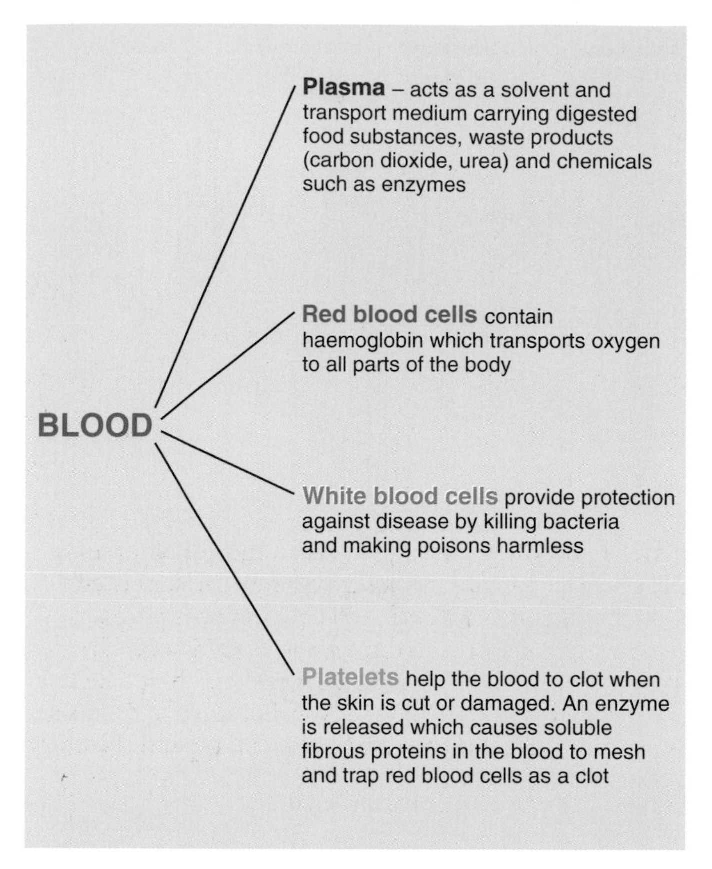

Figure 4.3 The constituents of blood and their main functions

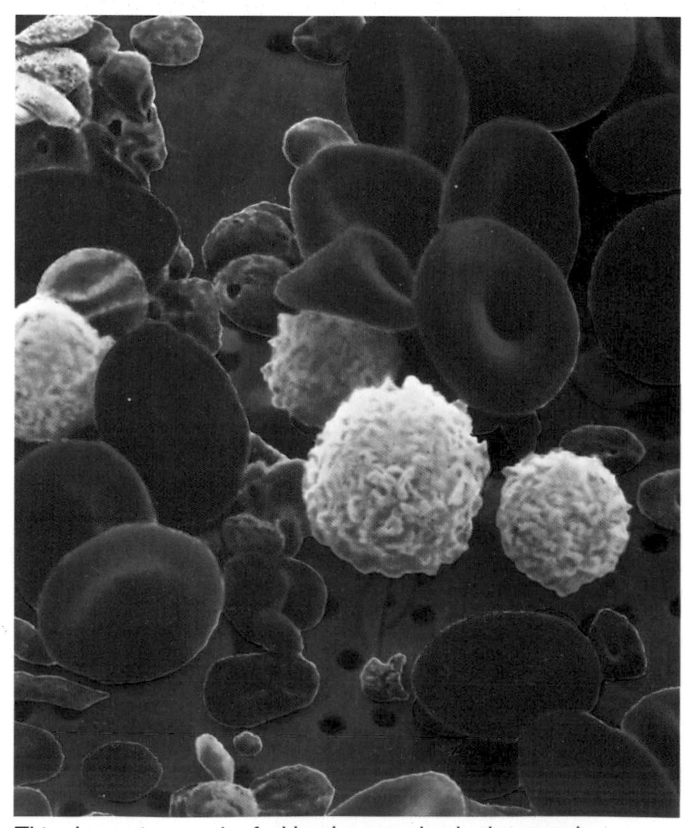

This photomicrograph of a blood smear clearly shows red blood cells, white blood cells and the smaller platelets

Plasma

Plasma is a straw-coloured liquid. It is about 90% water with 10% dissolved substances. The plasma transports:

- **digested foods** (nutrients), such as glucose, amino acids and minerals from the small intestine to all the body cells,
- **carbon dioxide** from all the body cells to the lungs where it is excreted,
- **urea** from the liver to the kidneys where it is excreted as urine,
- **hormones** from the endocrine glands to target organs.

Red blood cells

There are millions of red blood cells in every cubic centimetre of blood. Their main function is to transport oxygen to cells for respiration.

Red blood cells contain a red pigment called **haemoglobin**. This combines with oxygen in the lungs forming **oxyhaemoglobin**. The combination of oxygen with haemoglobin is relatively weak. As soon as the oxyhaemoglobin reaches the tissues where oxygen is required, oxygen is released.

$$\text{oxygen} + \text{haemoglobin} \underset{\text{requirement for oxygen}}{\overset{\text{high oxygen concentration}}{\rightleftarrows}} \text{oxyhaemoglobin}$$

The structure of red blood cells (Figure 4.4) is specially adapted for their function as oxygen transporters.

- They have no nucleus, thus enabling the whole cell to be filled with haemoglobin.
- The cell membrane is very thin and therefore readily permeable to oxygen.
- They are biconcave in shape (like pressed-in discs). This increases the surface area through which oxygen can diffuse.
- They are relatively small and flexible which allows them to pass through even the narrowest capillaries.

White blood cells

There are two kinds of white blood cells – **phagocytes** and **lymphocytes** (Figure 4.4). Both of these protect us against disease, but in different ways.

Phagocytes remove harmful bacteria from the body by engulfing them and then secreting an enzyme which kills them. When there is an infection, such as a boil on the skin, phagocytes, when loaded with dead bacteria, die and yellow 'pus' forms.

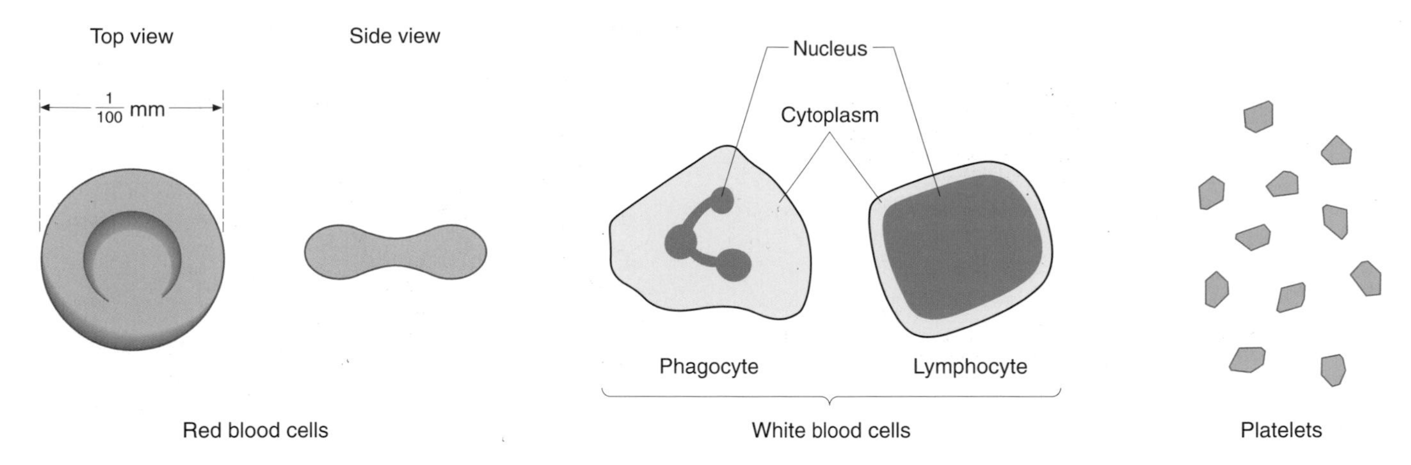

Figure 4.4 Blood cells and platelets

Lymphocytes become active when 'foreign materials' called **pathogens** invade the body. Bacteria, viruses and chemical poisons are all pathogens and the lymphocytes react by producing chemicals called **antibodies**. The antibodies dissolve in blood plasma, killing bacteria and viruses and neutralizing poisonous chemicals.

Platelets

Platelets are small fragments of cells without a nucleus. They provide protection by helping the blood to clot when the skin is cut or damaged. The clot prevents further loss of blood and acts as a barrier against infection.

Paul Ince, England's captain, was cut badly on the head when his team drew with Italy in the World Cup qualifier in October 1997. Platelets helped his blood to clot

Platelets produce an enzyme called thrombokinase on exposure to the air, for example when the skin is cut. This enzyme promotes the conversion of fibrinogen (soluble fibrous protein) in the blood to **fibrin**. The fibrin forms a mesh of fibres in which red blood cells are trapped to form a clot. As the clot dries, it forms a **scab** which prevents further loss of blood and the entry of harmful bacteria.

Functions of the blood

The blood has three main functions – **transport**, **protection** and the **regulation of body temperature**.

Transport

- The plasma transports nutrients, waste products and chemicals.
- The red blood cells transport oxygen.

Protection

- Phagocytes protect against harmful bacteria.
- Lymphocytes protect against all kinds of pathogens.
- Platelets, which help blood clots to form when the skin is damaged, ensure a barrier against infection.

Regulation of body temperature

The blood helps to keep our body temperature constant by allowing us to retain heat or lose it.

In cold weather, blood vessels and capillaries close to the surface of our skin contract. Less blood flows through them and therefore less heat is lost to the surrounding air. In warm weather or during energetic pursuits, the surface blood vessels and capillaries expand. This allows more blood to flow through them and heat can be lost from the body.

4.4 The effects of exercise, diet, smoking and stress on the circulatory system

Your heart beats about 70 times every minute, about 37 million times in one year. To do this, the heart muscles must have a good supply of nutrients and oxygen. These are supplied through the **coronary arteries** to many capillaries spread all over the heart wall.

Like any other muscular tissue, regular exercise improves the fibres so that they are stronger, larger and more efficient. The heart then may pump a larger amount of blood with each beat and the person is said to be very fit. If, however, the heart muscles become weak due to lack of exercise, poor diet or disease, many health problems may develop.

There are three main factors which contribute to heart and circulatory disorders:

- **high blood pressure**,
- **poor diet** and
- **smoking**.

If someone has high blood pressure, their heart must work harder to pump blood around the body. The causes of high blood pressure are not fully understood. It is, however, quite common in middle aged and elderly people and may be worsened by smoking, lack of exercise and too much stress.

As we get older, the walls of the arteries become thicker and harder. A layer of fat forms on the inside causing narrowing. So the flow of blood begins to be more difficult and high blood pressure results. It is advisable to eat a diet low in animal fat as a preventative measure to reduce this narrowing.

If we eat an excessive amount of carbohydrates and fatty foods and become overweight, the heart is put under strain. In addition, drinking too much alcohol is detrimental to the circulatory system.

Tobacco smoke contains a number of poisonous substances including nicotine and carbon monoxide. Carbon monoxide affects the blood pressure because it combines with haemoglobin in red blood cells more readily than oxygen. This means that the blood cannot carry as much oxygen to our body tissues. In order to improve the supply of oxygen, the heart must pump the blood faster, putting strain on itself and increasing the blood pressure.

So, we must take regular physical exercise, eat a balanced diet and refrain from smoking in order to maintain a healthy heart and circulatory system.

QUESTIONS

1 The diagram shows a section through the human heart.

a) Copy the diagram and label:
   i) a ventricle,
   ii) an atrium,
   iii) the artery which takes blood to the lungs.

b) The heart contains several valves.
   i) On your diagram, label **one** of these valves with the letter X.
   ii) Describe the function of the valve you have labelled. **NEAB**

2 Diseases can be caused when infective microbes such as bacteria and viruses enter the body.

a) Explain why infective microbes may make us feel ill.

b) Give **three** ways in which white blood cells help to defend us from infective microbes. **NEAB**

3 The diagram below represents the heart, some of the major blood vessels of the human circulatory system, and the organs they supply.

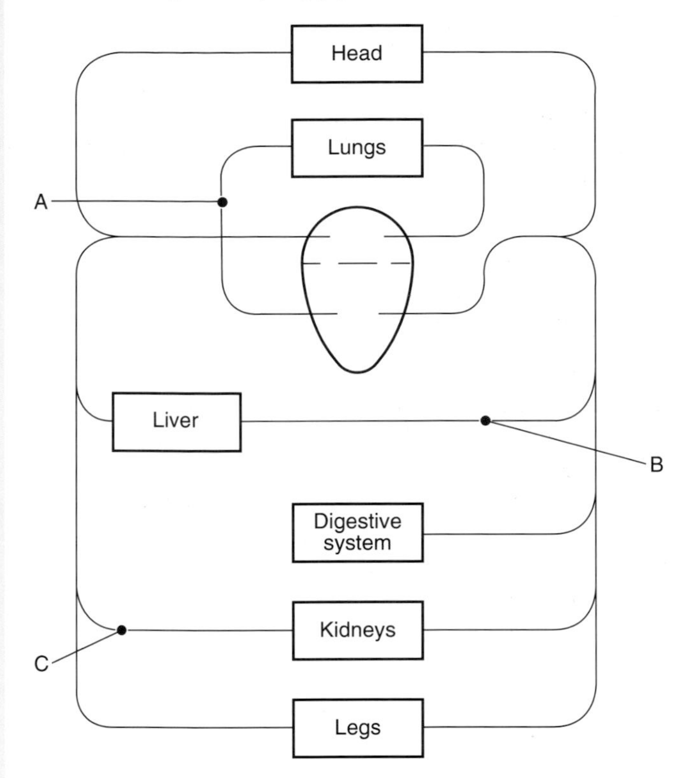

a) Name the blood vessels labelled, **A**, **B** and **C**.

b) Which direction would blood in vessel **A** be flowing, towards, or away from the lungs?

c) Where does a line need to go to complete the circulation of blood for the **digestive system**? In which direction would the blood flow? **NICCEA**

4 a) Describe how white blood cells defend the body against disease bacteria.

b) Study the graph below which shows the number of bacteria in the body of a child suffering from whooping cough.

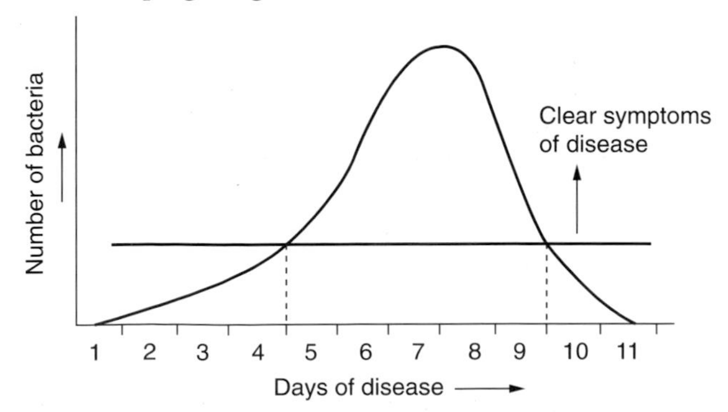

i) On what day did the child start to show symptoms?
ii) For how many days was the child ill?
iii) On what day did the biggest increase in bacteria occur? **WJEC**

5 a) What are the three main functions of the blood?

b) What part do blood platelets play in the clotting of blood?

c) In some people, blood clots without being exposed to the air. Why is this very dangerous?

d) The average speed of the blood in arteries is 45 cm/s, but the average speed in capillaries is only 1 mm/s.
i) What causes this difference in speed?
ii) What is the advantage of slow blood flow through capillaries?

6 The figure below shows a very simple diagram of the human blood system and the routes which blood takes during its circulation.

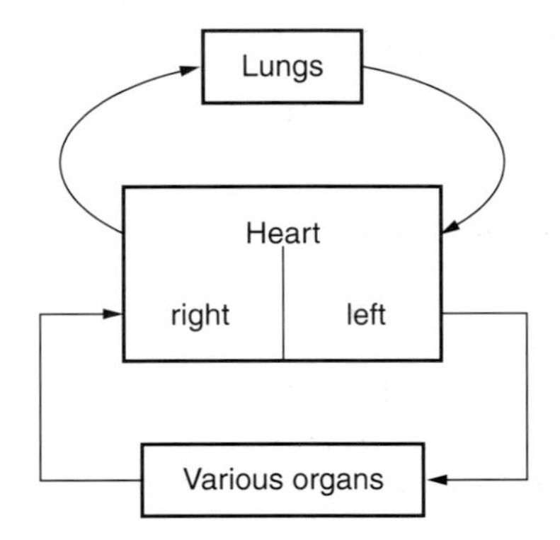

a) Describe the route which the blood takes from the right ventricle until it returns to the heart via the main vein (vena cava) into the right atrium. In your account use the terms aorta (main artery), pulmonary artery, vena cava (main vein), pulmonary vein and left and right atrium and ventricles.

b) Why is the blood system an example of double circulation?

c) Suppose you live to be eighty. Estimate the number of times your heart will beat in your lifetime.

7 a) How do red blood cells differ from white blood cells in their structure?

b) How are red blood cells specially adapted for transporting oxygen?

c) Experiments showed that people living at sea level had about 5 million red blood cells per mm^3 of blood, whereas those living in a nearby mountainous region had 7 million red blood cells per mm^3. Why do you think that there is such a difference?

CHAPTER

Photosynthesis

5.1 Reactants and products in photosynthesis

Plants do not feed like animals. One of the most important differences between plants and animals is that plants make their own food. Animals cannot do this. Animals must obtain their food by eating plants or by eating other animals. So you can see that humans and other animals rely on plants for their food. The process by which plants make their own food is called **photosynthesis**.

Conditions needed for photosynthesis

Light for photosynthesis

Most plants cannot grow well in dark and shady places where there is little light. Farmers and gardeners know how important light is for plants. Crops need plenty of sunshine because plants need light to photosynthesize.

We can show this using two similar geranium plants that have been *de-starched*. De-starching involves keeping the plant in the dark for several days. This uses up all the starch stored in the leaves.

When the experiment starts, one plant is put back in the dark to act as a *control* and the other is placed in the light. After a week or so, a leaf is taken from each plant and tested for starch (Figure 5.1).

The leaf which has been in the light turns dark blue with iodine indicating that starch is now present. This shows that **starch is a product of photosynthesis**. The presence of starch in plant leaves can be used as evidence for photosynthesis. The leaf which was kept in the dark is only stained brown from the iodine solution. There is no starch in this leaf showing that:

Light is necessary for photosynthesis.

Figure 5.1 Testing for starch using iodine

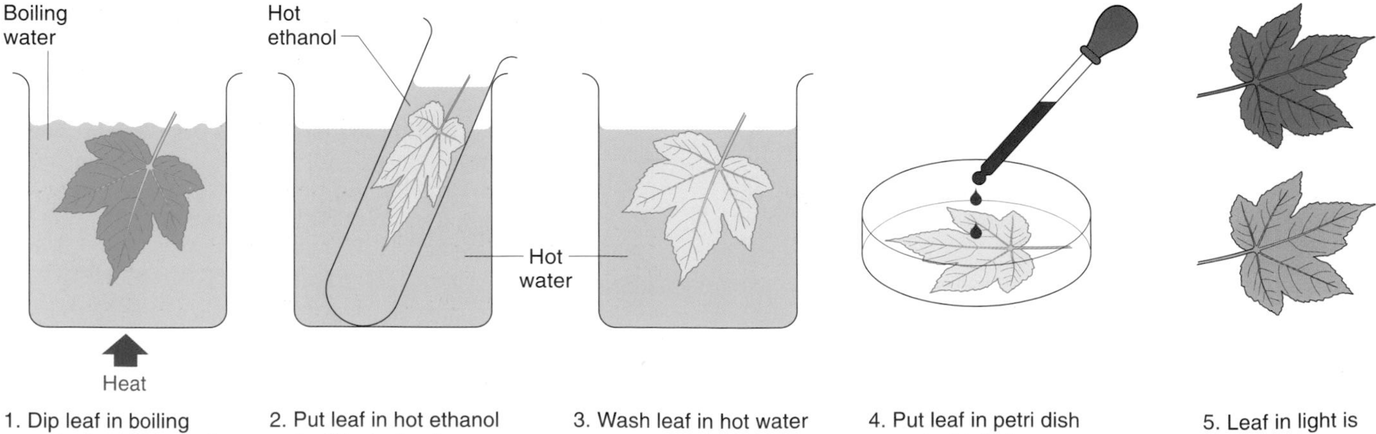

Where in the plant does photosynthesis happen?
Figure 5.2 shows what happens when a variegated leaf (one with white and green patches) is tested for starch. Only those parts of the leaf which were green turn dark blue with iodine. This shows that starch is present in the green area of the leaf, but not in the white area. This suggests that:

The **green pigment in leaves** is necessary for photosynthesis. This green pigment is called **chlorophyll**.

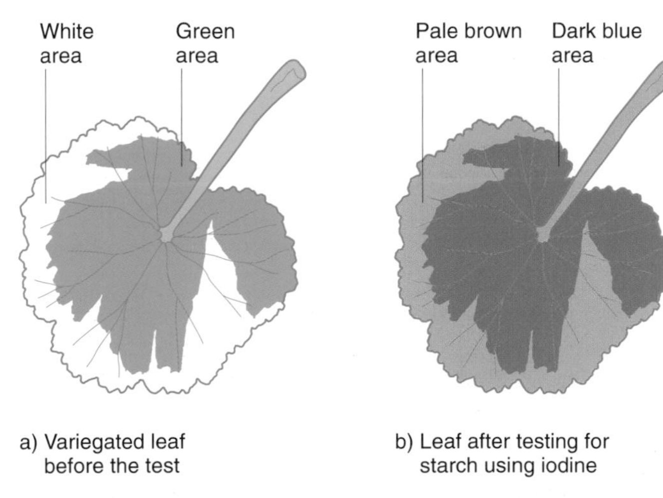

Figure 5.2 Testing a variegated leaf for starch using iodine

The products of photosynthesis

We know from the experiment shown in Figure 5.1 that **starch** is one of the products of photosynthesis. The other product of photosynthesis is **oxygen**. This can be shown very simply using the apparatus in Figure 5.3.

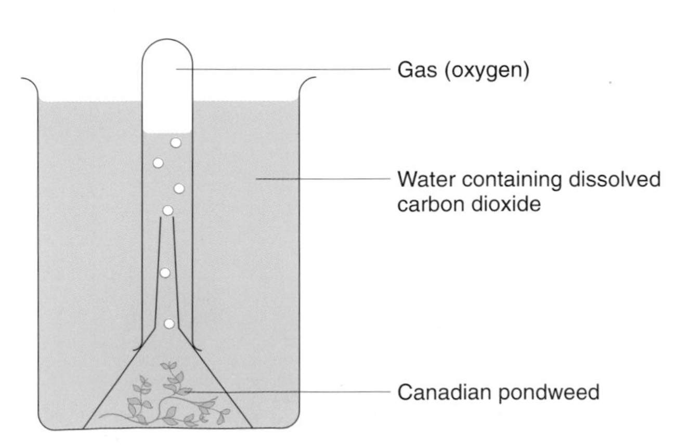

Figure 5.3 A test to show that oxygen is another product of photosynthesis

When the apparatus is kept in the light, bubbles of gas slowly collect in the test tube. After a few days, there is sufficient gas in the test tube to show that it is oxygen.

The reactants of photosynthesis

We know already from the experiments earlier in this section, summarised in Figures 5.1 and 5.2 that both **light** and **chlorophyll** are needed for the production of **starch** and **oxygen** during photosynthesis, but what are the reactants?

Starch is a carbohydrate. Therefore it contains carbon, hydrogen and oxygen. The most likely substances available to plants from which they might produce starch are carbon dioxide and water.

There are no simple experiments to show that plants produce starch from carbon dioxide and water. The reactions of water and carbon dioxide in starch production have been investigated by scientists using radioactive isotopes (carbon-14) in carbon dioxide and the heavy isotope of oxygen (oxygen-18) in water.

If plants are fed with water containing the heavy isotope of oxygen, oxygen-18 soon appears in the starch of the plants' leaves.

Similar experiments can be carried out using radioactive carbon dioxide containing carbon-14. If plants are grown in an atmosphere containing radioactive CO_2, the radioactive carbon-14 soon appears in the starch in their leaves. These experiments suggest that:

Carbon dioxide and **water** are the reactants for photosynthesis

5.2 A summary of photosynthesis

The experiments described in this chapter tell us that:

- carbon dioxide and water are the reactants for photosynthesis,
- starch and oxygen are the products of photosynthesis,
- light and chlorophyll are needed for photosynthesis.

Experiments show that the first product of photosynthesis is *not* starch, but glucose ($C_6H_{12}O_6$). The glucose molecules then link together forming starch.

Although photosynthesis is a complicated process, it can be summarised in a word equation as:

$$\text{carbon dioxide} + \text{water} \xrightarrow[\text{chlorophyll}]{\text{light}} \text{glucose} + \text{oxygen}$$

Using chemical symbols this can be written as:

$$6CO_2 + 6H_2O \xrightarrow[\text{chlorophyll}]{\text{light}} C_6H_{12}O_6 + 6O_2$$

Plants obtain their carbon dioxide for photosynthesis from two sources:

- from the carbon dioxide produced in the plant cells during respiration and
- from the carbon dioxide in the air and dissolved in the water around their roots.

The water for photosynthesis is absorbed through the roots of plants and then transported to the leaves (section 6.2).

5.3 Light and chlorophyll in photosynthesis

During photosynthesis, chlorophyll in plants absorbs light energy. The plant then uses this energy to turn carbon dioxide and water into glucose (sugar). So, the overall result of photosynthesis is to turn light energy from the Sun into chemical energy in the molecules of glucose.

It is easy to extract chlorophyll from leaves or grass by dissolving the chlorophyll in ethanol (Figure 5.4).

Figure 5.4 Extracting chlorophyll from grass

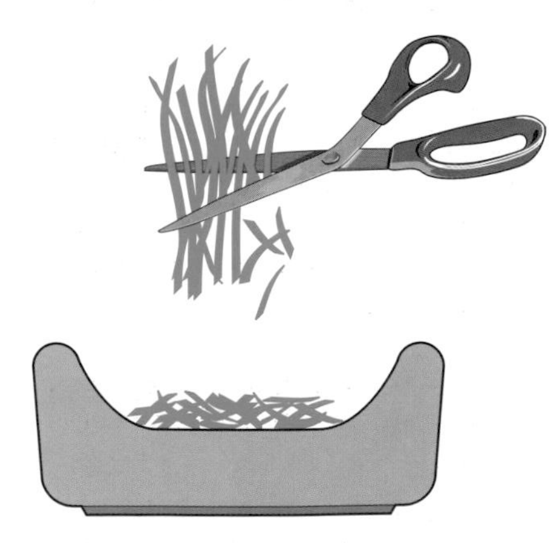

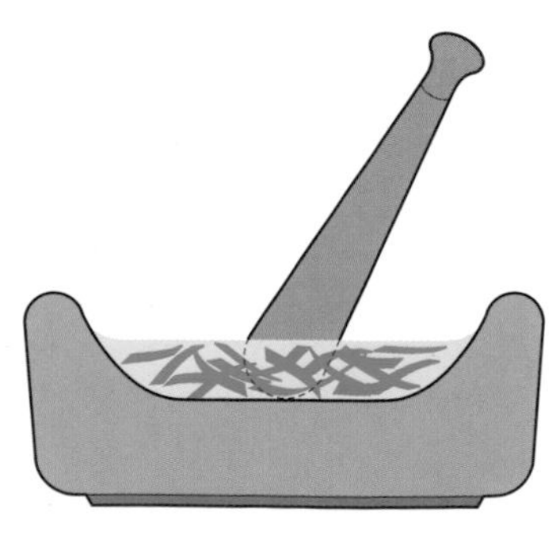

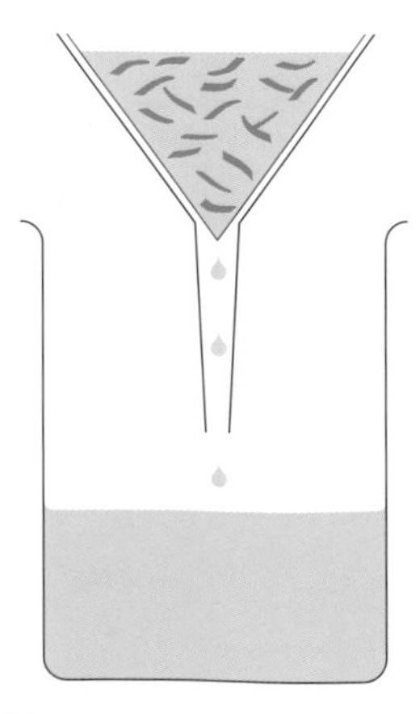

If light of different colours is shone on this solution of chlorophyll, blue and red light are mostly absorbed, whilst green light is mostly transmitted or reflected. This shows that plants do not absorb green light, so do not use the energy from it. In fact, plants appear green because they are reflecting green light.

Accurate experiments show that if a plant does not receive light in the blue or red regions of the spectrum, it cannot photosynthesize very well and its growth is slow.

Figure 5.5 shows the effect of different colours of light on the rate of photosynthesis. Notice that the rate of photosynthesis is highest in violet, indigo and blue light, effective in orange light but very poor in green and yellow light.

Figure 5.5 The effect of different colours of light on the rate of photosynthesis

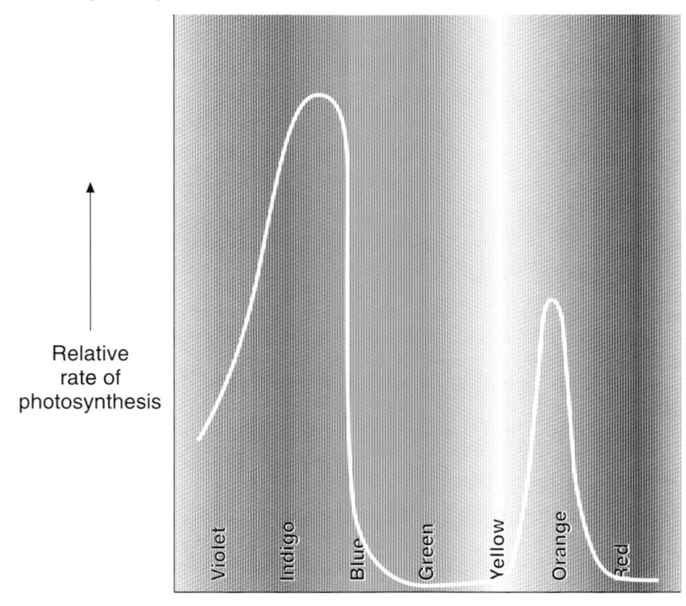

5.4 Leaves and photosynthesis

All plant cells which contain chlorophyll can photosynthesize. These cells are mainly in the leaves.

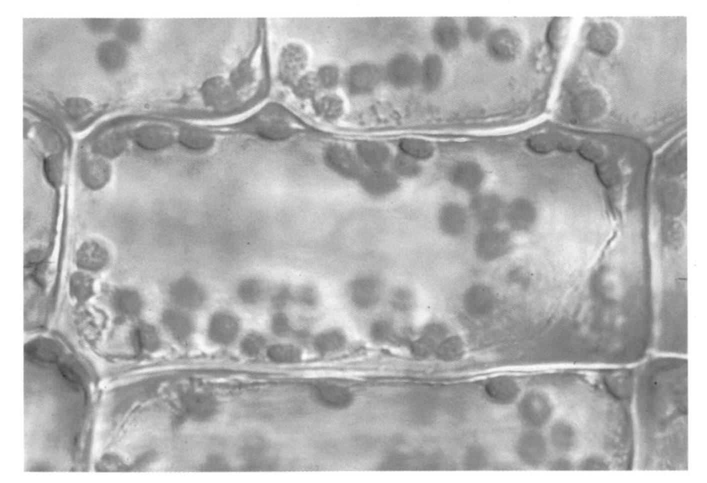

A magnified photo of leaf cells. The dark green blobs are chloroplasts containing chlorophyll

Leaves are usually flat and thin so they are well adapted for photosynthesis. Being flat gives them a large surface area. This helps them to absorb carbon dioxide from the air and reduces the distance that the carbon dioxide has to travel to reach the cells once it has been absorbed. The large surface area also helps the absorption of light from the Sun.

Look closely at a leaf (Figure 5.6). Notice the stalk or **petiole** which joins it to a stem or branch. The petiole divides into **veins**. These veins act as a kind of skeleton which supports the leaf, preventing it from drooping. The water required for photosynthesis travels up the plant from the roots and into the leaves via these veins.

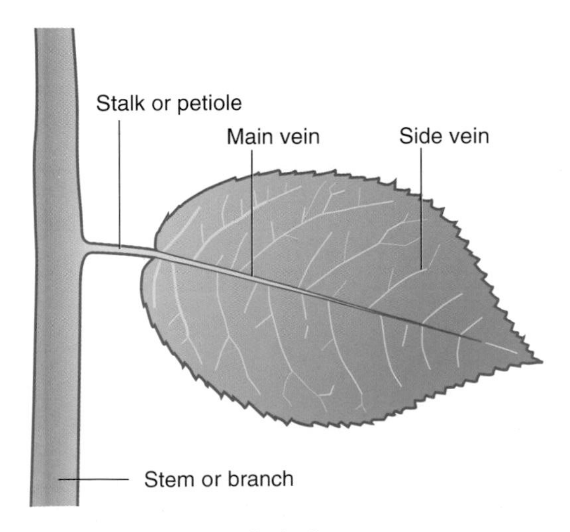

Figure 5.6 The structure of a leaf

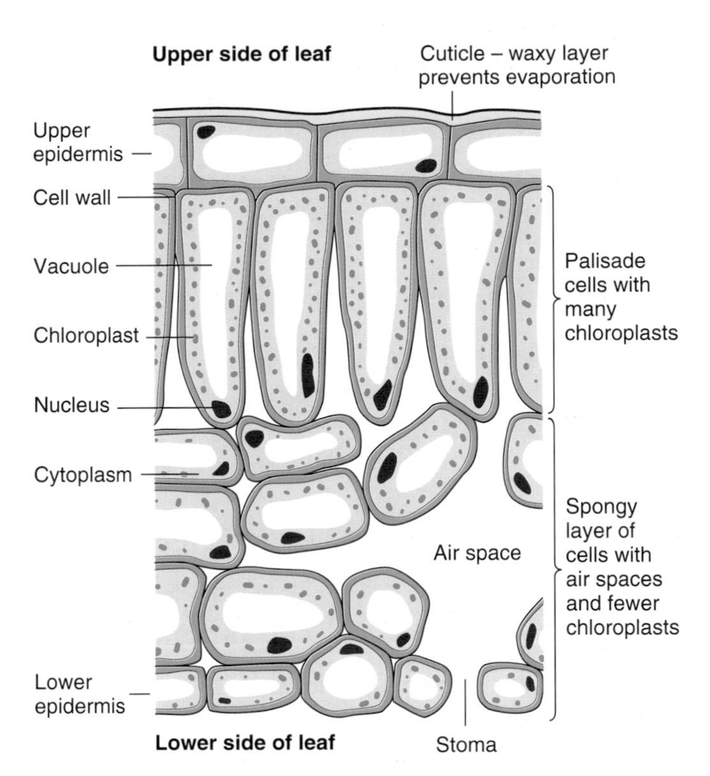

Figure 5.7 Cells in the cross section of a leaf

The carbon dioxide needed for photosynthesis enters through tiny holes called **stomata** (singular, stoma). Stomata are mainly on the underside of leaves (Figure 5.7).

Very often, the top side of a leaf has a smooth, shiny appearance. This is due to a thin waxy layer called the **cuticle** which reduces evaporation (water loss). Below the cuticle and also on the underside of the leaf, there is a single layer of tightly fitting cells. This is called the **epidermis** ('upper' on the top and 'lower' on the underside). Leaves are usually a deeper green on their upper surface. This is because the cells near the upper surface contain more chlorophyll than those below. As Figure 5.7 shows, there are two distinct areas of cells between the layers of epidermis.

- There are **palisade cells** near the upper surface, containing lots of chloroplasts filled with dark green chlorophyll. These palisade cells, as their name suggests, are arranged neatly like the vertical stakes in a fence. The large number of chloroplasts near the leaf surface allow maximum absorption of sunlight.
- Near the lower surface are more **rounded, spongy cells** containing fewer chloroplasts. These cells are arranged irregularly with air spaces between them. The air spaces allow a larger surface area of the spongy cells to be exposed for gas exchange during photosynthesis. Carbon dioxide diffuses into the cells whilst oxygen diffuses outwards.

In addition to long palisade cells and rounded spongy cells, leaves also contain a small number of phloem cells and xylem cells. The roles of these cells are covered in section 6.4.

5.5 Factors affecting photosynthesis

The word equation for photosynthesis is:

$$\text{carbon dioxide} + \text{water} \xrightarrow[\text{chlorophyll}]{\text{light}} \text{glucose} + \text{oxygen}$$

So, the rate of photosynthesis can be measured by recording either;

1 the rate at which carbon dioxide or water are used up, or
2 the rate at which glucose or oxygen are produced.

The rate of photosynthesis is usually measured by finding the rate of uptake of carbon dioxide and Figure 5.8 compares the rate of photosynthesis on a sunny, hot day with that on a cold, sunny day.

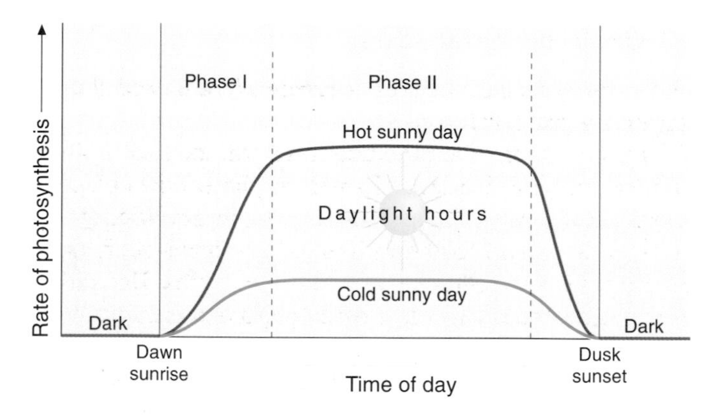

Figure 5.8 Comparing the rates of photosynthesis on a hot, sunny day and a cold, sunny day

Plants require light in order to photosynthesize. If there is no light at all, then the process cannot occur. Thus, in Figure 5.8, the rate of photosynthesis is zero before sunrise and after sunset. Look closely at the curve showing the rate of photosynthesis on a hot, sunny day. At dawn, the light appears and photosynthesis starts. As the light intensity increases during the morning, the rate of photosynthesis also increases. This is labelled as phase I on the graph.

After a while, probably about mid-morning, the rate of photosynthesis reaches a maximum value. Even though the light intensity continues to increase until mid-day, the rate of photosynthesis does not change. The rate of photosynthesis stays constant (phase II) until late in the afternoon when it begins to fall.

During phase I, photosynthesis increases as the light intensity increases. This means that the rate of photosynthesis is being determined by or limited by the level of light intensity. In this case, light intensity is described as the **limiting factor** for the rate of photosynthesis.

In phase II, light intensity cannot be limiting the rate of photosynthesis. Even though the light intensity has increased and then has started to decrease, the rate of photosynthesis stays constant at a plateau value. Some other factor must now be limiting the rate – possibly the temperature or the concentration of carbon dioxide.

Now, compare the rate of photosynthesis on a hot, sunny day in Figure 5.8 with that on a cold, sunny day. The reduced maximum rate of photosynthesis during the middle of the cold, sunny day suggests that the limiting factor in this case is temperature.

The major limiting factors for the rate of photosynthesis are:

- low temperature,
- shortage of carbon dioxide,
- poor light intensity.

These three factors interact and, in practise, any one of them may be the limiting factor for photosynthesis.

Commercial growers are very adept at controlling the conditions in which flowers, fruit and vegetables grow. They manage the environment in order to get the maximum yield from their crops by growing them at the maximum photosynthetic rate. In the greenhouses, they avoid the natural limiting factors of poor and variable light intensity, cold temperatures and poor carbon dioxide concentration in the air. Extra lighting is installed, there is a warm environment and the air may be enriched with extra carbon dioxide.

Plants being grown in a commercial greenhouse

Roughly speaking, the rate of photosynthesis doubles if the temperature rises by 10°C. However, although plants thrive in a sheltered garden or a greenhouse, there is a limit to the temperature at which they can survive. The rate of photosynthesis increases until the temperature reaches about 40°C. Above 40°C, photosynthesis slows down and then stops. This is because important chemicals in plants called **enzymes** are destroyed above 40°C. Enzymes are proteins which act as catalysts (section 18.8). They control the reactions in all living things.

5.6 How do plants use the products of photosynthesis?

The first products of photosynthesis are glucose and oxygen. Figure 5.9 summarises the four main uses to which plants put this glucose.

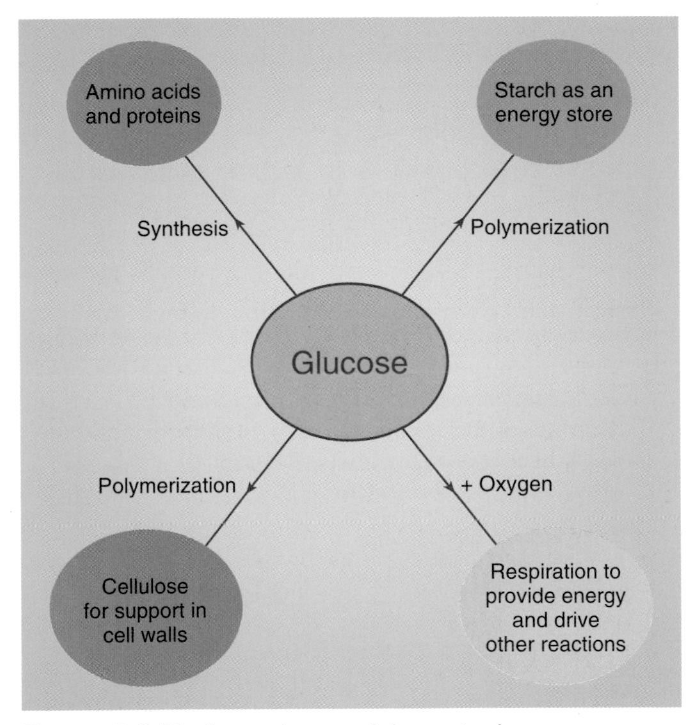

Figure 5.9 The four main uses of glucose in plants

i) Glucose and oxygen can be used as raw materials for respiration

Respiration is a **redox** process in which glucose reacts with oxygen to form carbon dioxide and water. The equation for the reaction is

$$\underset{\text{glucose}}{C_6H_{12}O_6} + \underset{\text{oxygen}}{6O_2} \longrightarrow \underset{\text{carbon dioxide}}{6CO_2} + \underset{\text{water}}{6H_2O} + \text{energy}$$

The **energy** produced in respiration is used to drive the other reactions which occur in plants. These reactions include the conversions of glucose (sugars) to starch, to cellulose and to proteins as shown in Figure 5.9.

Notice from the last equation that the overall process of respiration is the *reverse* of photosynthesis.

ii) Glucose molecules can be polymerized by joining with one another (section 20.8) to produce starch

The starch is then **stored** for future use. This is why the leaves of plants usually contain starch. Root vegetables also store starch in their roots. Potatoes store it in large root tubers and carrots store it as a swollen tap root. Extra starch and carbohydrates produced by onion plants are stored as bulbs at the bottom of their long stems.

Starch (carbohydrate) is stored until the plant requires it for other processes (e.g. for respiration, growth and seed production). It is an advantage for plants to store carbohydrates as starch rather than glucose. This is because starch is insoluble and therefore does not cause large amounts of water to accumulate in storage cells as a result of osmosis. In many cases, the starch stored in plants provides a very useful source of food for animals, including ourselves.

iii) Glucose can be converted to cellulose which is needed to support the plant as it grows

This process, like the production of starch, also involves polymerization. Growth in plants takes place mainly at the tips of roots and shoots, in developing buds, flowers and fruit and in storage organs such as potato tubers, swollen roots and onion bulbs. Because of this, glucose must reach these regions to supply both the materials and energy for growth. Some of the glucose is converted to cellulose and used to build cell walls which help support the plant. The glucose is transported up and down stems and into and out of leaves and roots via the phloem tissue (section 6.4).

iv) Glucose can be used together with nitrogen from nitrates in the soil to synthesize amino acids and proteins

Photosynthesis is the single most important process in plants. New and larger molecules can be synthesized from carbon dioxide and water. The first product is glucose and this is the starting material for all other carbohydrates, fats and proteins. Glucose is the starting material for almost all the other compounds containing carbon in plants. It is right at the centre of plant metabolism. Chemical pathways exist in plants which allow them to convert glucose to substances as simple as carbon dioxide and molecules as complex as proteins and nucleic acids.

1 a) Copy and complete the equation for photosynthesis:

carbon dioxide + ________ + light energy
→ sugar + ________

b) A plant with variegated (two coloured) leaves was left in sunlight for several hours. Pieces of one of its leaves were then removed and tested for sugar. The diagram below shows the results of the test.

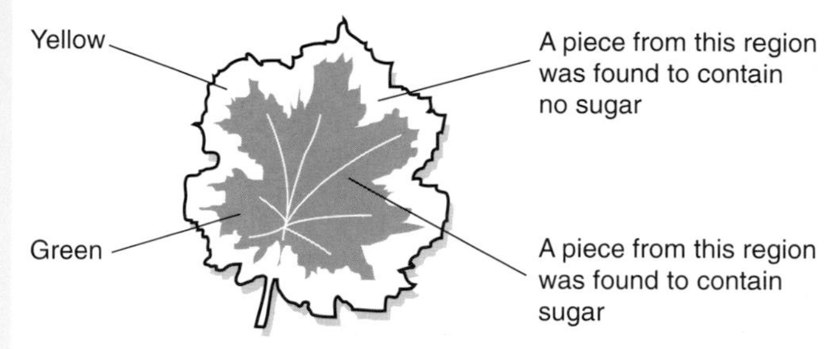

Explain, as fully as you can why the yellow regions of the leaf had not produced sugar.

a) Explain why plants need i) nitrate and ii) potassium for healthy growth. **NEAB**

2 The table below shows the results of an experiment to study the effects of carbon dioxide concentration on photosynthesis in tomato plants.

Carbon dioxide concentration (percentage)	Rate of photosynthesis (arbitrary units)
0.00	0
0.02	20
0.04	29
0.06	35
0.08	39
0.10	43
0.12	46
0.14	46
0.16	46

a) Use the results in the table to draw a graph, with Carbon dioxide concentration on the horizontal axes and Rate of photosynthesis on the vertical axes.

b) What is the rate of photosynthesis at 0.03% carbon dioxide?

c) Why was the light intensity kept constant during the experiment?

d) Give **one** other factor which must be kept constant.

e) A market gardener buys bottled carbon dioxide to add to his greenhouse. Use the results from the experiment above to advise him at what level he should maintain the carbon dioxide in his greenhouse. Give a reason for your answer.

f) Photosynthesis in plants produces sugars. Name the process where sugars are moved from the leaves to other parts of the plant.

g) Sugar can be stored in plants as starch. Name **one** other storage product that plants can make from sugars.

h) In growing regions of plants, sugars are respired to provide energy. Copy and complete the equation for respiration.

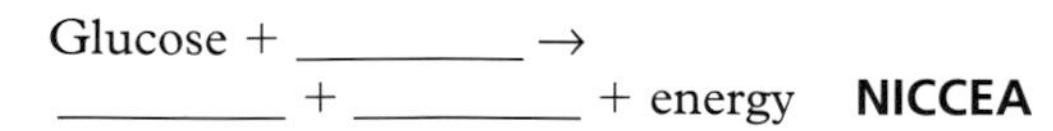

Glucose + ________ →
________ + ________ + energy **NICCEA**

3 A gardener grew tomato plants in a small greenhouse. The door and window were covered by fly screens and a small electric heater warmed the greenhouse. He grew 30 plants and was disappointed in a crop of only 1 kg of tomatoes per plant.

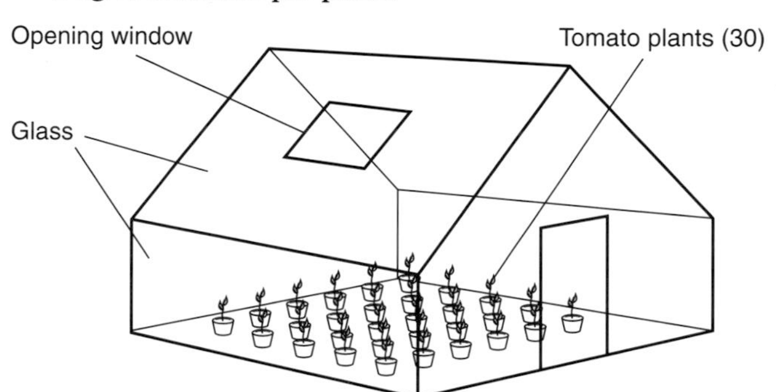

His neighbour said he should have used a gas heater instead of an electric heater. Explain how and why this could have improved the crop. **MEG**

4 a) Plants make their food by photosynthesis. Sarah used the apparatus below to show that carbon dioxide is needed for photosynthesis.

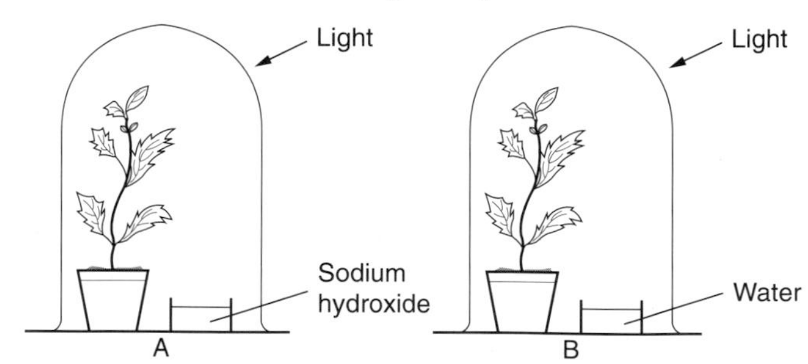

At the end of the experiment she tested a leaf from each plant to see if starch had been made.
The starch test had three parts, shown below.
i) The leaf was dipped in boiling water. Why?
ii) The leaf was then placed in boiling ethanol. Why?
iii) Iodine solution was then added to the leaf. Why?

b) Explain the results that Sarah would see after adding iodine solution to a leaf from A and to a leaf from B.

c) Explain which set of apparatus (A or B) would contain the most oxygen at the end of the experiment. **Edexcel**

CHAPTER

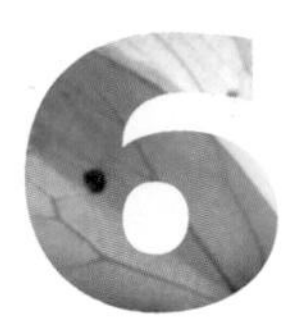

Water uptake and transport in plants

6.1 Introduction

In order to survive and grow, plants need to transport materials from one part of their structure to another. In particular:

- **Water** must be transported from the soil into the roots and up to the leaves for photosynthesis.
- **Glucose and other sugars** which are produced during photosynthesis must be transported from the leaves to other parts of the plant for growth and storage.
- **Mineral salts** containing nitrates and phosphates and magnesium ions must be absorbed from the soil into the roots and then transported to other parts of the plant.

Materials such as glucose and mineral salts can only be moved around a plant if they are in solution. The movement of these solutions through the plant involves a complex system of tubes connecting the roots, the stem and leaves. This transport system is called the plant's **vascular system** and can be compared to an animal's blood system. If you cut a plant stem, liquid oozes out. This liquid, escaping from the cut vascular system is called **sap**.

6.2 Water uptake by plants

Figure 6.1 shows an experiment to investigate water uptake by a small plant.

The volume of water in the cylinder was read at the start of the experiment and on each of the next four days. The results are shown in Table 6.1.

Figure 6.1 Investigating the uptake of water by a plant

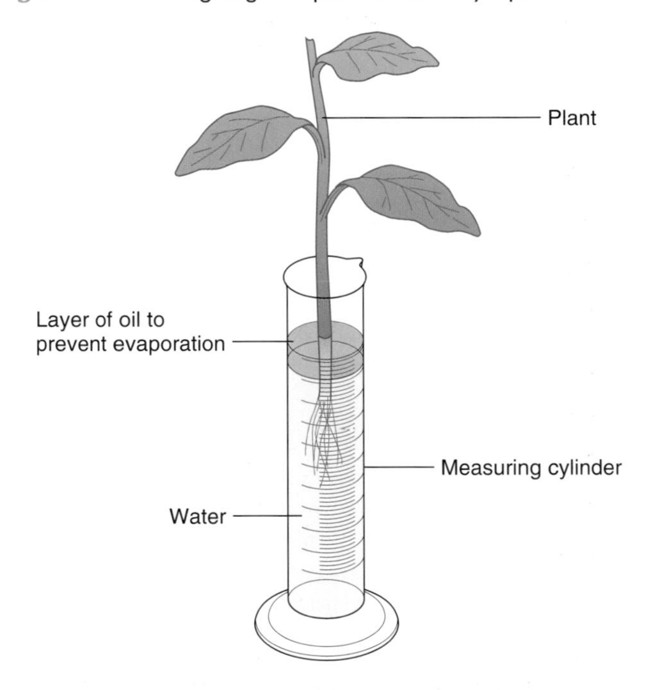

Table 6.1 Results for the uptake of water by a plant

Time from start of experiment (days)	Volume of water in cylinder (cm^3)
0	100
1	97
2	94
3	92
4	89

The results in Table 6.1 show that water was taken up by the plant.

Although **roots** are the organs which anchor plants firmly in the ground, they are also crucial in allowing plants to absorb water. Most water enters the roots of a plant through specially adapted surface cells known as **root hair cells** (Figure 6.2). The outer wall of each root hair cell has a long projection into the soil which increases the surface area for water absorption. In addition, the cell wall of this extension is relatively thin, allowing water to pass through easily.

The cytoplasm and cell sap in the vacuole of a root hair cell contain a large number of dissolved solutes. As the soil water contains lower concentrations of dissolved solutes, water enters the root hair cells by osmosis (section 1.5).

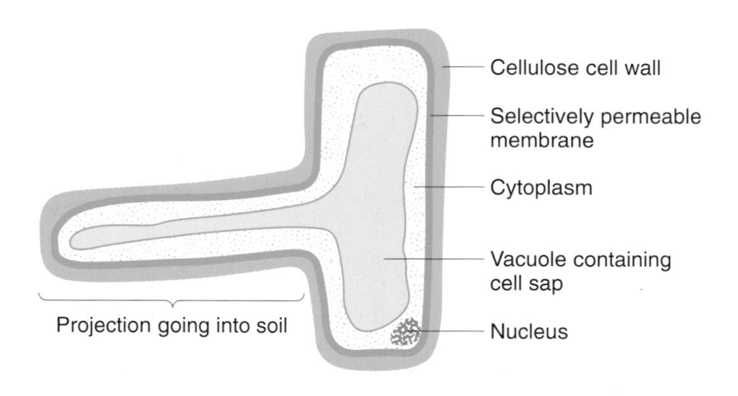

Figure 6.2 A root hair cell

Some of the water absorbed by a plant is used in photosynthesis. However, most of the water evaporates from cells in the leaves and escapes through the stomata (Figure 5.7). As water is lost from the leaves, it is replaced by more water drawn up from the roots through the vascular system.

The evaporation of water from the leaves of a plant is called **transpiration**. The continuous flow of water through the plant from roots to leaves is called the **transpiration stream**.

The rate at which transpiration occurs is largely determined by atmospheric conditions which affect the rate of evaporation via the stomata in the leaves.

It can be shown that transpiration increases:

- on hot days when the temperature is high causing greater evaporation;
- on dry days when evaporation of water vapour into the atmosphere is easier;
- on windy days as the moving air promotes evaporation;
- on bright days when the extra light keeps the stomata well open. The stomata open in response to light to allow more carbon dioxide to enter the leaves for photosynthesis to take place in the leaf cells.

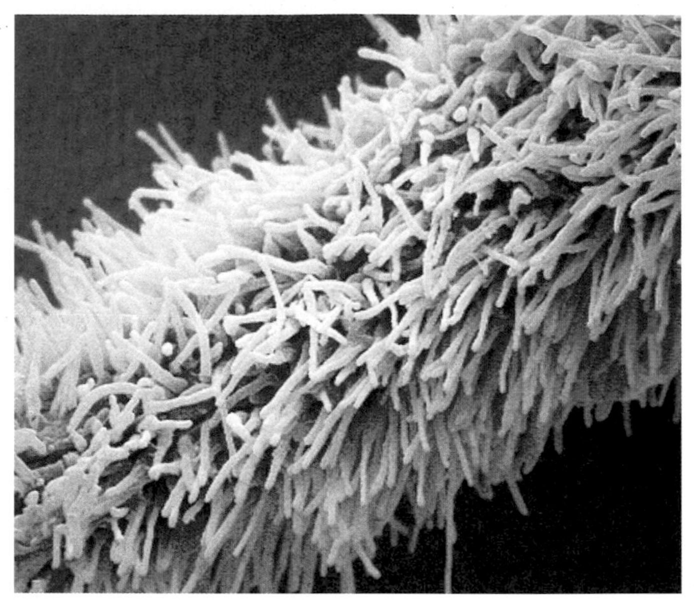

Hundreds of root hair cells allow this plant to absorb water from the soil

6.3 Water and support in plants

Plant cells differ from animal cells in having a strong, fairly rigid cell wall. This cell wall is completely permeable to water and solutes, but inside this wall, the cell membrane is selectively permeable.

When cells in the leaves and stem of a plant have a good supply of water, they fill up and pack tightly. Water moves into the cells by osmosis and increases the pressure inside the cells. The epidermis of the leaf or the stem holds the cells in place. The cells press against one another and make the leaf or stem firm, yet flexible (Figure 6.3a).

In this case, the leaves and the stem are stiffened by filling the cells with water in the same way that a bicycle tyre is stiffened by filling the inner tube with air.

When plant cells are swollen in this way they are described as **turgid**. The support which the swollen cells give to the plant is called **turgor**. Turgor is a major factor in the support of non-woody plants, like dandelions and daffodils. When plants, like these lose their water supply by cutting or drought, their cells are no longer stiffened by turgor. The leaves droop and the stem may wilt (Figure 6.3b).

This busy lizzie plant was wilting badly on a hot summer's day but has recovered after watering

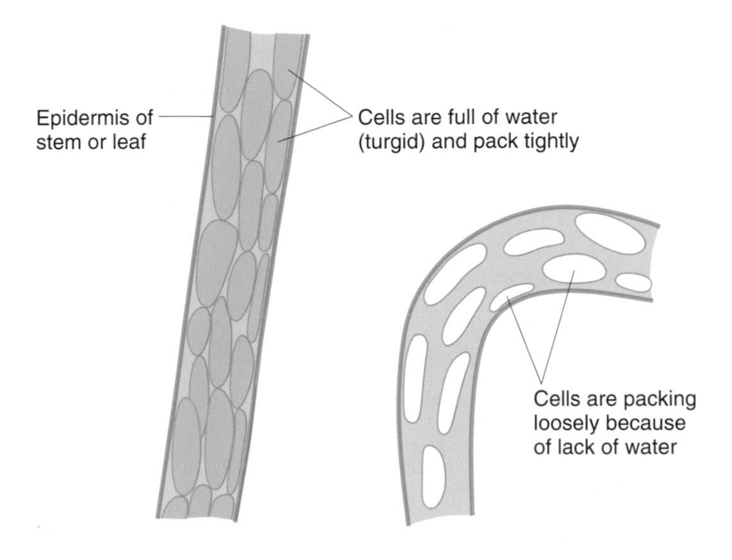

Figure 6.3 The importance of water in the support of plant tissue

Perennial flowering plants, shrubs and trees which continue from year to year, produce woody tissue to stiffen stems and do not rely so completely on turgor for support.

6.4 Transport in plants

The photograph in Figure 6.4 shows a section through the stem of a clematis plant. This shows the internal structure of a plant stem very neatly. The internal stem structure is composed of three tissues.

- The **epidermis** is an outermost layer of cells which is a kind of skin. On the leaves and stem the epidermis is covered with a waxy cuticle pierced by stomata.
- **Packing tissue** fills most of the stem. This consists of rounded cells packed close together.
- The **vascular system** is made up of **xylem**, (pronounced 'zylem') and **phloem**, (pronounced 'flowum'). The vascular system is responsible for the transport of materials around a plant. In the roots and stem, the vascular system is concentrated in regions known as **vascular bundles**. So, xylem vessels and phloem tubes are arranged side by side through the stem of the plant as bundles of tubes. This is illustrated in Figure 6.5 over the page.

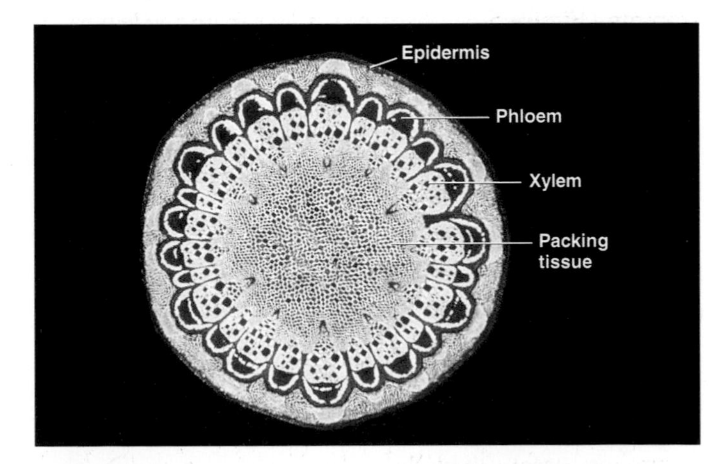

Figure 6.4 A photograph of a section through the stem of the climbing plant, clematis, showing the epidermis, packing tissue, phloem and xylem. The magnification is 15 times

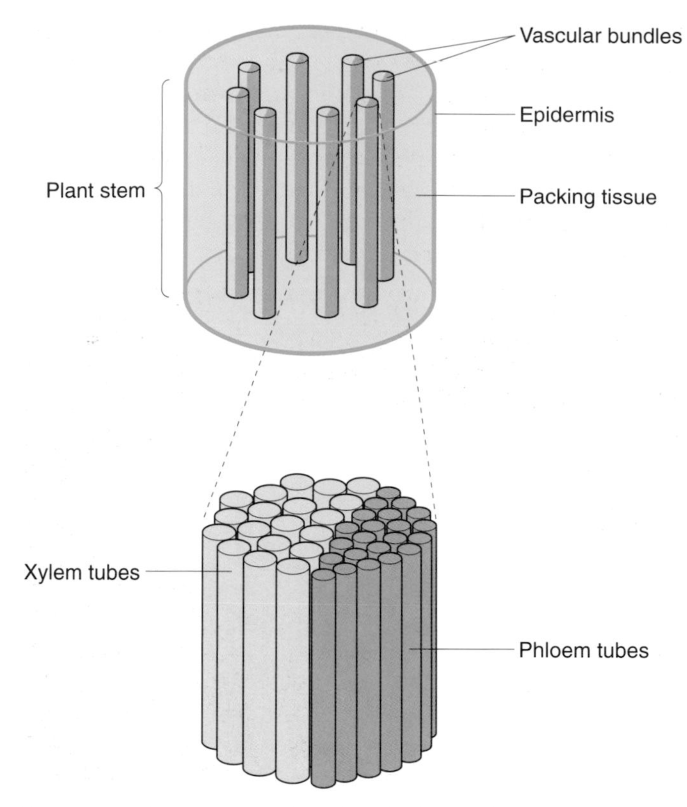

Figure 6.5 The vascular bundles of xylem and phloem tissues in a plant stem

The xylem and phloem tissues in the vascular bundles form a continuous system throughout a plant between roots, stem and leaves. This means that water taken in through the roots can be transported up the stem and to the leaves, whilst glucose, photosynthesized in the leaves, can pass to the stem and roots.

In each vascular bundle, the inner tissue is xylem. This is composed of long, thin, *dead* cells joined together like capillary tubes. The xylem tubes are arranged in series, one after the other. They form a continuous link from the roots to the leaves carrying water and dissolved mineral salts.

Xylem carries the transpiration stream in plants and plays a crucial part in the transport of **water**.

Transpiration causes water to be drawn up the xylem tubes from the roots and into the rest of the plant. The walls of xylem cells are made of cellulose, like the cell walls of all plant cells. As they get older, the cellulose is slowly converted to **lignin** which is less permeable and much harder. As a result of this, xylem plays an important part in supporting perennial and woody plants and eventually it becomes part of the packing tissue.

The outer tissue in each vascular bundle is phloem. Phloem contains very long, thin *living* cells joined end to end and running continuously from the roots, along the stem and into the leaves. The end walls between one phloem cell and the next are perforated by tiny holes like a sieve. Because of this, phloem is sometimes described as sieve tubes.

Phloem plays an important role in the transport of **glucose** and **other sugars** dissolved in water throughout the plant.

Glucose is produced by photosynthesis in the leaves. It is then distributed to the roots, the growing tips, the flowers and the fruit via the phloem. Notice that material in the xylem travels in only one direction along a plant stem – from the roots to the leaves. On the other hand, material in the phloem may travel in either direction – up the stem to the growing tips, fruit and flowers or down the stem to the roots.

In a tree trunk, the phloem is in the soft inner part of the bark. So, if a complete ring of bark is cut from a tree trunk, food cannot be transported to the roots and the tree will slowly die.

Table 6.2 summarises the important differences between the xylem and phloem in a plant.

Table 6.2 A summary of the important differences between xylem and phloem tissues

	Xylem	Phloem
State of being	dead	living
Material of cell wall	lignin	cellulose
Nature of cell wall	non-permeable	permeable
Materials transported	water and mineral salts	glucose and other sugars
Direction of transport	*up* the plant from the roots	*up and down* the plant from the leaves
Support provided	by lignin (woody tissue), can be very strong	by turgor pressure, much weaker

6.5 The mineral requirements of plants

Plants require certain elements if they are to flourish and grow strongly. These elements are therefore called **essential elements**. The most important essential elements for plants and those required in the largest amounts are carbon, hydrogen and oxygen. These are obtained from water in the soil and from oxygen and carbon dioxide in the air.

In addition to carbon, hydrogen and oxygen, the major elements required by plants are nitrogen, phosphorus and potassium.

Nitrogen is required by plants for the synthesis of amino acids, proteins and nucleic acid, deoxyribose nucleic acid (DNA) and ribose/nucleic acid (RNA).

Plants obtain their nitrogen from ammonium salts (containing NH_4^+ ions) and nitrates (containing NO_3^- ions) in the soil.

Phosphorus is required by plants for reactions involving the synthesis and interaction of adenosine diphosphate (ADP) and adenosine triphosphate (ATP). These two substances play an important part in the transfer of energy in cells. Phosphorus is also needed in the synthesis of nucleic acids and in the formation of cell membranes. Plants obtain their phosphorus from phosphates (containing PO_4^{3-} ions) in the soil.

Potassium is required by plants for the transfer of materials across cell membranes and for the synthesis of carbohydrates and proteins. Plants obtain their potassium from mineral salts in the soil containing potassium ions (K^+).

Other elements, such as **magnesium** and **iron** are needed in smaller amounts for healthy plant growth. Magnesium is an essential constituent of chlorophyll, whilst iron is a component of the enzymes which synthesize chlorophyll.

Deficiencies in mineral salts result in weak plants and the yield from any crop may be severely affected by the lack of any one essential element.

Different plants require different amounts of particular essential elements. If one particular crop is grown year after year in the same soil, rather than varying the crop, it is more likely that there may develop a deficiency in certain minerals as these are continually being taken from the soil. This situation of **monoculture** requires the application of fertilizers to prevent these deficiencies and ensure high crop yields.

A healthy tomato plant leaf

A tomato plant growing in soil lacking magnesium. Magnesium ions are needed for chlorophyll formation. Without magnesium the leaves turn yellow

1 a) The diagram below shows a section of a leaf.

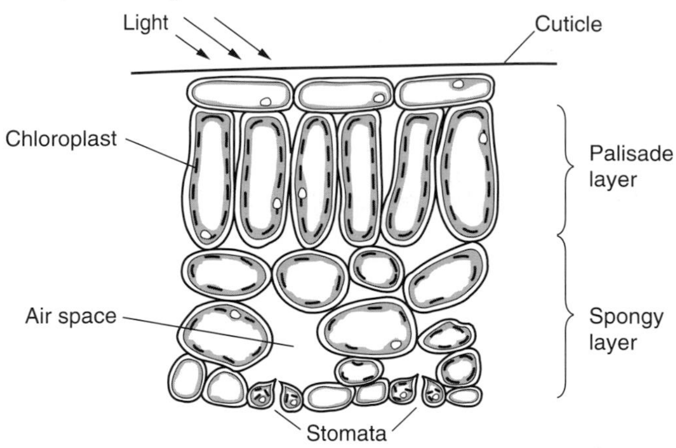

i) Name the process in which the leaf loses water.
ii) Name the structures through which this loss occurs.
iii) How is the leaf supplied with water for photosynthesis?

b) Hydrogencarbonate indicator is orange in colour. It turns yellow when carbon dioxide is present and purple when carbon dioxide is removed. When the experiment was set up, as shown in the following diagram, the indicator was orange in all tubes. The tubes were left in the light for four hours.

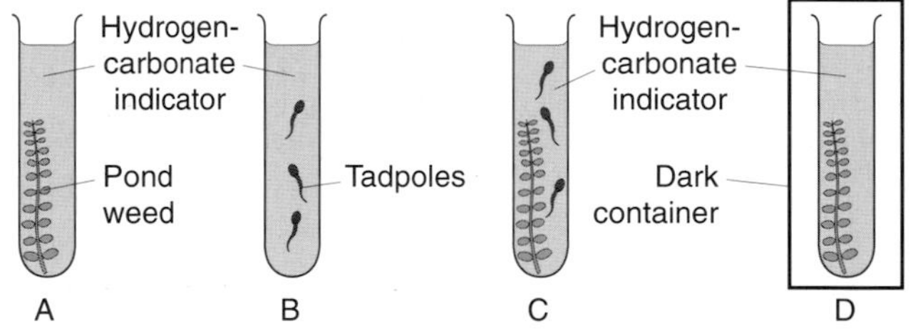

The following table shows the colour of the indicator in each tube after four hours.

Tube	Colour of the indicator
A	purple
B	yellow/orange
C	orange
D	yellow

Explain the results in each of the four tubes.

c) Plant growth is controlled by chemicals called plant growth hormones.
i) Explain why gardeners use rooting powders when growing plants from cuttings.
ii) Plant growth substances are the basis of some weedkillers (herbicides) such as 2,4-D. Explain how these herbicides kill weeds. **MEG**

2 The following graphs relate to the uptake of nitrate ions by the epidermal cells (piliferous layer) of a plant root when immersed in a culture solution.

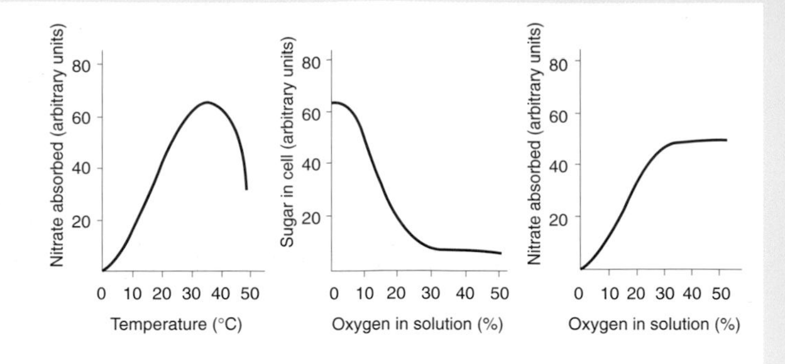

i) How is the uptake of nitrate ions affected by the environmental temperature?
ii) In what way does the temperature/uptake graph suggest that the absorption of ions is an active process?
iii) What additional evidence provided by the other graphs, supports the active uptake hypothesis? **NICCEA**

3 The drawing shows a root hair cell from near the tip of a young root.

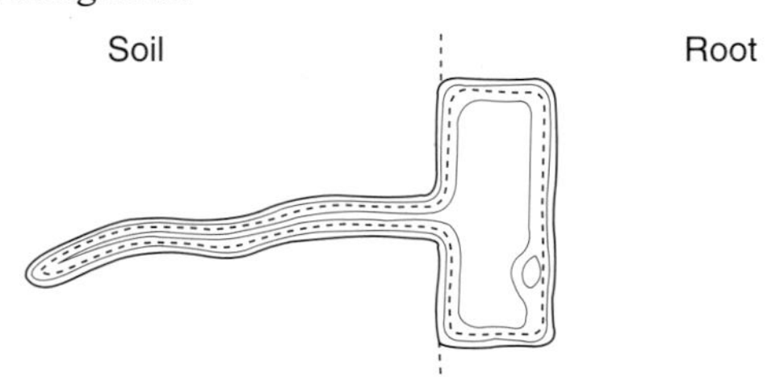

a) This cell needs oxygen. Name the process by which oxygen enters the cell from the air in the soil.

b) Describe the process by which water enters the root hair cells.

c) A seaside garden is flooded by the sea in an exceptionally severe storm. Several of the plants wilt and die. Explain why flooding with sea water caused the plants to wilt. **NEAB**

4 Many people like to keep pot plants inside their houses. These plants need watering regularly.

a) Copy the following diagram and draw arrows to show the path taken by the water from the time it enters the plant to the time it leaves as water vapour.

b) i) Name the process by which the water enters the plant from the soil.
ii) Explain how this process works.

c) i) Explain briefly how a plant loses water and how it controls the rate of loss.
ii) Suggest **two** reasons why it is important for a plant to be able to control its rate of water loss. **MEG**

CHAPTER

Communication and nerves

Stimuli and responses The nervous system Reflex actions	The structure of the eye and its response to light Focusing with our eyes

7.1 Stimuli and responses

Living things depend on their senses for survival. Our senses help us to avoid danger. They also enable us to find food, drink and light. If someone whistles in the street, you would probably look round. The sound of the whistle causes a change in your sense of hearing. Changes like this which can be detected by our sense organs are called **stimuli** (singular, stimulus). Our reactions to stimuli are called **responses**. So, your looking round is a response to the whistle.

Figure 7.1 shows our various senses, the sense organs involved and the stimuli detected by these organs.

Sense organs enable us to react to our surroundings. They are linked with the brain and our nervous system so that we can control and co-ordinate our behaviour.

All our sense organs have specialised cells called **receptors** which detect stimuli (changes in our environment).

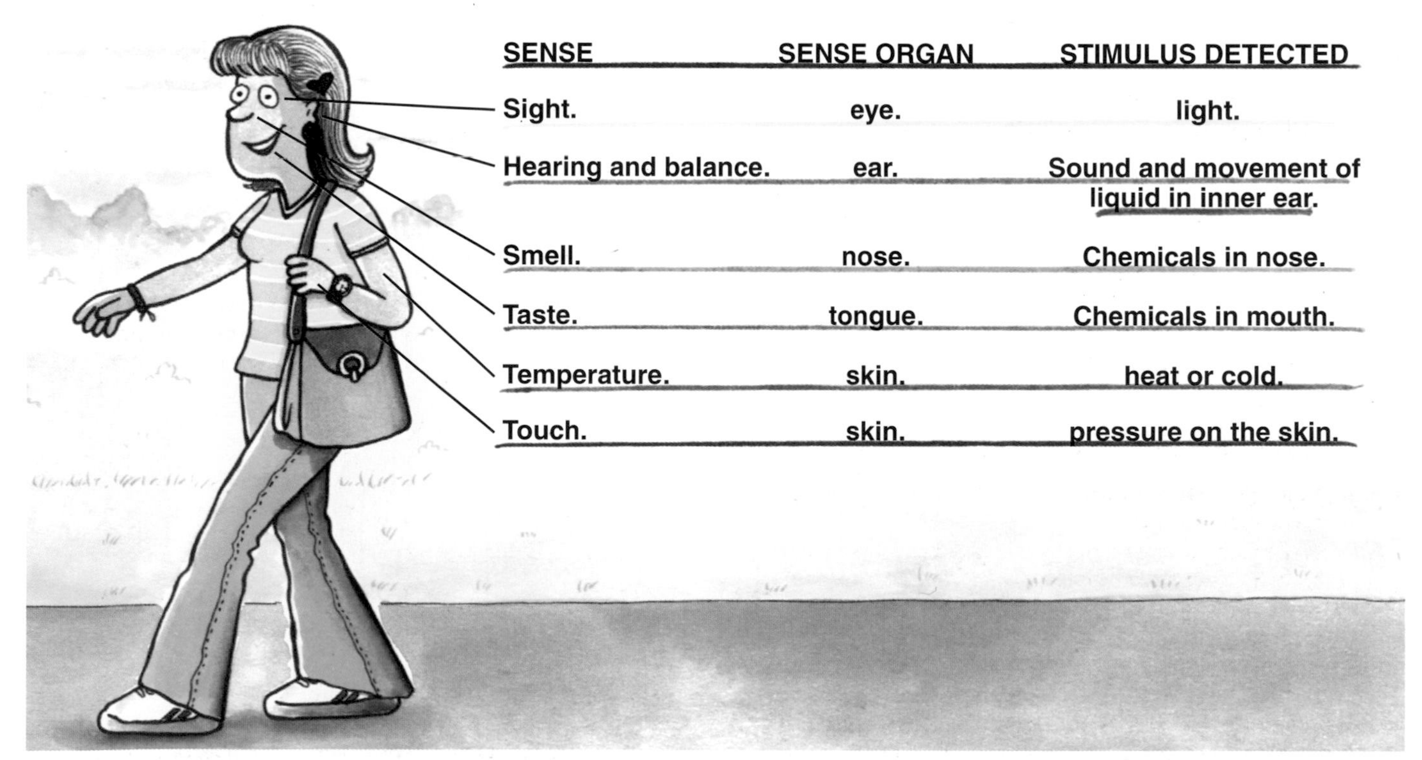

Figure 7.1 Our various senses, sense organs and the corresponding stimuli they detect

We all rely on our senses. Which senses are these students using in order to help them cross the road safely?

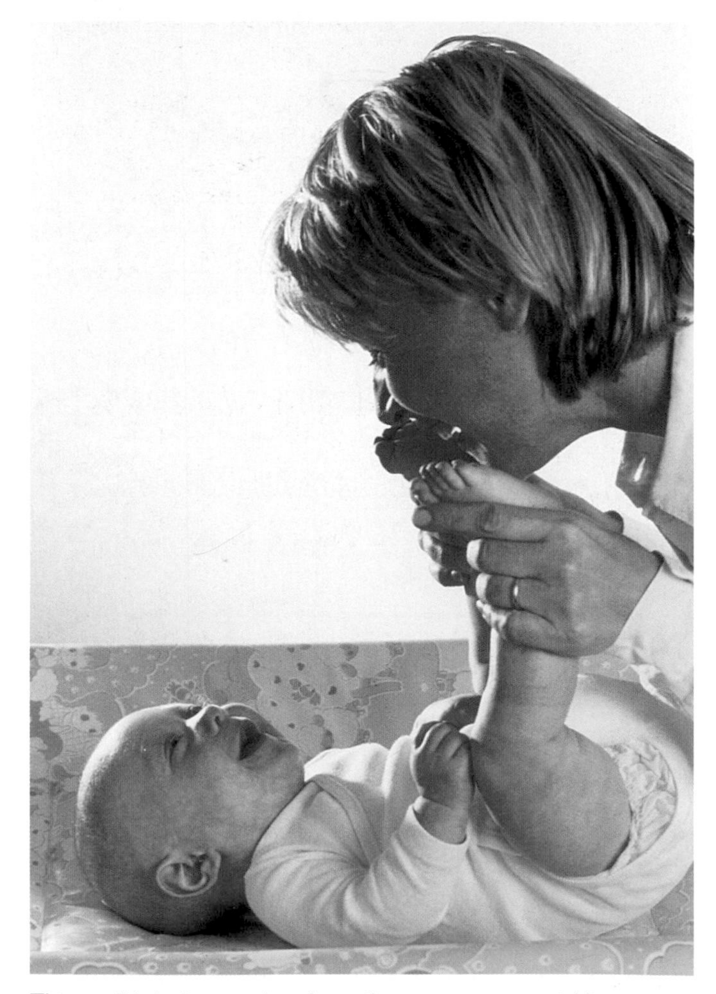

This small baby has produced a smiling response to a tickling stimulus

These include:

- receptors in the eyes which are sensitive to light,
- receptors in the ears which are sensitive to sound,
- other receptors in the ears which are sensitive to the movement of liquid in the inner ear to help us keep our balance,
- receptors in the nose and on the tongue which are sensitive to chemicals,
- receptors in the skin which are sensitive to touch and temperature.

Receptors can turn (transduce) the energy from stimuli into **impulses**. These tiny electric impulses pass along nerve fibres to and from the spinal cord and the brain. The spinal cord and the brain are the centres of our nervous system.

7.2 The nervous system

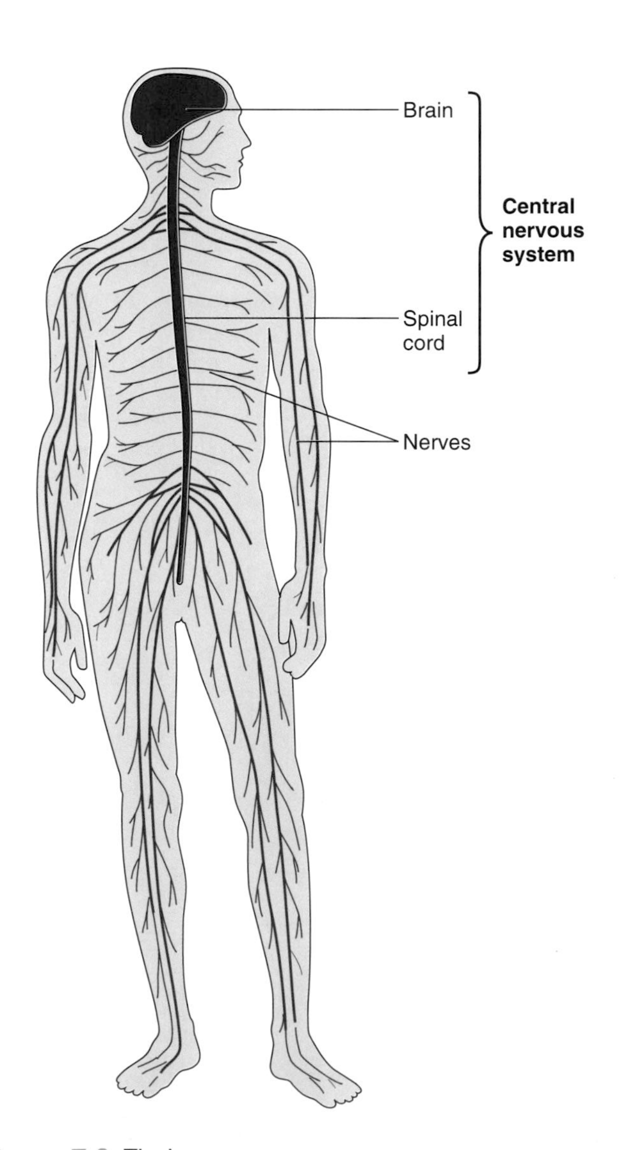

Figure 7.2 The human nervous system

The nervous system plays a crucial role in allowing us to respond to stimuli. The main function of the nervous system is to carry impulses (messages) from one part of the body to another. Experiments have shown that these impulses along the nerves consist of tiny electric pulses which travel rapidly through the nervous system.

> The nervous system consists of two parts:
>
> 1 the **central nervous system (CNS)** comprising the brain and spinal cord,
> 2 a **network of nerves** linking the CNS with the different organs in the body.

The brain is enclosed neatly within the cranium which is part of the skull. The spinal cord runs down the centre of the backbone. Thus, the entire CNS is well protected by bone. Some nerves come out of the brain and go to the sense organs and muscles in the head. Other nerves come out of the spinal cord and go to structures in the arms, legs and rest of the body.

Figure 7.3 shows the structure of a typical nerve cell, often called a **neurone**, linking a sense organ to the brain.

The main part of a neurone is the **cell body** containing the nucleus. Part of the cytoplasm of the cell body is extended into a long thin nerve fibre. A nerve fibre, called a **dendron,** links receptors in the sense organ to the cell body. A second nerve fibre called an **axon** then links the cell body to the brain. The electrical impulses (messages) which pass along the nerve fibre are produced by the movement of Na^+ and K^+ ions.

Nerve cells differ from other cells in having protruding branches called **dendrites**. Dendrites at the end of a dendron pick up stimuli in sense organs and then transmit impulses to the brain. Dendrites in the brain link up with other nerve cells to form a complex network.

There are four types of neurones (nerve cells) in the body.

1 **Sensory neurones** like that in Figure 7.3 which carry impulses from sense organs to the spinal cord.
2 **Motor neurones** which carry impulses from the spinal cord to muscles and other organs.
3 **Relay** (**connector**) **neurones** in the spinal cord and the brain. In the spinal cord, relay neurones *connect* sensory neurones to motor neurones and pass on (relay) messages. In the brain, relay neurones *connect* to and from pyramidal neurones. The main function of relay neurones is to allow a large number of cross connections, rather like a telephone exchange.
4 **Pyramidal neurones** *connect* the relay neurones and other pyramidal neurones in the brain. They have a vast network with thousands of branches to other cells like the inter-connections in a computer.

Although relay neurones are relatively small, individual sensory and motor neurones can be very long indeed. For example, a sensory neurone, more than half a metre in length, could link the tip of your little finger and the top of your spinal cord.

Figure 7.3 The structure of a typical nerve cell (neurone) linking a sense organ to the brain

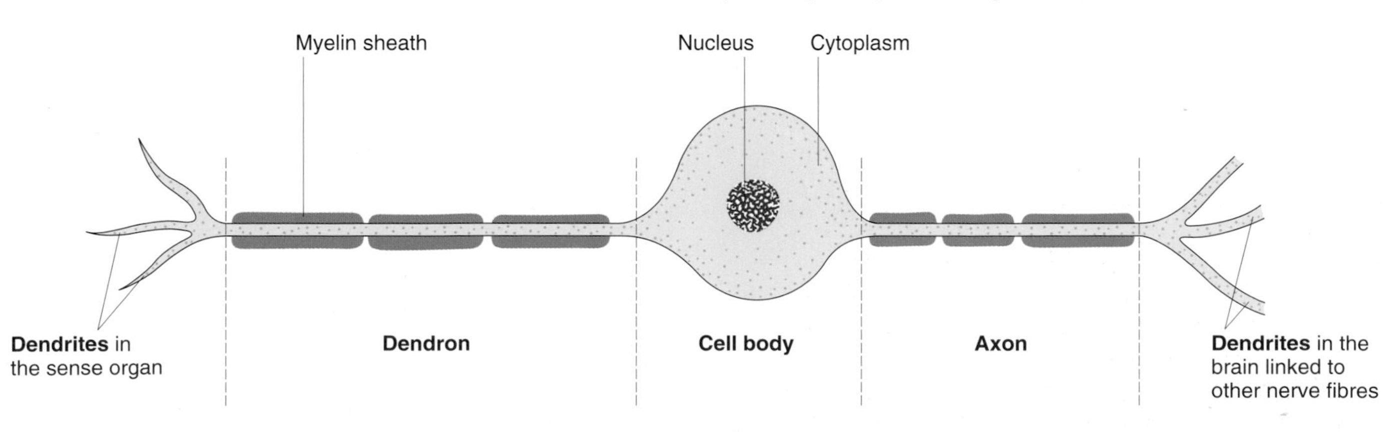

Sometimes, nerve fibres are grouped into bundles which are simply called **nerves** (Figure 7.4). The **myelin sheath** of fatty tissue around each nerve fibre insulates it from neighbouring fibres and prevents impulses crossing from one fibre to another.

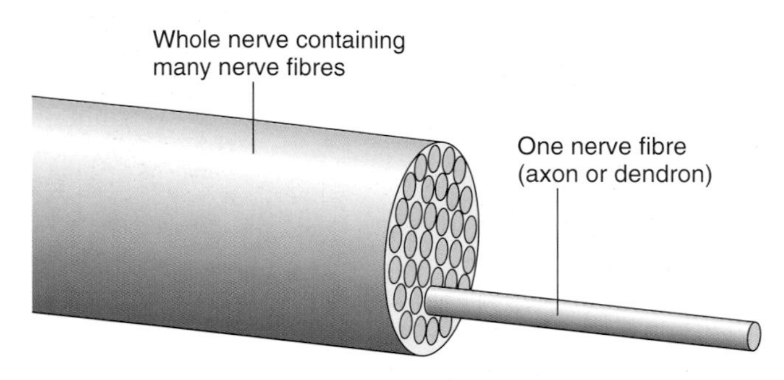

Figure 7.4 Bundles of nerve fibres in a nerve

7.3 Reflex actions

If someone whistles, you look around. If you accidentally touch a hot kettle, you pull your finger away immediately. If your nose tickles, then you sneeze. You don't have to think. Each of these actions happen automatically. The responses are spontaneous and out of your control.

Rapid and automatic responses to stimuli are described as **reflex actions**. Reflex actions are rapid because the nervous impulses travel by the shortest route in your body. These shortest routes are called **reflex arcs**. The reflex arc involved in responding to a whistle is shown as a flow diagram in Figure 7.5.

- The sound of the whistle acts as a **stimulus**.
- **Receptors** in the ear detect the stimulus.
- Impulses pass along a **sensory neurone** from the ear to the central nervous system.
- Impulses pass via a **relay neurone** in the CNS to a motor neurone.
- Impulses pass along **motor neurones** to muscles (**effectors**) which allow you to turn round.
- The final **response** is to look around.

Figure 7.5 A flow diagram showing the pathway for a simple nervous response (reflex arc)

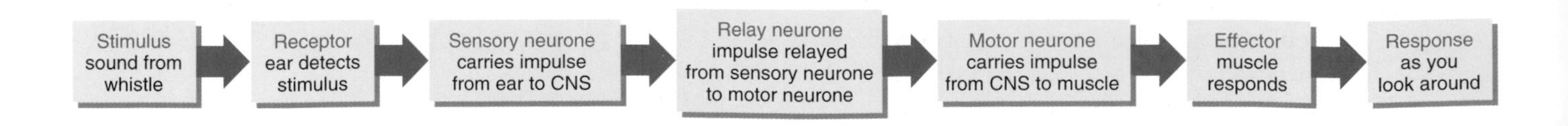

Effectors which bring about responses are either muscles or glands. Muscles usually respond by contracting, whereas glands respond by secreting chemicals (hormones). The actions of hormones are described in Chapter 8.

The reflex arc which operates when you pull your finger away from a hot kettle and the transfer of impulses between different neurones is shown in detail in Figure 7.6.

As soon as you touch the hot kettle, heat and pain receptors in your finger create nervous impulses (Figure 7.6a). These pass along sensory nerve fibres to the spinal cord. Here, the nerve fibres make junctions or **synapses** with relay neurones (Figure 7.6b).

The relay neurones then transfer the impulses through further synapses, to motor neurones controlling muscles in the arm. Finally, impulses carried by the motor nerve fibres make the muscle contract, pulling your finger away from the kettle.

Nervous impulses are produced by the flow of Na^+ ions into a neurone and K^+ ions out of a neurone. This movement of positive ions creates a similar effect to the movement of negative electrons in an electric current. The nervous impulse, with its movement of ions, causes the secretion of a chemical at a synapse. This chemical triggers the movement of ions in the next neurone and so the impulse passes on. At the same time, the secreted chemical is destroyed by enzymes.

Notice that the brain is *not* involved in reflex actions. This makes the response much quicker because impulses have less distance to travel and no decisions have to be made in the brain.

Many actions are, however, not automatic reflexes. They are thoughtful, **conscious actions**.

Conscious actions involve the brain as well as the spinal cord. During conscious actions, impulses must travel to the spinal cord and then up to the brain which co-ordinates a response. Return impulses start at the brain and must travel down to the spinal column before moving out to the muscles. Therefore, impulses involving conscious actions cross more synapses and take longer routes.

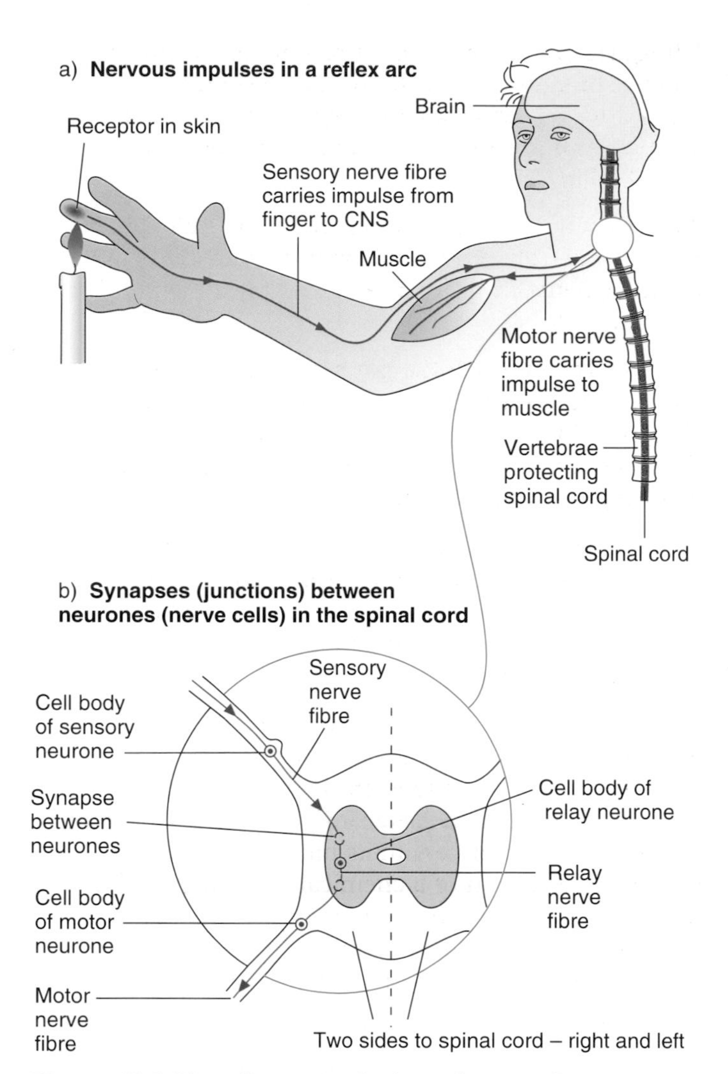

Figure 7.6 The reflex arc involved in pulling your finger away from a hot kettle

You may have had your reflexes tested like this patient. The doctor is tapping the patient's leg just below the knee. This test is a useful guide to the condition of the spinal cord

7.4 The structure of the eye and its response to light

Our eyes respond to light and enable us to see things. They are very sensitive to changes in light. Figure 7.7 shows the structure of an eye. Its structure is very complex. Five parts are crucial in our response to light – the cornea, iris, lens, retina and optic nerve.

- The **cornea** is a transparent portion at the front of the outer sclera. Light from an object enters the eye through the cornea. The cornea is curved and helps us to focus.
- The **iris** controls the size of the **pupil** and hence the amount of light entering the eye. The iris has two sets of muscles – circular muscles and radial muscles. Circular muscles are activated and contract when the light intensity on the retina increases. The pupil becomes smaller and less light enters the eye.

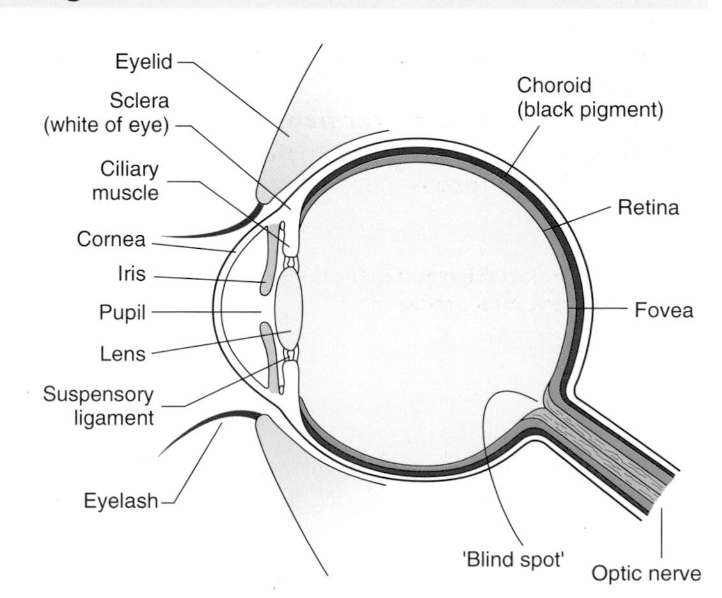

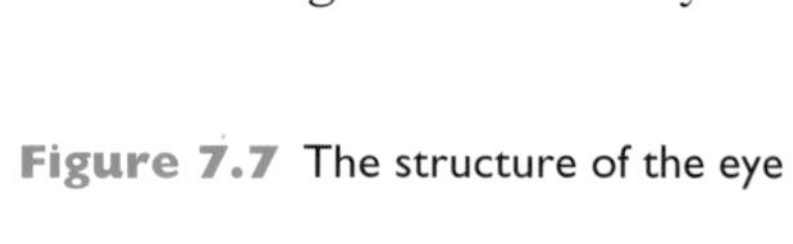
Figure 7.7 The structure of the eye

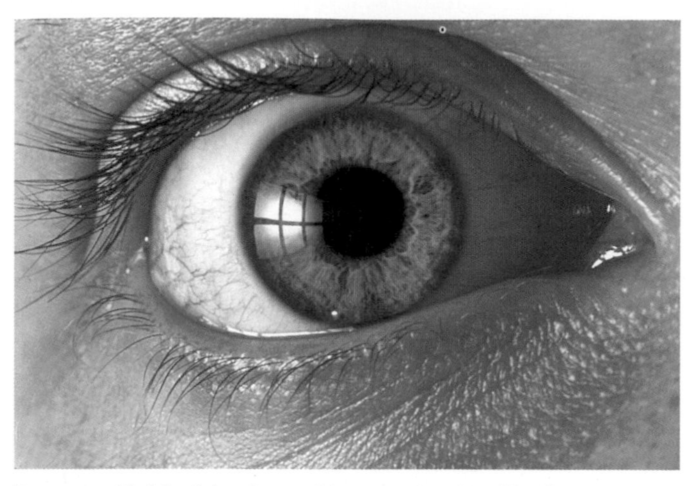

Because of bright light, the pupil is constricted in this photo.

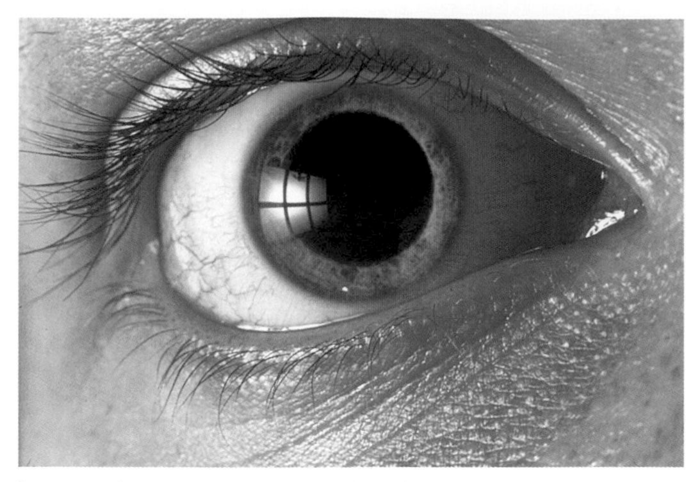

Because of poor light the pupil is dilated in this photo

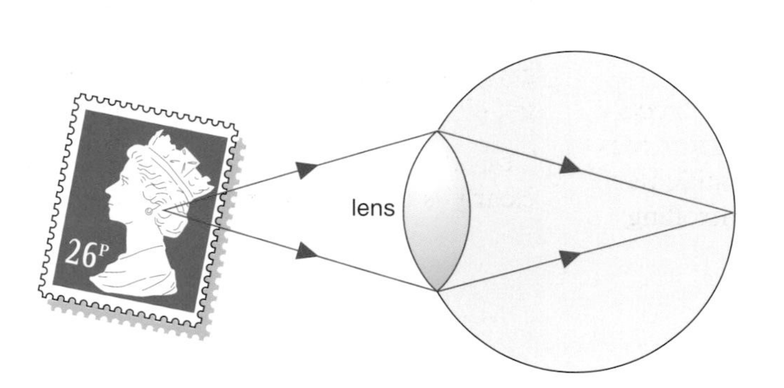

Ciliary muscle contracts

Suspensory ligament slack

Lens

a) **Looking at something close up**
Ciliary muscle contracts, suspensory ligament is slack and the lens is fatter and more powerful

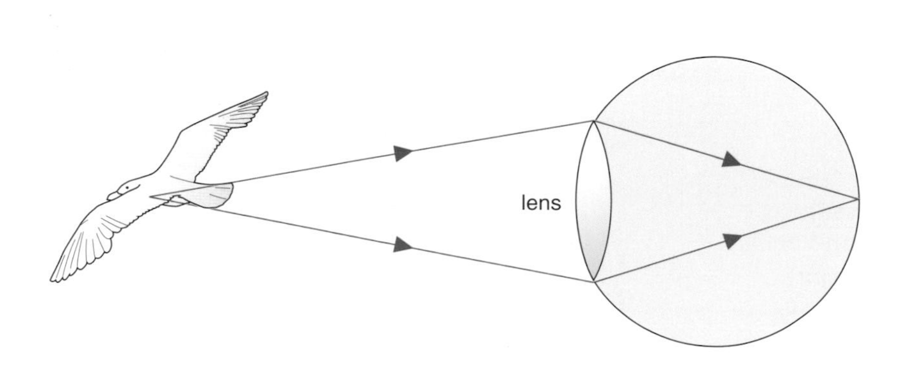

Ciliary muscle relaxed

Suspensory ligament stretched

Lens

b) **Looking at something in the distance**
Ciliary muscle relaxes, suspensory ligament is stretched and the lens is thinner and less powerful

Figure 7.8 The shape of the lens changes in order to focus either close up or far away

The radial muscles are activated and contract when the light intensity on the retina decreases. This draws back the centre of the iris making the pupil larger and allowing more light to enter the eye.

- The **lens** is held in position by the suspensory ligaments and ciliary muscles. The lens and **ciliary muscle** enable us to make fine adjustments to our focusing.
- The **retina** contains receptor cells sensitive to light.
- The **optic nerve** allows impulses from the retina to pass to the brain along sensory neurones.

The most sensitive area of the retina containing the highest density of light-sensitive receptor cells is the **fovea**. The fovea lies on the central axis of the eye. There are very few light-sensitive cells at the point where the optic nerve leaves the retina. Because of this, the area is called the **blind spot**.

7.5 Focusing with our eyes

Our eyes must be able to focus and cope with objects at different distances as well as coping with differences in light intensity. The iris, pupil and retina enable us to cope with differences in light intensity by controlling the amount of light entering the eye.

The parts of the eye which enable us to focus on different objects are the cornea, lens and ciliary muscles.

As light enters the eye it passes through the cornea. The cornea changes the direction of light rays, bending (diffracting) them towards the lens and onto the retina. Fine adjustments to our focusing on the retina can be made by altering the shape of the lens. It is the contraction or relaxation of the ciliary muscle, which forms a ring around the lens, that alters its shape.

When you look at something close up, the circular ciliary muscle contracts and its diameter gets smaller. This causes the suspensory ligament to go slack and the lens becomes fatter (Figure 7.8a). The lens now bends the light rays more and we can focus on the object such as the printed word on this page.

When you look at something in the distance, the ciliary muscle relaxes and its diameter gets larger. This causes the suspensory ligament to stretch and the lens becomes thinner (Figure 7.8b). We then see beyond the classroom window out and into the distance.

This ability of the lens to change its shape and focus clearly is called **accommodation**.

QUESTIONS

1 The drawing shows a light-sensitive receptor cell from the eye. The structures labelled A, B and C are found in most animal cells.

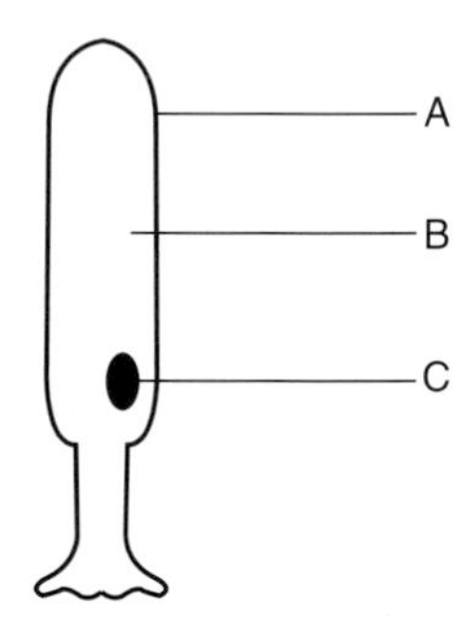

a) Name the structures labelled A, B and C.

b) Name the layer of the eye which contains light-sensitive receptor cells.

c) Name the **two** structures in the eye that produce an image on the light-sensitive receptor cells.

d) In the pupil reflex, information is passed from the light-sensitive receptor cells to the brain and then to the muscles of the iris. Describe, in as much detail as you can, how information is passed from the receptor cells to the iris muscles. **NEAB**

2 a) The diagram below shows five parts of the body, A, B, C, D and E.

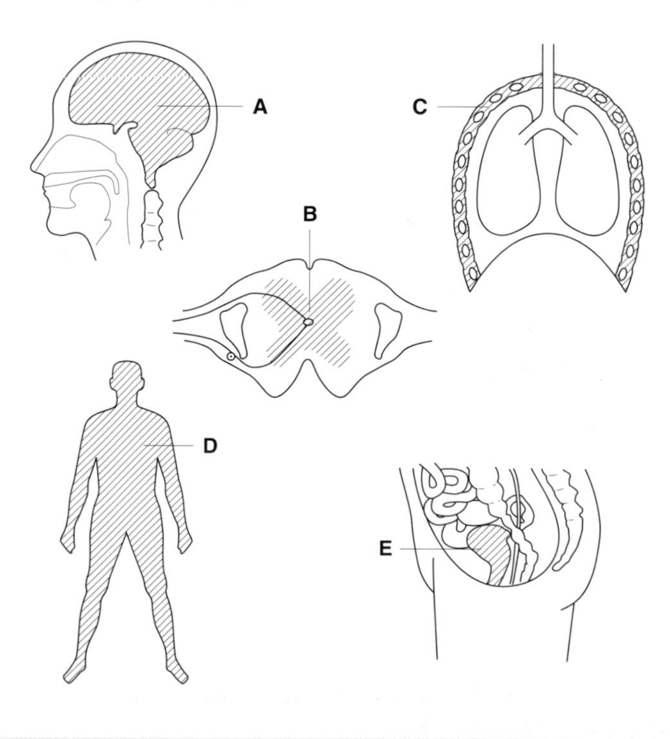

The following table contains five jobs that the parts shown on the previous page do. Copy the table and write the label of the correct part next to the job that it does.

Job	Organ
moves the rib cage	
stores urine	
senses the temperature of the surroundings	
controls voluntary actions	
controls the knee-jerk reflex actions	

b) The nervous system is made up of nerve cells. The diagram below shows a motor nerve cell.

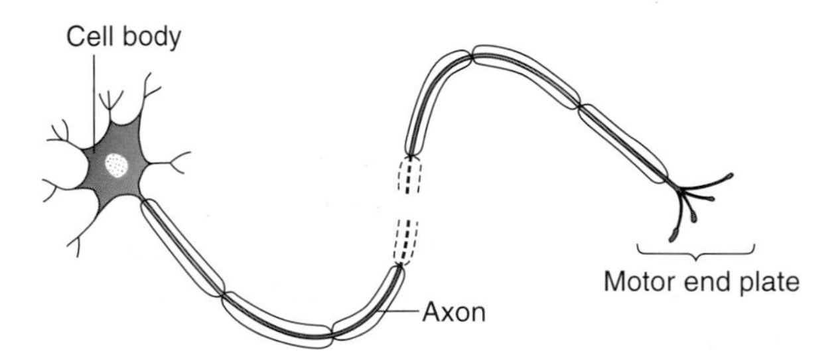

Copy the following sentences and finish them by choosing the **best** words from the list:

heart long muscle short spinal cord tendon thin

The cell body of a motor nerve cell is in the brain or in the ______________ .

The motor end plate is connected to a ______________ .

The axon is ______________ so that it reaches the organ it affects.

c) The following diagram represents a reflex arc.

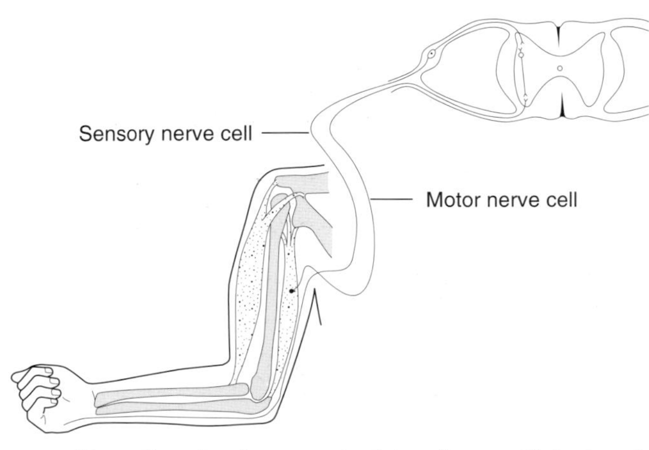

Describe what happens in this reflex arc if the hand touches a very hot object. **MEG**

3 When a reflex action takes place, it can be summarised as follows:
1. Stimulus → **2.** Receptor → **3.** Co-ordinator → **4.** Effector → **5.** Response.
Give an example of a reflex action and describe it in terms of **1**–**5** above. **WJEC**

4 A section through the eye and two light rays are shown below.

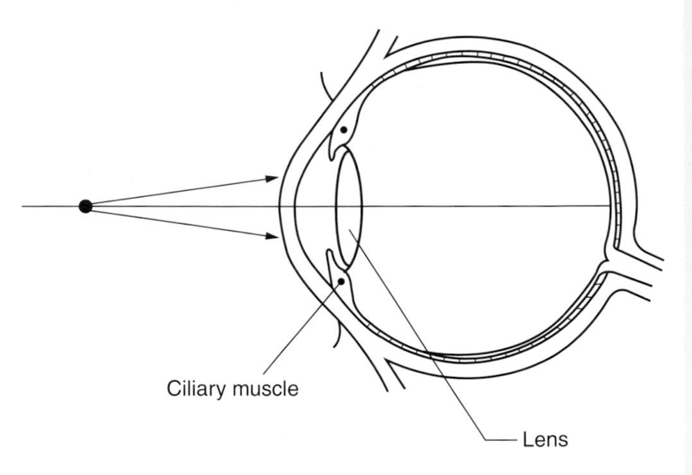

a) Copy the diagram and:
i) draw an arrow labelled **X** to show the iris;
ii) draw an arrow labelled **Y** to show the retina;
iii) continue the two light rays to show them focused on the retina.

b) Describe the change in the shape of the lens when the eye is focused on near objects. **SEG**

5 a) What type of receptors are stimulated when you:
i) read a book,
ii) remove your coat,
iii) suck a sweet,
iv) cut your finger?

b) Hundreds of volunteers were tested to find the temperature that caused them to feel pain as the water in which they had their fingers was heated up. The results are shown below. What conclusions can you draw from the results in the graph?

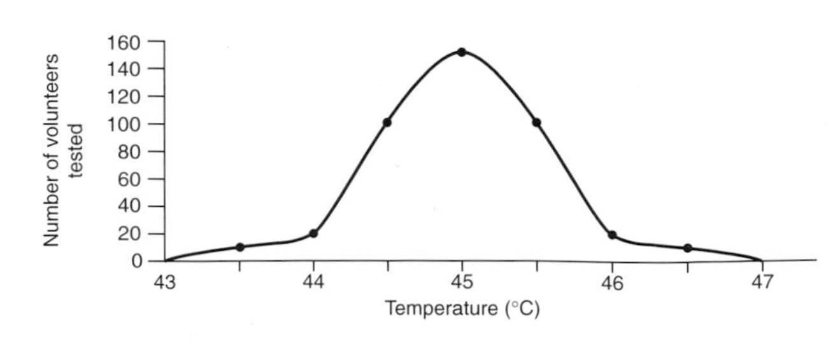

c) Which structure(s) in the eye is (are) responsible for:
i) detecting light,
ii) moving the eyeball,
iii) focusing an image,
iv) controlling the amount of light entering,
v) sending messages to the brain?

CHAPTER

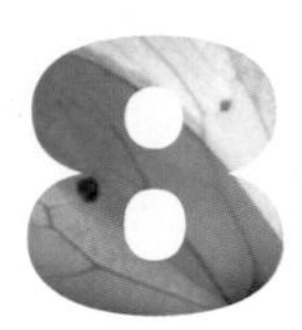

Hormones and control

8.1 Our hormones and hormonal control

Many processes in our bodies are controlled and co-ordinated by chemicals called **hormones**. Hormones are produced in our bodies by organs called **glands** and then released (secreted) into the bloodstream.

When a particular hormone is needed, it is released into the bloodstream in tiny amounts and then transported to all parts of the body.

Usually, however, the hormone affects only certain cells which are described as its **target cells**. So, hormones carry messages from one part of the body to another and they are sometimes called **chemical messengers**.

The hormone system consists of a number of glands. The positions of these glands, the hormones which they release and the effects of these hormones are shown in Figure 8.1.

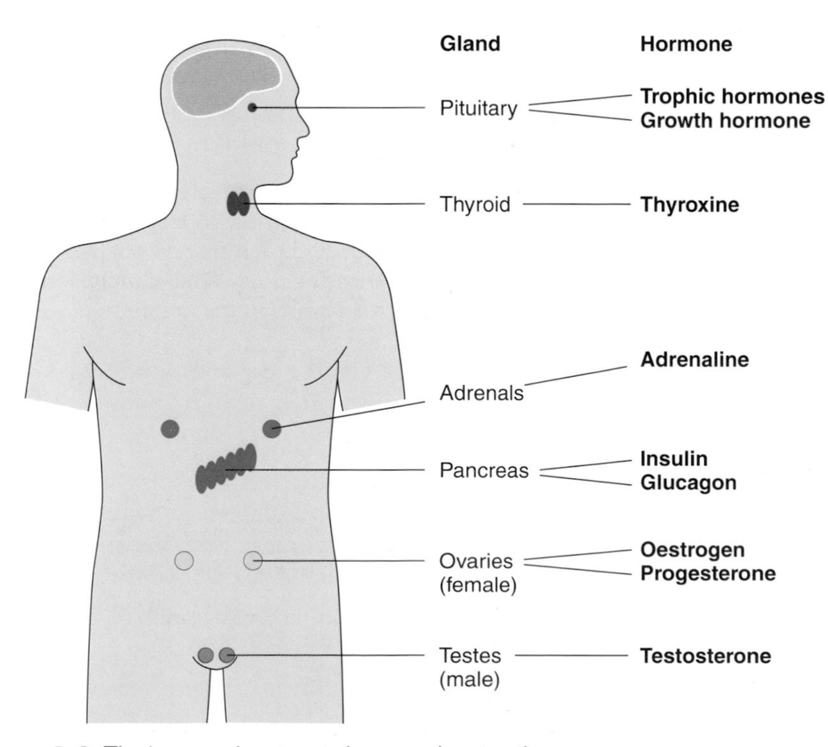

Figure 8.1 The hormonal system in humans showing the hormones released and their effects

Notice the following points in Figure 8.1:

- The pituitary gland produces the **growth hormone** and also **trophic hormones** which stimulate other glands to produce their hormones. Because of this it is described as the 'master gland'.
- **Adrenaline** has different effects on different organs. It causes cells to respire faster and produce extra energy. This causes the heart to beat faster and increases the blood supply to our muscles so that they are ready for action. The extra blood supply to our muscles is also obtained by cutting down the flow of blood to other organs like the gut. All this leads to the way we feel in an emergency. You can tell that your heart is racing and there is a sinking feeling in your stomach.
- The concentration of glucose (sugar) in the blood is controlled by two hormones produced by the pancreas – **insulin** and **glucagon** (see section 8.2).
- **Oestrogen** and **progesterone** are female sex hormones which are produced by the ovaries. **Testosterone** is the male sex hormone produced by the testes. These hormones promote the development of secondary sexual characteristics during puberty such as changes in body proportions, the development of body hair, breasts in girls and a lowering in the pitch of boys' voices. These hormones continue to control sperm or egg production and the female menstrual cycle throughout adult life.
- Hormones play a major part in controlling body processes. Like the central nervous system, they allow 'messages' to be sent from one part of the body to another. Whereas the central nervous system controls our actions and responses very rapidly, second by second, the hormone system exerts its control much more slowly. This happens over hours, days or even years as in the case of growth or the development of secondary sexual characteristics.

8.2 Insulin and diabetes

When we eat carbohydrates, such as starchy foods like bread, the enzymes in saliva and in the stomach break down the starch to produce glucose. The glucose is then absorbed into the blood and broken down still further in our cells producing energy, carbon dioxide and water.

Cells will only metabolise effectively if the glucose concentration stays at about the same level. This means that our bodies must control the level of glucose available to cells and in the blood. This **regulation of glucose levels** is effected by the **pancreas** and the **liver**. There are two mechanisms in this process:

1. If the *glucose concentration in the blood is too high*, the pancreas releases insulin into the bloodstream. The insulin causes liver cells to convert glucose into insoluble glycogen. The glycogen remains stored in the liver and the blood glucose level falls.
2. If the *blood glucose level becomes too low*, this is also registered by the pancreas and its release of insulin is reduced. Glucose is no longer converted to glycogen and the concentration of blood glucose rises (Figure 8.2).

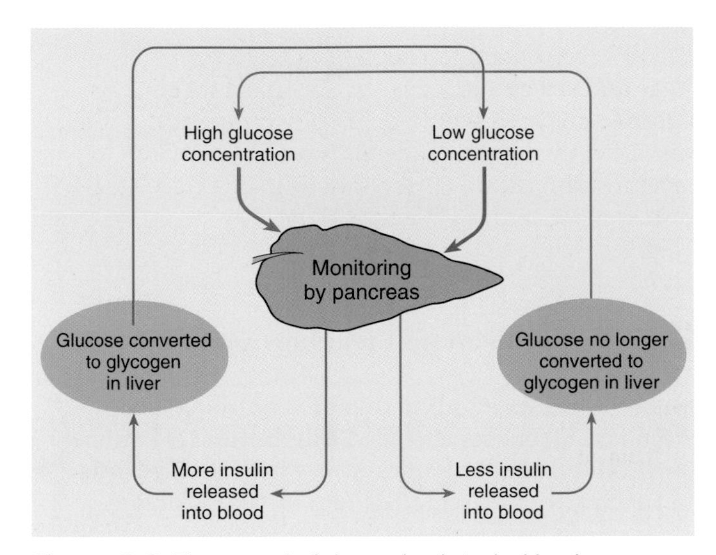

Figure 8.2 The control of glucose levels in the blood

This control of insulin output and hence glucose levels by the feeding back of information to the pancreas about glucose concentration in the blood, is known as **feedback control**. It is an example of **homeostasis** (section 8.4).

When the concentration of glucose in the blood becomes too low, cells in the pancreas can also produce a second hormone, glucagon, which causes the liver to convert glycogen into glucose and release it into the blood.

This maintenance of glucose concentration at a steady level using the two hormones, insulin and glucagon is also an example of homeostasis (Figure 8.3).

Some people cannot produce enough insulin to keep their concentration of blood glucose at a steady level. Without treatment, their glucose level becomes too high. They become weak and dizzy, eventually going into a coma. These people are called diabetics and their condition is known as **diabetes**.

In order to follow normal lives, diabetics must control the level of glucose in their blood. It is very important for them to pay attention to dietary habits, eating regular meals which do not include sugary foods.

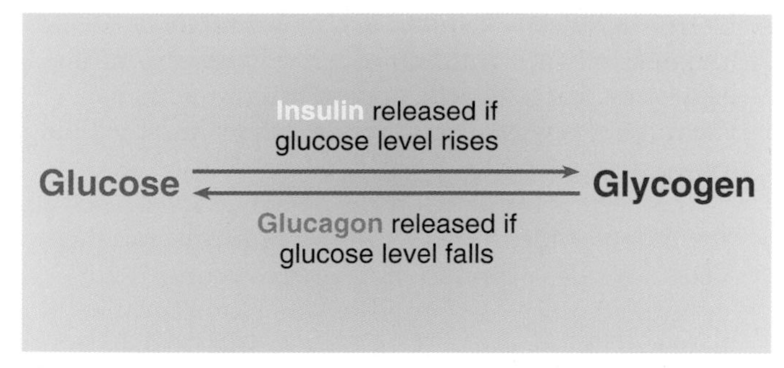

Figure 8.3 Maintaining glucose concentration at a steady level using insulin and glucagon

It may be necessary to take tablets to stabilise blood sugar and severe diabetics have to inject insulin. As digestive enzymes would break down the insulin if it was taken by mouth, diabetics have to learn how to inject insulin just under the skin after testing a sample of their urine to determine the dose.

8.3 Sex hormones and fertility

Important hormones are involved in the menstrual cycle of a woman and in the control and promotion of fertility. The menstrual cycle lasts approximately 28 days. There are three important aspects of the cycle:

1 egg and follicle development,

2 changes to the lining of the uterus and

3 changes in the concentration of hormones.

The cycle starts with **menstruation** when the lining of the uterus breaks down. During this time, tissues and blood seep out of the vagina over 4 to 6 days. This is usually known as a **period**.

Figure 8.4 Stages of the menstrual cycle showing changes in the ovary, in the uterus and in the concentration of hormones

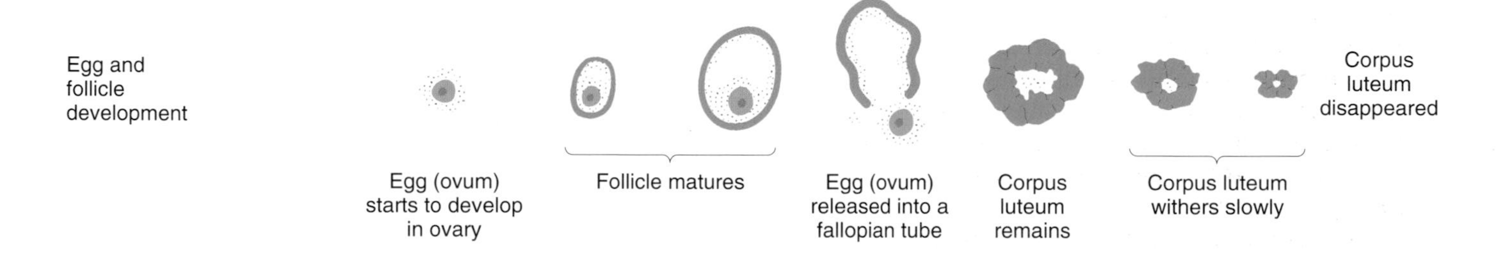

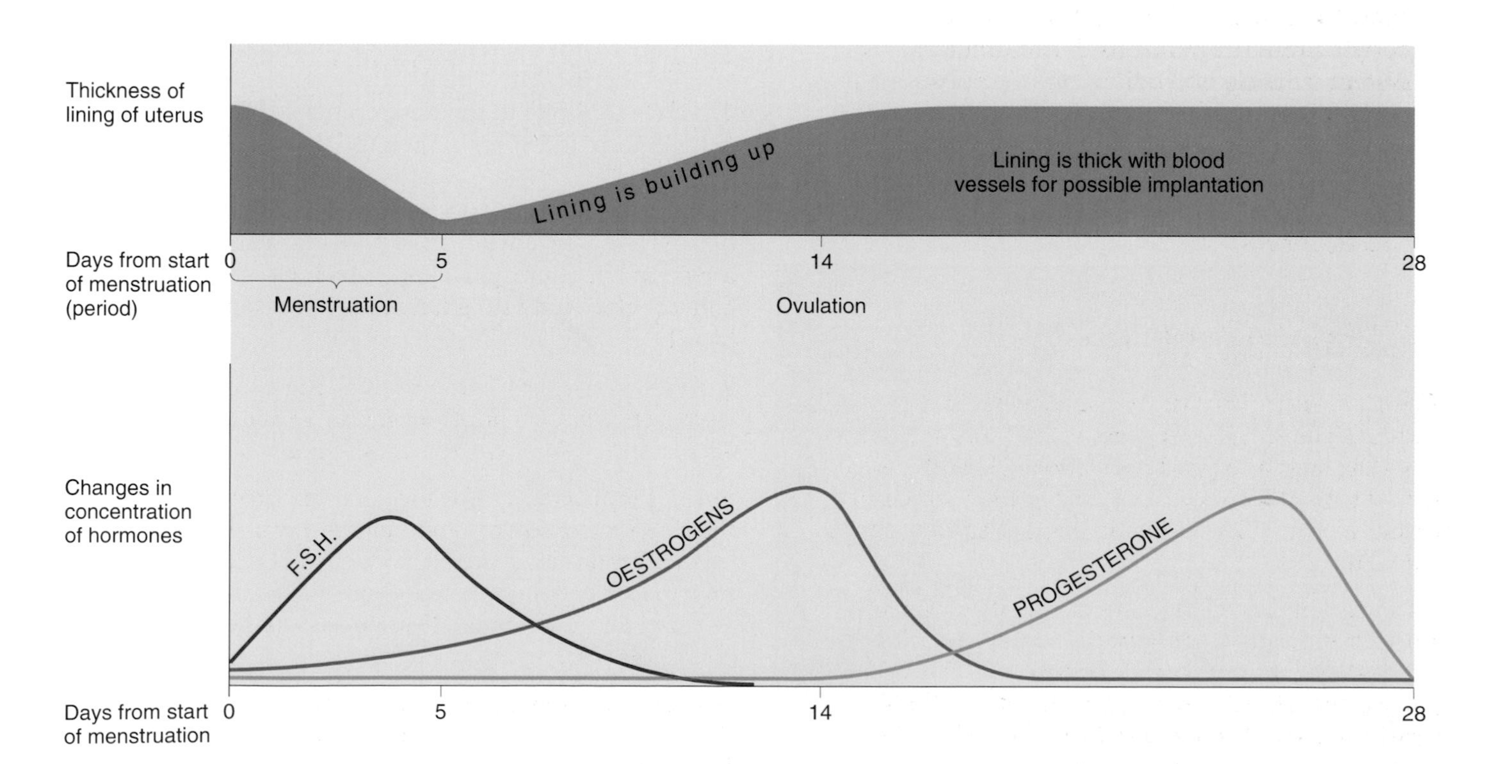

Thousands of immature eggs (**ova**) are present in the two ovaries. Immediately after menstruation, one of these eggs starts to develop inside a protective sheath called a **follicle**. About two weeks later, the follicle is mature and its egg (ovum) is released into one of the Fallopian tubes. This process is called **ovulation**. The follicle remains in the ovary and over the next two weeks becomes a **corpus luteum** (yellow body) and withers away.

As the follicle has been developing in the ovary, the lining of the uterus has been building up ready to receive a fertilized egg. If during sexual intercourse a sperm fertilizes this egg, the woman becomes pregnant and hormones prevent menstruation from taking place.

If conception has not occurred, about two weeks after ovulation when the corpus luteum has disappeared, the lining of the uterus breaks down. Menstruation starts and the whole cycle begins again.

Controlling the menstrual cycle

The menstrual cycle is controlled by hormones.

- At the beginning of the cycle, **follicle stimulating hormone (FSH)** is secreted by the pituitary gland. This causes the follicle to develop in the ovary.
- FSH also stimulates the ovary to produce **oestrogens**. Oestrogens have two controlling functions.
  (i) They promote the repair and build up of the lining of the uterus after the period has finished.
  (ii) They inhibit the production of FSH and therefore prevent further ovulation.
- **Progesterone** is produced by the corpus luteum after ovulation has occurred, two weeks into the cycle. This hormone maintains the lining of the uterus and keeps it thick with blood ready to receive the possible implantation of a fertilized egg. If the egg is not fertilized, the corpus luteum withers away, progesterone production ceases and the lining of the uterus breaks down once more after 28 days.

  If an egg is fertilized, the embryo attaches itself to the lining of the uterus. A placenta forms and this takes over the production of progesterone. The continued release of progesterone maintains the lining of the uterus and inhibits further ovulation.

So, female sex hormones control egg development and the whole menstrual cycle. They can therefore be prescribed in order to increase or reduce a woman's fertility.

- FSH is sometimes used as a **fertility drug** where it is found that a woman's level of FSH is too low to stimulate the development of eggs and follicles.
- Oestrogens are taken as a **contraceptive** measure (the 'pill') to inhibit FSH production and prevent eggs maturing.

8.4 Maintaining control – homeostasis

Living things cannot survive unless the conditions inside their bodies, such as temperature and pH are fairly constant and under control. This means that our body systems must work together to provide the most suitable (optimum) conditions for cells to function efficiently.

Temperature control in warm blooded animals can be compared with the temperature control of a room using a thermostat. The thermostat acts as a **sensor** which is sensitive to the temperature. If the temperature is too high, then the thermostat switches

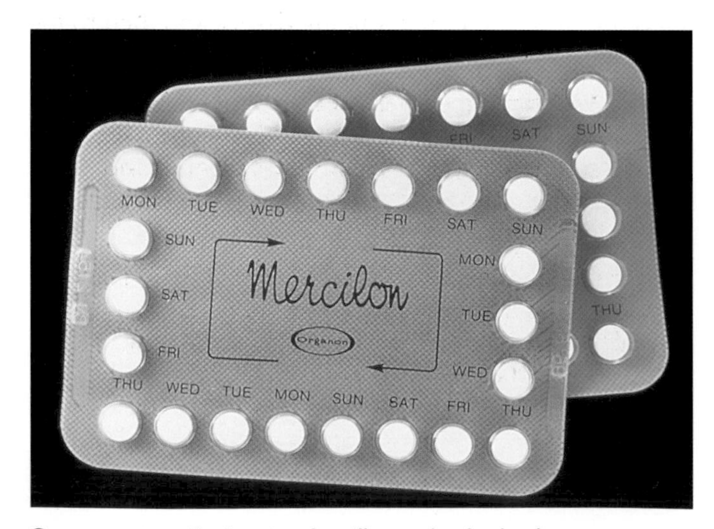

Contraceptive pills showing the pills to take day by day

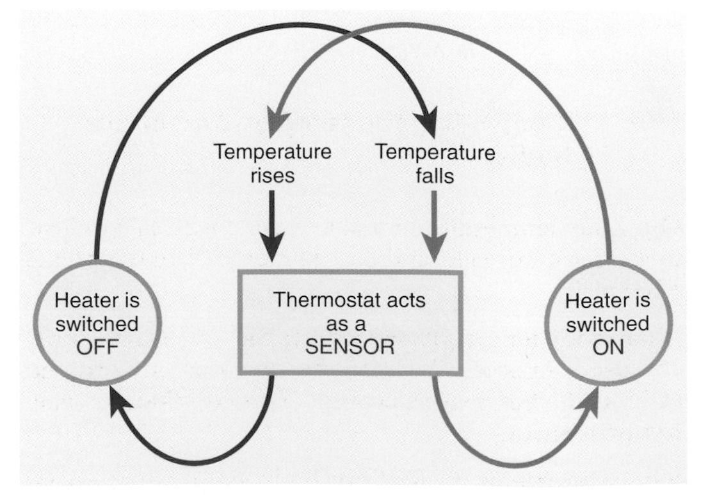

Figure 8.5 Feedback control in maintaining the temperature in a room using a heater and thermostat

the room heater off and the temperature falls (Figure 8.5). If the temperature now becomes too low, the thermostat switches the heater on and the temperature rises (Figure 8.5). This 'feedback' of information to control the heater is called **'feedback control'** or **'negative feedback'** as the feedback of information always tries to change the conditions.

Our blood and skin can act as sensors of temperature rather like a thermostat. If the temperature rises or falls, then our blood and skin 'feedback' information to the brain. The brain then passes messages to our muscles, hairs, arteries and sweat glands to put things right.

This 'feedback' control of machines and organisms which tries to keep conditions more or less the same is called **homeostasis**. This word comes from two Greek words – *homeo* meaning the same and *status* meaning state or condition.

There are many examples of homeostasis and some of these are shown in Table 8.1.

Table 8.1 A summary of some important homeostatic systems

Variable being controlled	**Sensor** which 'senses' (monitors) the variable	**Actuator** which changes conditions (puts things right)
room temperature	thermostat	heater
body temperature	skin and blood	muscles, hairs, sweat glands
glucose in blood	brain	insulin and glucagon from pancreas
water content of body	brain	kidneys (and to a lesser extent lungs and skin)
traffic flow	TV cameras, police HQ	diversions, traffic lights

8.5 Controlling the temperature of our bodies

Our body temperature must be maintained fairly close to 37°C for our cells and overall metabolism to operate effectively.

If our body temperature becomes too low, we become confused. At about 27°C, we become unconscious and at 25°C the heart stops beating. This condition is called **hypothermia**.

Similar problems occur if our body temperature is too high. In this case, death occurs because the heart is working too fast.

There are three main ways in which our bodies gain heat and three in which they lose heat.

Gain of heat

1. Respiration, which happens in the cells, is an exothermic reaction (gives out energy).
2. We are warmed by the Sun, especially in the summer months. When it is cold outside, heating systems in buildings will make us feel warmer.
3. A hot meal or a hot drink may help to warm us up.

Loss of heat

1. When we breathe out, we are losing a warm gas.
2. Heat is lost when excreting warm urine and warm faeces.
3. Conduction and radiation to our surroundings causes us to lose heat.

Our bodies have developed important control systems to counteract these gains and losses of heat and maintain our internal body temperature at 37°C.

> The skin is our main contact with the environment and it plays an important part in maintaining a constant body temperature.

Temperature receptors in the skin send impulses to the heat regulating (thermoregulatory) centre of the brain which then controls the responses of the skin. The skin can respond and control body temperature in four ways.

1 Variable blood flow

Blood flows through the capillaries in the skin which are close to the surface. Heat from the blood is therefore lost to the surrounding tissues and then to the air (Figure 8.6 on the following page).

In cold weather, blood vessels and capillaries close to the surface of our skin contract. This is called **vasoconstriction**. Less blood flows through the narrower (constricted) capillaries and therefore less heat is lost from the blood.

In warm weather, our surface blood vessels and capillaries expand and this process is called **vasodilation**. This enables more blood to flow through them and excess heat can be lost from the body.

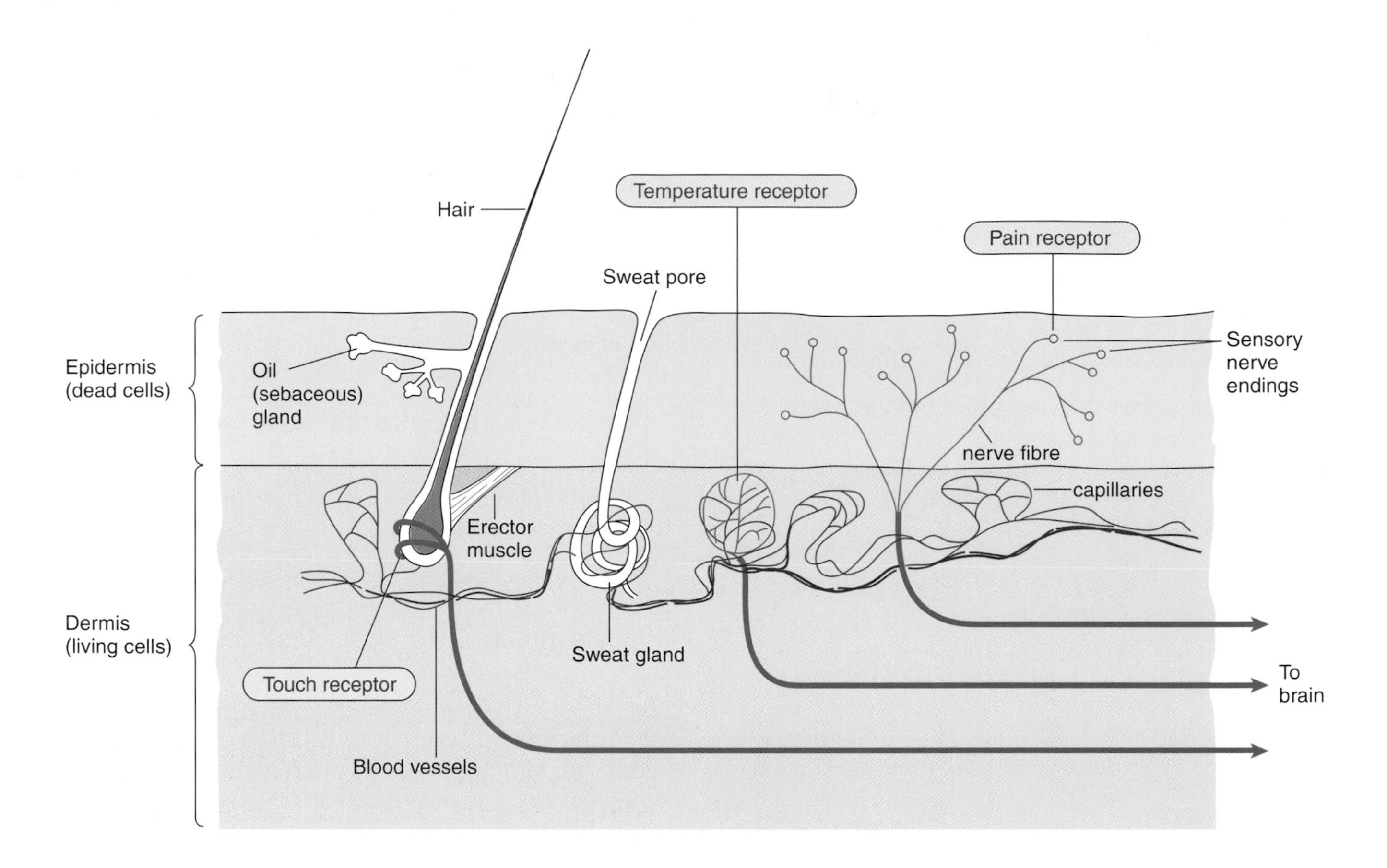

Figure 8.6 The structure of human skin

2 Shivering

When we shiver, our muscles contract. This activity results from and stimulates respiration in our muscle cells. Respiration is an exothermic process, so heat is produced to warm us up.

Our body temperature rises when we exercise vigorously. This makes us sweat. As sweat evaporates, it takes heat from the skin and so cools us down

3 Sweating

Sweat glands secrete sweat (which is mainly water) onto the surface of our skin. As soon as our blood temperature rises above normal, we begin to sweat. As the sweat evaporates, it takes the heat needed for evaporation from the skin and so cools us down.

4 Hairs and muscle activity

Each hair in our skin can be raised or lowered by its own tiny **erector muscle** (Figure 8.6). The erector muscle attaches the base of the hair to the top of the dermis.

In cold weather, the muscle contracts, pulling the hair away from the surface of the skin. The raised hairs therefore trap more air which is a very poor conductor of heat. The layer of air can now act as an insulator and heat loss from the body is reduced.

In warm weather, the erector muscles relax, less air is trapped close to the body and heat loss is less restricted.

The clothes we wear also play an important part in helping us to control body temperature.

- Clothing traps air which acts as an insulator and prevents heat loss. Garments made of wool are particularly effective.
- Light, thin clothing reduces the amount of trapped air and allows for the loss of heat in hot weather.
- Light coloured (particularly white) clothing reflects more radiation and lowers the amount of heat absorbed. This is why cricketers and tennis players generally wear white.

8.6 Getting rid of waste

In Chapter 2, we learnt how food is broken down. Then, in Chapters 3 and 4 we learnt how this digested food plus oxygen is made available to the cells in our bodies via the blood system.

After using the digested food and oxygen, our bodies produce waste products. These waste products resulting from metabolism must be removed from our bodies in order to keep the internal environment in and around our cells relatively constant.

> Most of the waste products from body processes are removed by the lungs and the kidneys.

- Carbon dioxide and water are produced during respiration in our cells. These diffuse into the blood and then through the lung alveoli where they are breathed out (see section 3.3).
- Urea, excess water and excess salt are removed from the blood by the kidneys and excreted in the urine. This is discussed in more detail in the next section.
- Water and salts are also lost via the skin when we sweat in order to get cooler.

8.7 Controlling the water and salt content of our bodies – the kidneys

The kidneys play a major role in controlling the water and salt content of our bodies. We take in water and salts when we eat and drink. We lose water and salts from our bodies in urine and sweat.

Urine contains mainly water. It is sometimes almost colourless, but when it contains greater concentrations of other waste products, it looks yellow.

Besides water and salts, the other major waste product in urine is **urea**. Urea is produced when proteins are metabolised. Proteins are first broken down into amino acids. Most of these amino acids are used for growth and the repair of body tissues. Excess amino acids are broken down in the liver to form urea (Figure 8.7).

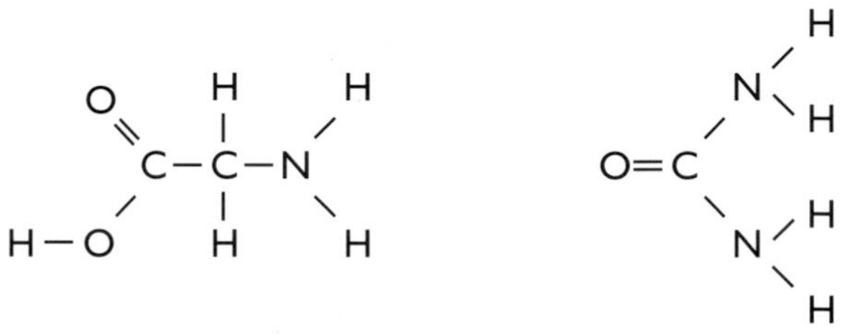

glycine

urea

Figure 8.7 The chemical structures of glycine (the simplest amino acid) and urea

Urea is poisonous and therefore it must be removed from the body. It dissolves in the blood surrounding the liver cells and is then removed from the blood by the kidneys. Figure 8.8 over the page shows how the kidneys are connected to the blood system and the **urinary excretory system**.

The kidneys

We have two kidneys at the back of the body, just above the waist. The kidneys have an excellent supply of blood through the renal arteries coming from the aorta (Figure 8.8). Blood is carried back to the heart via the renal veins and the vena cava.

As branches of the renal arteries pass into the kidneys, they divide into an intricate network of blood capillaries (Figure 8.9 over the page).

These capillaries are wrapped around thousands of tiny, convoluted and looped **tubules**. As the blood flows over the tubules, water, salt, urea and other waste products are first filtered out. Water and other substances, including sugars and salts, are then re-adsorbed to meet the body's needs. This leaves urea, excess salt and excess water. The tubules later join up to form connecting ducts which lead into the **ureters**. At the same time, the blood capillaries join up again to form a single vein leading to the renal vein.

The ureters (one from each kidney) carry urine to the **bladder** (Figure 8.8). The bladder expands as urine collects in it. Urine is excreted from the bladder via the **urethra**. The urethra runs to an opening in front of the vagina in a female and down the centre of the penis in a male.

The kidneys produce dilute urine if there is too much water in the blood and concentrated urine if there is too little water in the blood. Cells in the brain called **osmoreceptors** act as sensors. These monitor the water content of the blood. When the water content falls below normal, **antidiuretic hormone (ADH)** is released from the pituitary gland into the bloodstream. This causes the capillaries to re-adsorb more water and results in concentrated urine being passed. The osmoregulators sense the increased water content of the blood and negative feedback causes less ADH to be released.

If the water content of the blood is too high, ADH is no longer released. Less water is re-adsorbed into the blood capillaries of the kidneys and the urine passed is dilute.

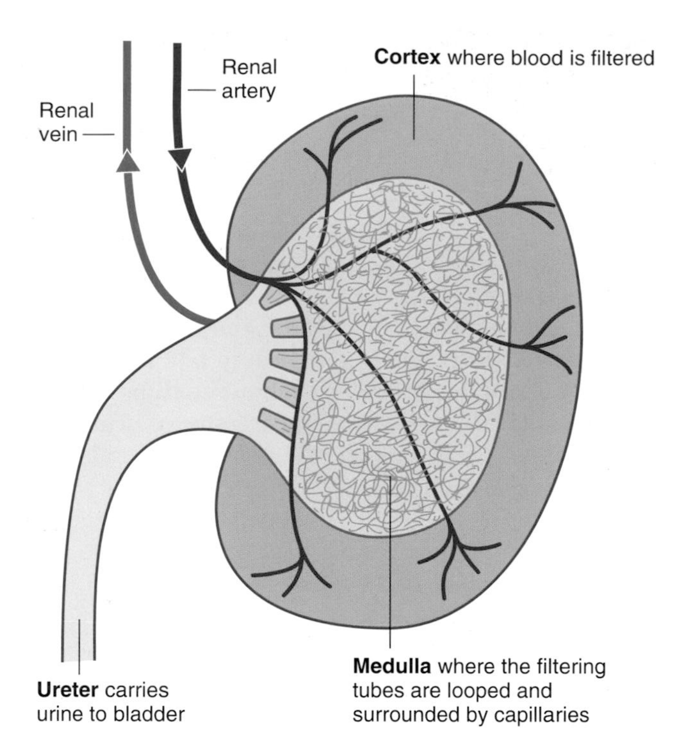

Figure 8.9 A cross section through a kidney

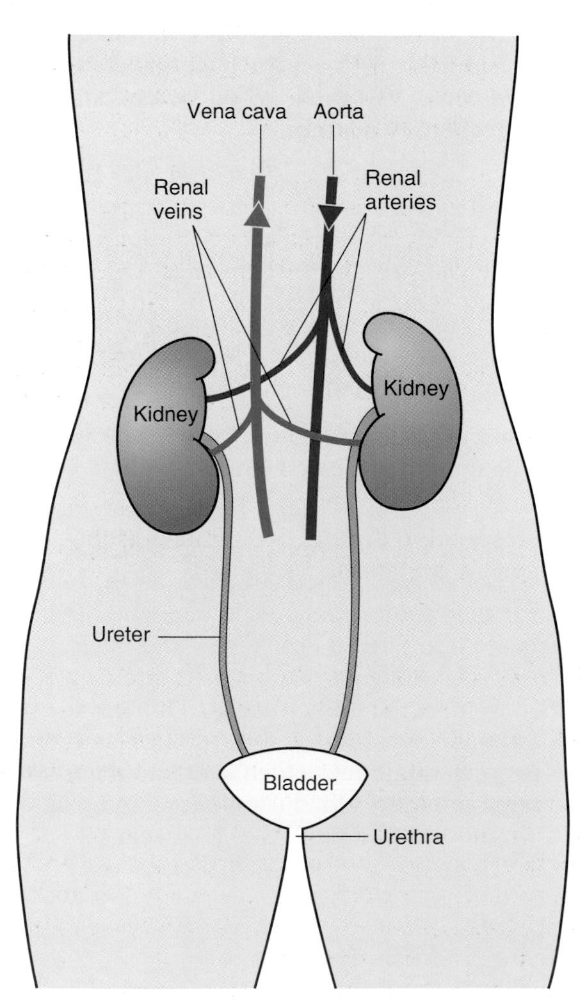

Figure 8.8 The kidneys and the urinary excretory system

8.8 Drugs – beneficial and harmful

Chemicals which affect our metabolism and behaviour are often referred to as drugs. Almost all drugs influence the way we feel and behave and the correct use of drugs under medical supervision can have many beneficial results.

Drugs which affect the central nervous system can be divided into four different types.

1. **Sedatives** (**depressants**) slow down the brain and nervous system and can cause drowsiness. The drug valium is prescribed to help people suffering from anxiety. Alcohol is another example and it does not take much alcohol for a slowing down of mental faculties to take place and this is why you must not drink and drive.
2. **Stimulants** speed up the nervous system and make you more alert. They include caffeine in tea and coffee, 'pep pills' and cocaine.
3. **Analgesics**, such as paracetamol and aspirin are used to reduce the feeling of pain such as tooth-ache. Morphine is prescribed to alleviate intense pain, while heroin is used illegally to dull the senses.
4. **Hallucinogens**, such as LSD and cannabis (pot, dope, joints) cause weird and unusual experiences and sensations, and may cause the person under the influence of these drugs to be irresponsible and act dangerously.

Don't drink and drive!

Table 8.2 The harmful effects of certain drugs

Drug	Effect on behaviour and health
Nicotine in tobacco	Tobacco smoke inhibits the action of cilia and destroys cells which line the air passages and the lungs. This causes bronchitis and emphysema and may lead to lung cancer. A strain is put on the heart and disease of the circulatory system ensues because of lowered levels of oxyhaemoglobin in the blood (see section 3.4).
Alcohol	Taking alcoholic drinks affects the nervous system. Reactions are slowed down, speech slurred, walking becomes unsteady and a really drunk person may become unconscious. There is a lack of self-control. Long-term abuse causes liver and brain damage.
Organic **solvents** in glues, sprays, polishes, etc.	Inhaling these substances affects the nervous system and key areas of metabolism. This causes damage to the brain, heart, lungs and liver.
Cannabis	This drug affects the brain and the nervous system. Long-term abuse can cause damage to the brain and the reproductive system.

8.9 Hormones in plants

Which one of these cress plants has been grown with light from only one side?

Have you ever noticed how plants on a window ledge grow towards the light? The stems bend towards the window and the leaves and any flowers turn to the Sun. Responses like this, where plants are responding to stimuli, are called **tropisms**. Tropisms occur much more slowly than responses in animals, however.

Scientists now know that the responses of plant roots and shoots to light, gravity and moisture are caused by chemicals which control cell growth and development. These chemicals are **plant hormones** which act in a similar way to hormones in animals. So plants have a slowish *hormonal* response system rather like animals (but, animals have a rapid response system in addition, involving the central nervous system and muscles).

> Plants respond to three major stimuli – **light**, **gravity** and **moisture**.

- **Phototropism** is the name for a plant's response to light. Plant shoots grow towards the stimulus of light and are described as *positively phototropic*. On the other hand, plant roots grow away from the light so they are *negatively phototropic*.
- **Geotropism** is a plant's response to gravity. Plant roots grow towards the stimulus of gravity and so they are described as *positively geotropic*. On the other hand, plant shoots are *negatively geotropic* (Figure 8.10 over the page).
- Plant roots also grow towards moisture. This is another example of positive tropism.

The responses of plants to these stimuli are very beneficial. Roots growing towards gravity and moisture will anchor the plant firmly and provide the all important water for photosynthesis. Shoots growing towards the light will photosynthesize more effectively and therefore grow more successfully.

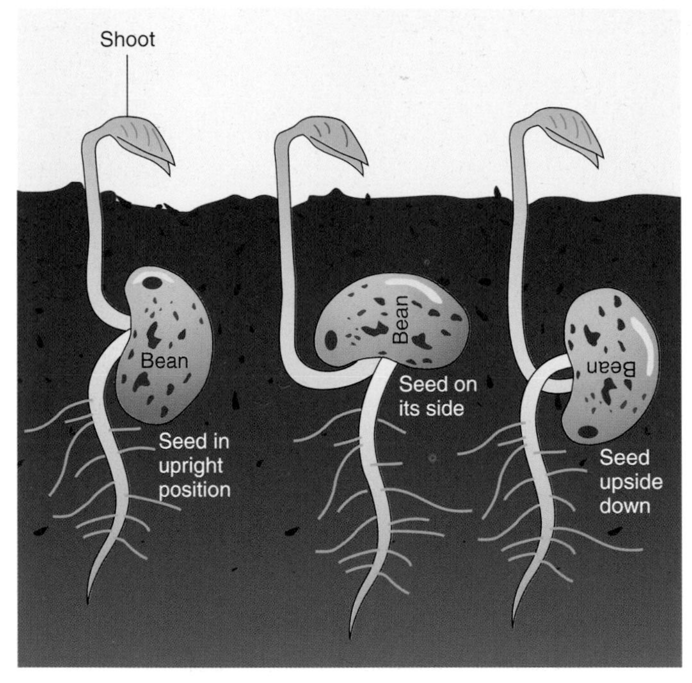

Figure 8.10 When a bean seed grows, the shoots are always negatively geotropic and the roots are always positively geotropic

8.10 The control of the way a plant grows

The growth in plant roots and shoots is controlled by a group of plant hormones called **auxins**. For example, phototropism occurs because one particular auxin collects on the side of the shoot furthest away from the light. This auxin promotes growth so the shoot grows faster on the shaded side and it bends towards the light (Figure 8.11).

Shoot tips also produce hormones that inhibit side shoot growth. Removal of the tip removes this inhibition and side growth occurs much faster. This explains why regular hedge clipping promotes bushier and thicker hedges. Other auxins are produced in response to gravity, causing opposite effects in roots to that in shoots.

Auxins and other plant hormones are now produced commercially and used by gardeners and horticulturists.

- Hormone rooting powders are chemicals which promote the growth of roots in plant cuttings.
- Selective weedkillers are synthetic growth substances which make broad leaved weeds like dandelions and daisies grow so fast that they use up all their food supply and then die.
- Growth substances can be sprayed on unpollinated flowers to make them grow and develop into fruits without seeds. This is how seedless grapes and oranges are produced.

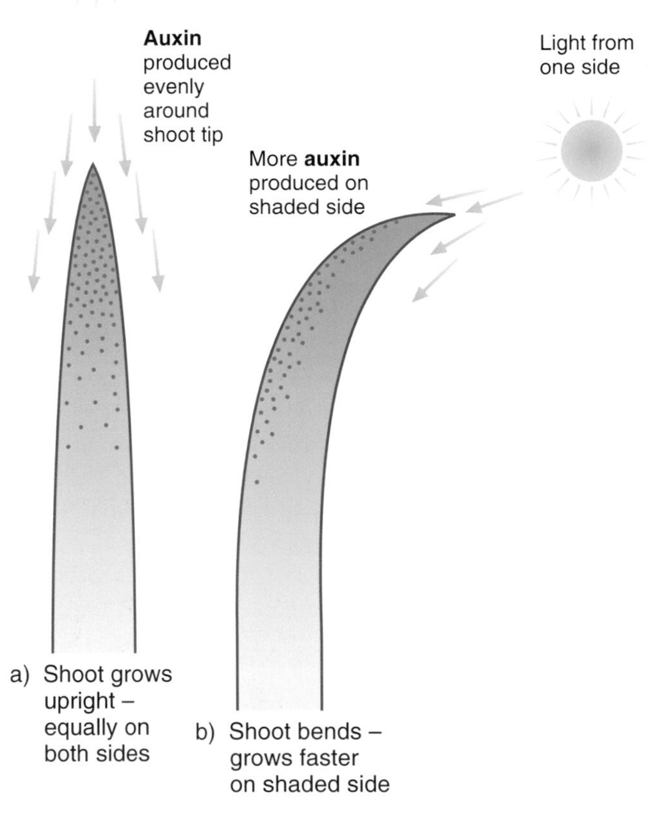

Figure 8.11 Auxins control the way in which a plant grows

These two cuttings were taken at the same time from the same plant. The cutting on the left has been treated with hormone rooting powder and the other was untreated

The dock plant in the bottom photo has been treated with selective weedkiller. The hormones speed up the plant's metabolism so that it grows too fast for its supply of water and nutrients and then dies

QUESTIONS

1 a) Diagrams A, B, C and D show blood vessels and a sweat gland in the skin.

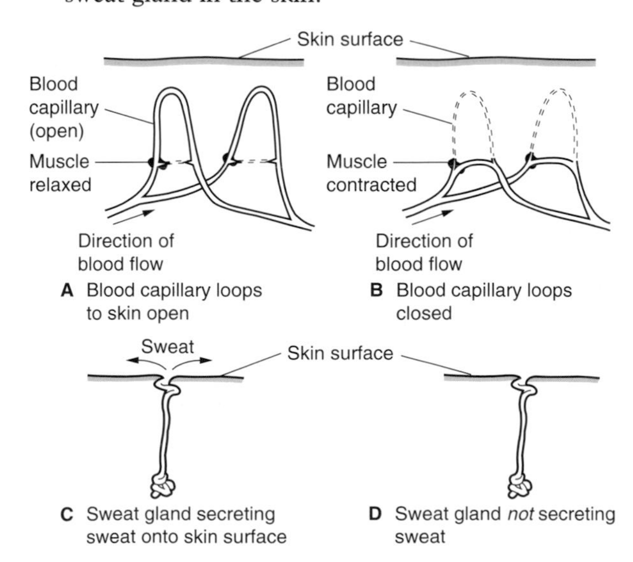

i) Which **two** diagrams show the blood vessels and the sweat gland when the body is reducing heat loss from the skin?
ii) Explain how sweating helps to cool the skin.

b) In an investigation into temperature control a person was kept in a room maintained at a temperature of 45 °C. The graph shows the effect on the person eating a mouthful of ice on the temperature of the skin and of the blood.

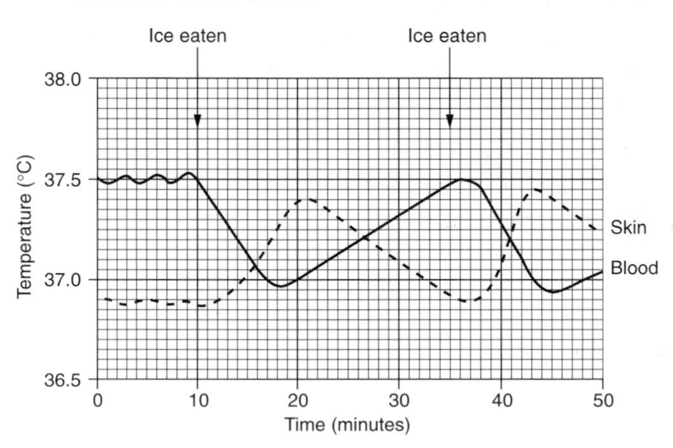

i) Give **two** changes, other than skin temperature, which will occur in the body in response to the change in blood temperature after eating the ice.
ii) Explain why the skin temperature changes after eating the ice. **NEAB**

2 The diagram shows the sequence of events that occur in the body of a person not suffering from diabetes.

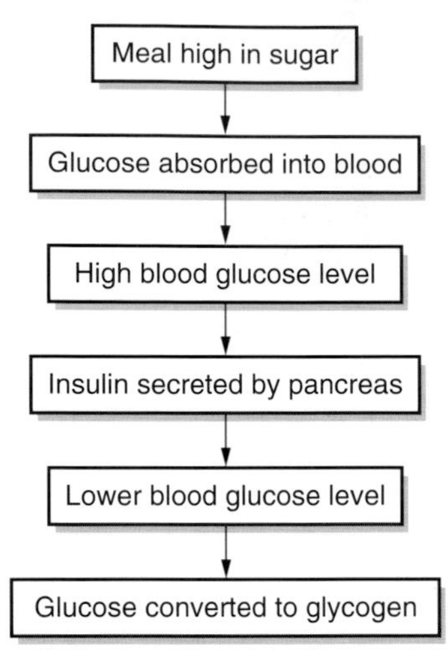

Use the diagram and your knowledge to answer the following questions.

a) What influences the conversion of glucose to glycogen?

b) Name **one** organ in which glycogen is stored.

c) Name the mechanism that controls blood glucose.

d) Explain how this mechanism controls blood glucose. **NICCEA**

3 Your skin is covered by an invisible protective barrier. It protects the skin from possible irritation and harmful bacteria. This barrier has a pH value of 5.5.

- Most soaps have a pH value of approximately 10.
- Soaps with added moisturisers have a pH of 7.3.
- Facial Wash is a soap-free cleaner which has a pH of 5.5.

a) Copy and complete the following chart using two crosses (X) to show the pH value of soap with moisturisers and Facial Wash.

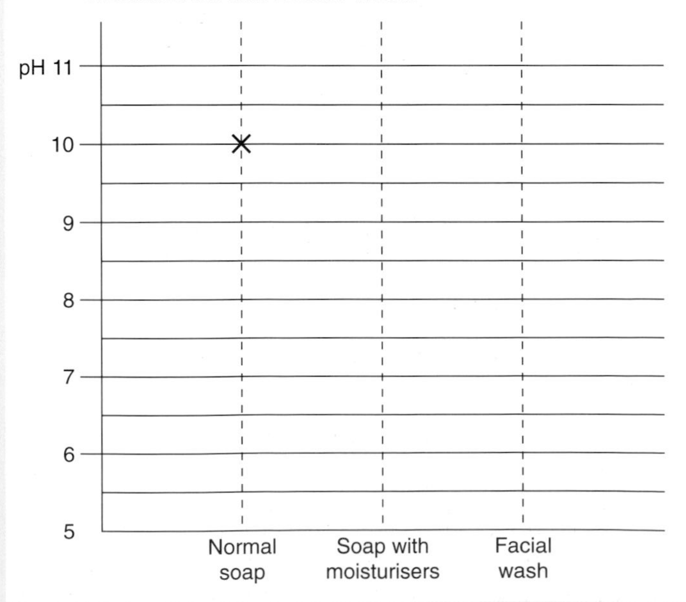

b) Copy the following conclusions from these results and use the words from the following list to complete them.

neutral substance strong acid strong alkali weak acid weak alkali

i) Normal soap is a ______________ .
ii) The protective barrier on the skin is a ______________ .

c) Why is the Facial Wash better than the soaps for washing delicate skin? **MEG**

4 Co-ordination in animals can be brought about by hormones. Explain how the body reacts when the blood sugar level rises. **WJEC**

5 Humans can survive in environments as hot as 50 °C and as cold as −50 °C. They are able to maintain a blood temperature between 36 °C and 38 °C. The table shows some of the possible responses to cold and heat.

Responses to cold	Responses to heat
increased voluntary movement	decreased voluntary movement
increased metabolic rate	reduced metabolic rate
shivering	sweating
putting on more clothing	taking off clothing
moving to a warmer place	moving to a colder place
vasoconstriction	vasodilation

a) Explain how an increased metabolic rate can help to maintain blood temperature in a cold environment.

b) Suggest why sweating has a greater cooling effect in a dry desert than in the humid tropics.

c) Explain how vasoconstriction helps to maintain blood temperature in a cold environment.

d) The figure shows the principle of a homeostatic process.

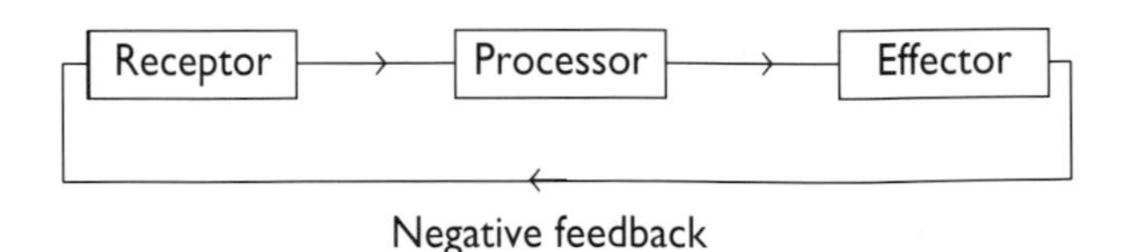

Use this figure to explain how blood temperature can be maintained by shivering. **MEG**

CHAPTER

Chains and cycles

9.1 Ecosystems – habitats for communities

Look at a field, or a garden or even a single rose bush. Notice that organisms live together in groups rather than as individuals. There might be several rabbits in the field, blackbirds in the garden or ladybirds on the rose bush.

These groups of organisms of the same species which live together in one particular area are called **populations**.

Usually there are populations of many different species in a particular area. Just think of the dozens of different populations in even the smallest garden – daisies, roses, daffodils, blackbirds, squirrels, etc.

Different populations living together are called a **community** and the area in which the community lives is called a **habitat** (Figure 9.1).

Different habitats support very different communities of plants and animals. So, the community in a pond will be very different to a country garden community.

The combination of a habitat and a community is called an **ecosystem** and the study of ecosystems is known as **ecology**.

For example:

ecosystem =	habitat +	community
(pond)	(water, mud, stones)	(tadpoles, frogs, beetles, sticklebacks, spirogyra, pondweed)

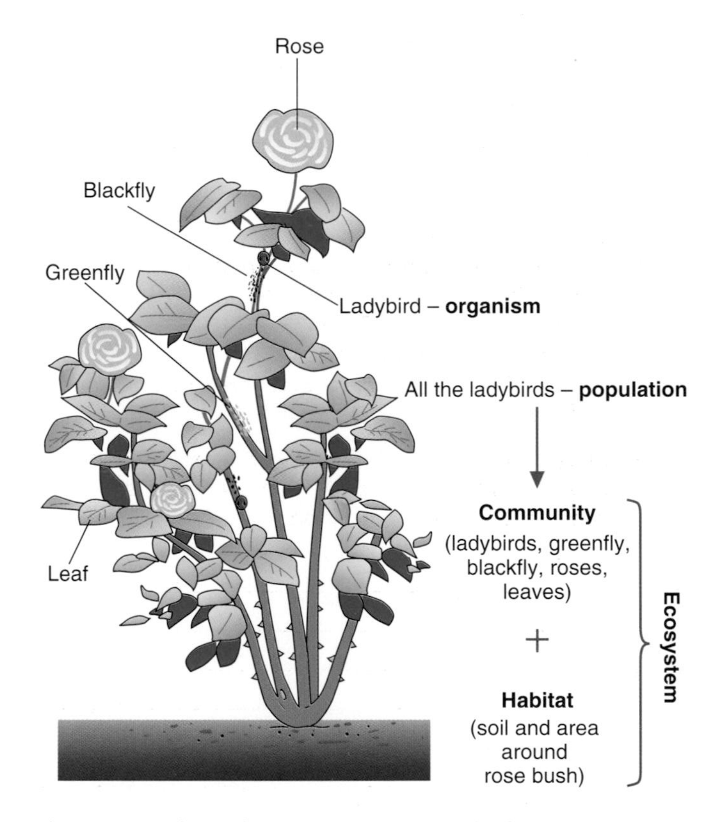

Figure 9.1 A simple ecosystem on a rose bush

The simple ecosystem on a rose bush is shown in Figure 9.1.

Although different ecosystems can be studied individually, they are not usually isolated. For example, the ladybirds, greenfly and blackfly in the ecosystem in Figure 9.1 might well fly to other nearby ecosystems. On a much larger scale, ecosystems in different continents can be linked by migrating birds, fish and other animals.

This swallow links ecosystems in Europe and Africa

9.2 Interactions and energy transfer in ecosystems

In any ecosystem there are different interactions between the animals and plants in the community and with the habitat itself. One of the most important of these interactions is that based on feeding.

Feeding interactions involve the **transfer of energy** from one organism to another. The ultimate source of energy for almost every ecosystem is sunlight.

Energy enters an ecosystem as light which is captured by green plants so that they can photosynthesize and produce their own food. Plants are the only organisms which can produce food in this way. Because of this, plants are described as the **producers** in an ecosystem. All other living things depend on green plants for their food and energy. They consume plants or animals which have eaten plants and are known as **consumers**. So all animals are consumers because they rely on plants and other animals for food.

Figure 9.2 shows one example of the feeding interactions between producers and consumers in an ecosystem. Sequences of producers to consumers such as grass and seeds → rabbit → fox are called **food chains**. Food chains show which organisms eat other organisms.

Three more examples of food chains are shown in Table 9.1.

Table 9.1 Examples of food chains

Producer	First consumer	Second consumer	Third consumer
grass	snail	thrush	cat
rose bush	greenfly	ladybird	
seaweed	crab	seagull	

Figure 9.2 Producers and consumers in a food chain

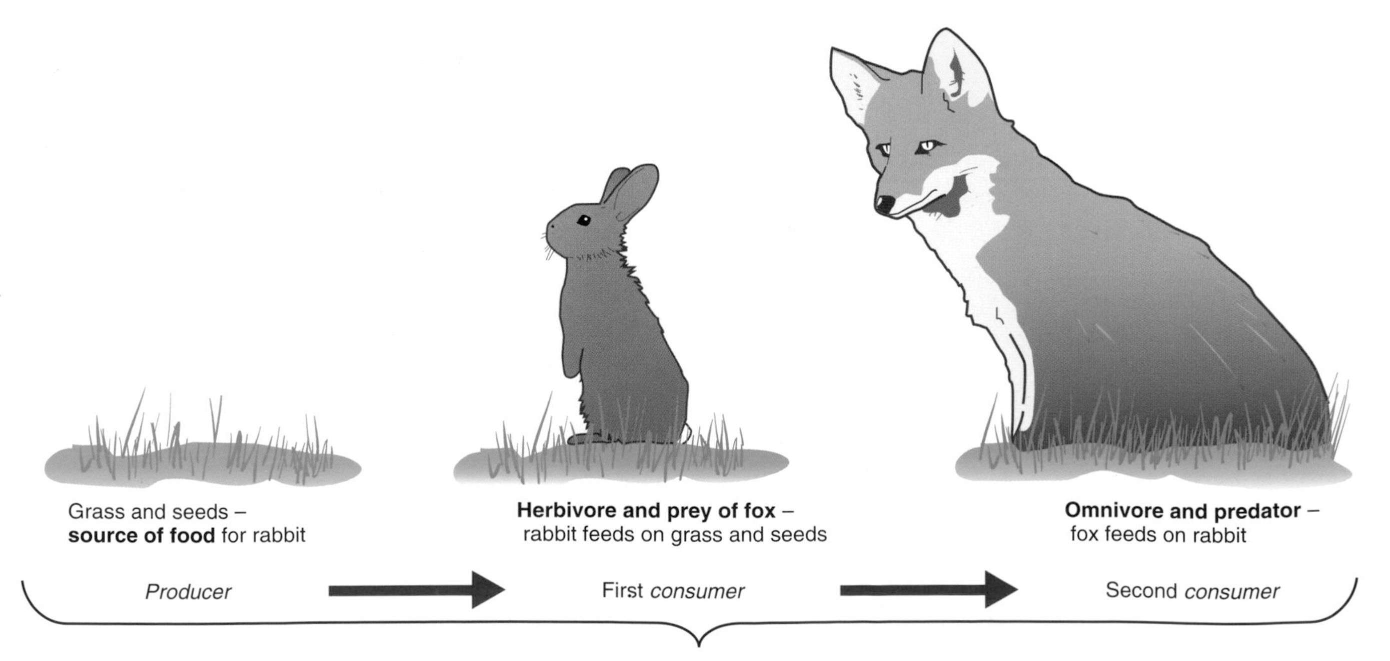

In any food chain:

- energy and materials are transferred along the chain from one organism to another;
- the producers are always plants;
- the first consumers are often **herbivores** – animals that eat only plants;
- the second and subsequent consumers are either **carnivores** – animals that feed only on other animals, or **omnivores** – animals that eat both plants and animals.

Carnivores, such as foxes, which hunt other animals are called **predators**. The animals, like rabbits, which they eat are the predator's **prey**.

9.3 Food chains and food webs

In an ecosystem, feeding patterns are rarely as simple as those in Figure 9.2 or Table 9.1. For example, thrushes will eat greenfly and ladybirds as well as snails. Crabs will consume snails as well as seaweed. In any community of plants and animals, several food chains will probably exist and these will be interlinked with some organisms in more than one food chain. These interconnected food chains are known as **food webs**.

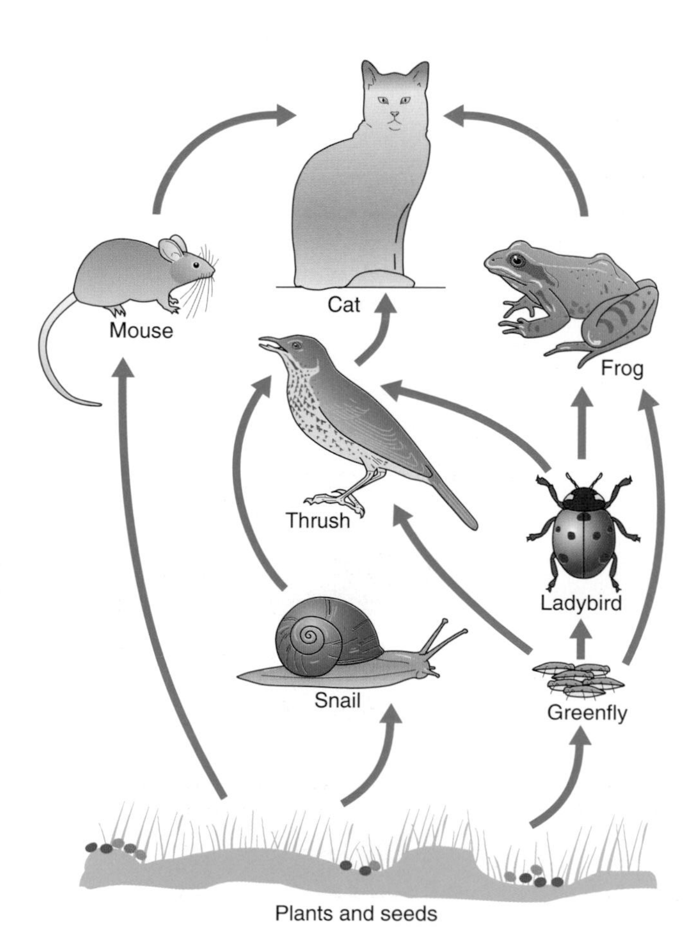

Figure 9.3 A food web for a garden community

A food web for a garden community is shown in Figure 9.3.

As energy is transferred along a food chain from producers to consumers, most of it is lost to the environment as heat. Plants and animals must respire in order to live. Respiration converts the chemical energy of food into other forms of energy including heat. Most of the heat is lost to the environment and there is less to pass on to the next consumers. In time, very little energy is left and this limits the number of steps in the food chain.

Pyramids of numbers

When biologists first studied ecosystems, they looked at the number of organisms in the populations at each step in a food chain. Information from these studies was presented in diagrams called **pyramids of numbers** and Figure 9.4 shows a pyramid of numbers for one food chain in a small garden. Each level of the pyramid is represented by a block and the width of the block represents the number of organisms at that level.

As you pass along a food chain, the number of organisms at each level usually gets less and the pyramidal shape becomes clear. Millions of plant producers provide food for fewer (perhaps thousands) of herbivores which are the primary consumers. Finally, the omnivores and carnivores are usually fewer again in number but often larger in size.

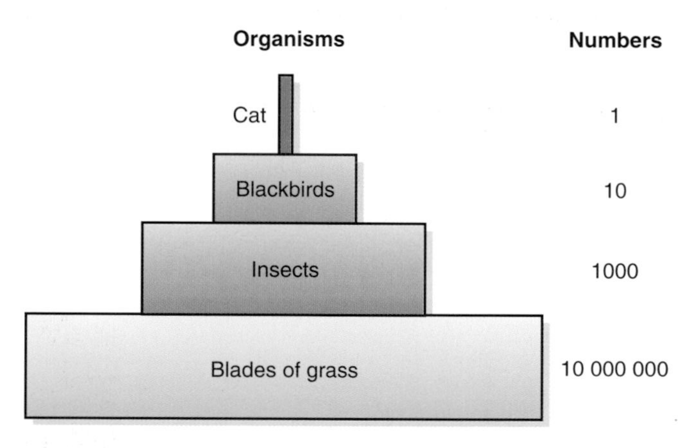

Figure 9.4 A pyramid of numbers for one food chain in a small garden

Occasionally, a food chain does not produce a neat pyramid of numbers like that in Figure 9.4. This can happen when one relatively large producer, like a rose bush supports a huge population of smaller herbivores and carnivores (Figure 9.5).

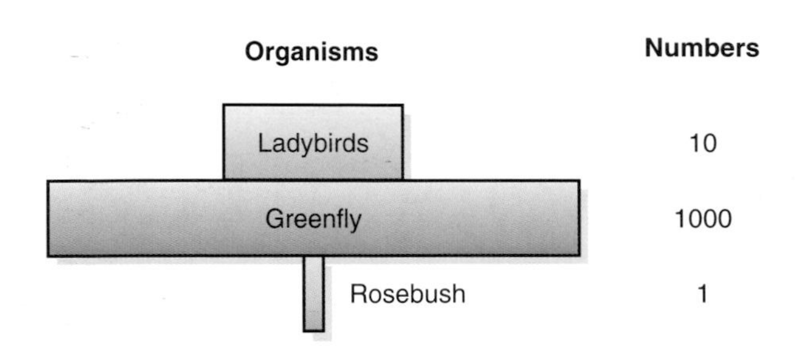

Figure 9.5 A pyramid of numbers for greenfly and ladybirds on a rose bush

9.4 Pyramids of biomass and energy transfer

A more accurate description of the food and energy transfer in a food chain is shown by a **pyramid of biomass** rather than a pyramid of numbers.

The different levels of feeding in a food chain are sometimes called **trophic levels**. Pyramids of biomass are based not on the number of organisms at each trophic level but on their total dry mass or biomass. Figure 9.6 shows a pyramid of biomass for the food chain in Figure 9.5. When this is done, the biomasses of the ladybirds, greenfly and the rose bush produce a neat pyramid.

Pyramids of biomass are almost always neat pyramids with the biomass of organisms decreasing from one trophic level to the next along the food chain.

This decrease in biomass along the food chain shows very clearly that only a fraction of the total food, and therefore the total energy, passes from one trophic level to the next.

The loss of energy from one trophic level to the next occurs because:

- Respiration uses up food (energy) in maintaining the living processes in organisms. These include energy for movement and warmth. Energy is also lost as heat to the surroundings. Warm blooded animals, particularly mammals and birds, whose body temperature is often higher than that of the atmosphere around them, need extra energy to maintain their body heat.
- Energy and materials are lost in an organism's waste products. Carbon dioxide is breathed out as a warm gas. There is heat lost to the body when urine or faeces are passed out.

Careful measurements show that only 10% (or even less) of the available energy gets passed on from one trophic level to the next. So, a herbivore (trophic level 2) has access to one tenth of a plant's energy (trophic level 1), but a carnivore (trophic level 3) has access to only one hundredth.

This loss of energy explains why food chains rarely include more than four trophic levels. With four trophic levels, only one thousandth (0.1%) of the original producer's energy is available to the third consumer.

The loss of energy along a food chain also means that it is far more energy efficient to eat plants than to eat animals. As humans, we receive far more energy from cereals by eating them directly than by feeding the cereals to cattle and then eating the beef (Figure 9.7).

Figure 9.7 also explains why an area of agricultural land can be used more effectively and can feed more people if it is used to cultivate crops rather than to rear

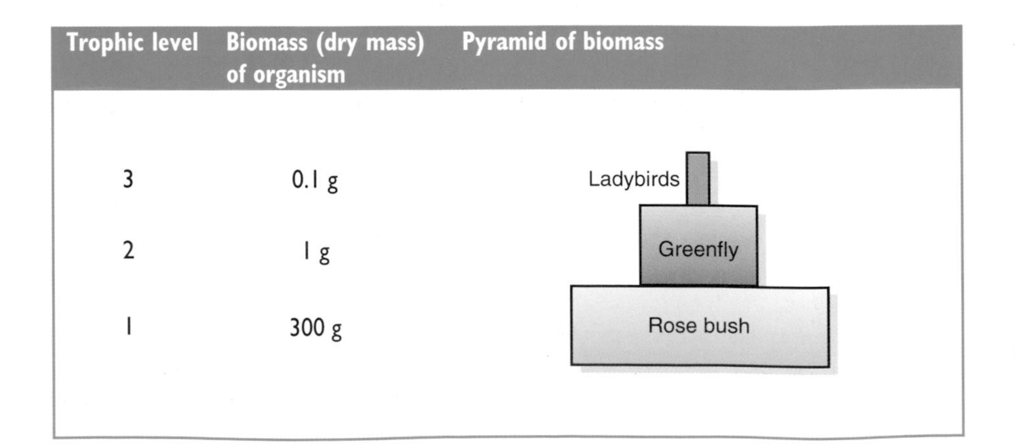

Figure 9.6 A pyramid of biomass for greenfly and ladybirds on a rose bush

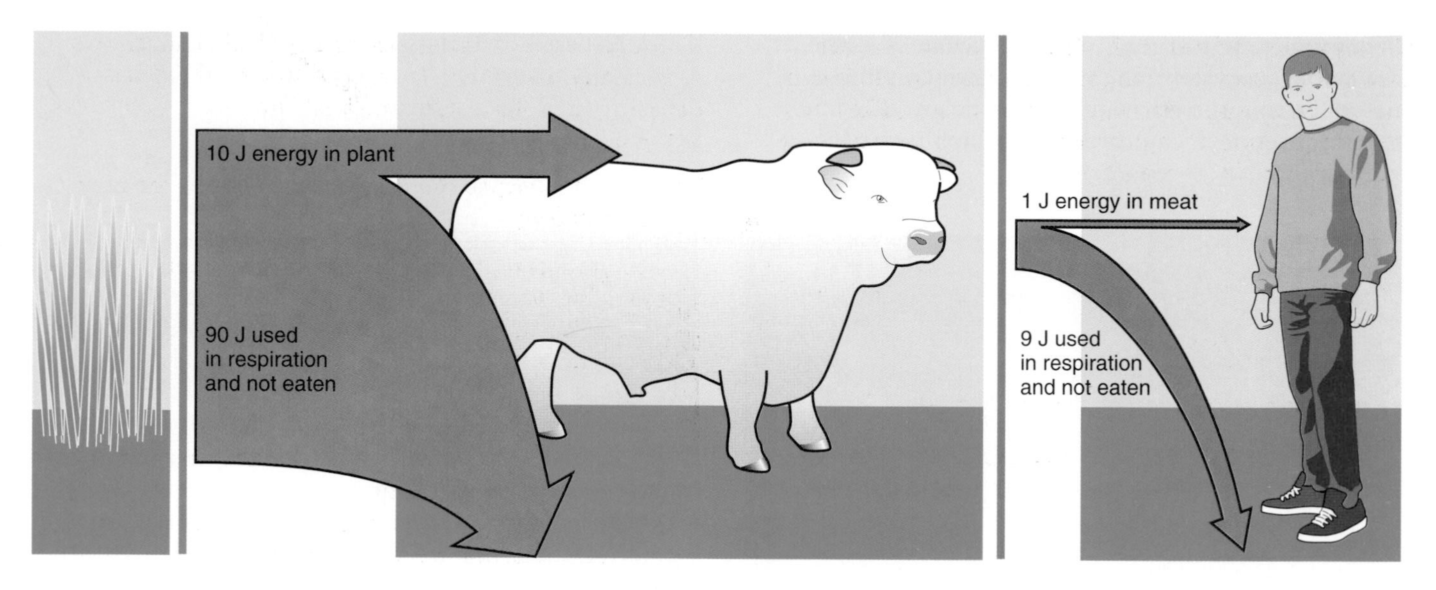

Figure 9.7 Energy efficiency in a human food chain

animals for meat. This argument also applies to animals that are reared intensively, such as pigs and chickens. These animals are fed on cereals so they require 'land' indirectly and also put another trophic level in our food chain.

So, the efficiency of food production can be greatly improved by:

- reducing the number of stages in our food chains,
- restricting the energy loss from animals reared for their meat. This can be done by limiting their movements, which may seem cruel, and controlling the temperature of their surroundings.

9.5 Microbes and decomposers in ecosystems

Living things remove materials from the environment in order to live and grow. For example, plants remove carbon dioxide from the air and water from the soil in order to photosynthesize. As a result of photosynthesis, plants produce carbohydrates containing carbon, hydrogen and oxygen.

Plants also take water and minerals containing nitrogen from the soil in order to grow. The growth of a plant involves the synthesis of carbohydrates and proteins. Proteins contain carbon, hydrogen, oxygen and nitrogen. These chemicals, like carbohydrates and proteins containing carbon and hydrogen, together with oxygen and possibly nitrogen, are called **organic compounds**.

When plants and animals respire, the elements in some of these organic compounds are returned to the environment as waste materials in compounds such as carbon dioxide, water and urea. The elements in animal and plant compounds are also returned to the environment when living things die and decay.

Materials decay because they are broken down by **decomposers** like fungi, crows, earthworms and other organisms which, in turn, are broken down by **microbes** (micro-organisms) such as bacteria and viruses. Decomposers feed on the animal and plant material, releasing carbon dioxide, water and nitrogen.

Microbes decompose materials faster when:

- there are moist conditions,
- the temperature is close to 37°C,
- there is a plentiful supply of oxygen.

Microbes are active at sewage works where they are used to decompose human waste. Microbes are responsible for breaking down plant material in a compost heap. These decay processes produce carbon dioxide, water and nitrogen compounds. These are precisely the materials which plants need for photosynthesis and growth.

Notice that:

> Elements such as carbon, hydrogen, oxygen and nitrogen are recycled in ecosystems.

Decomposition is crucial in this recycling of elements. In a stable ecosystem, the processes which remove materials from the environment are balanced by the decomposition (decay) processes which return materials to the environment.

Carrion crows play their part as decomposers by eating dead animals

Without decomposers, recycling would be impossible and the Earth would be cluttered with dead plants and animals. Fungi are important decomposers. Look at this huge fungus growing on the remains of a dead tree

Decomposers recycling faeces. These flies are busy eating (decomposing) and recycling the elements in cow dung

9.6 The carbon cycle

Plants and animals need a steady supply of carbon to make important compounds like carbohydrates, proteins and nucleic acids, all of which they need to live. Plants can synthesize most of their carbon compounds by photosynthesis from carbon dioxide in the air. Animals obtain their carbon compounds by eating plants and other animals. As the plants and animals carry out respiration, carbon is returned to the air as carbon dioxide. Carbon dioxide is also returned to the air when decaying plants and animals are decomposed by bacteria and when this material is burnt. These processes in which carbon is continually being 'fixed' in animals and plants and then returned to the air in CO_2 make up the **carbon cycle** (Figure 9.8).

Notice the following key stages in Figure 9.8:

1. Carbon dioxide is removed from the atmosphere by green plants during photosynthesis. The carbon in this carbon dioxide is used by plants to synthesize carbohydrates, proteins, fats and nucleic acid which make up plant cells.
2. When plants are eaten by animals, these carbon compounds become part of the carbohydrates, proteins, fats and nucleic acid which make up their bodies.

Grass and leaves in this compost heap are decaying to carbon dioxide and water. In this case, the reaction produces so much heat that the water comes off as steam

Large amounts of energy were trapped in fossil fuels such as coal, oil and gas after animals and plants had died millions of years ago. A similar process occurs when peat forms in bogs

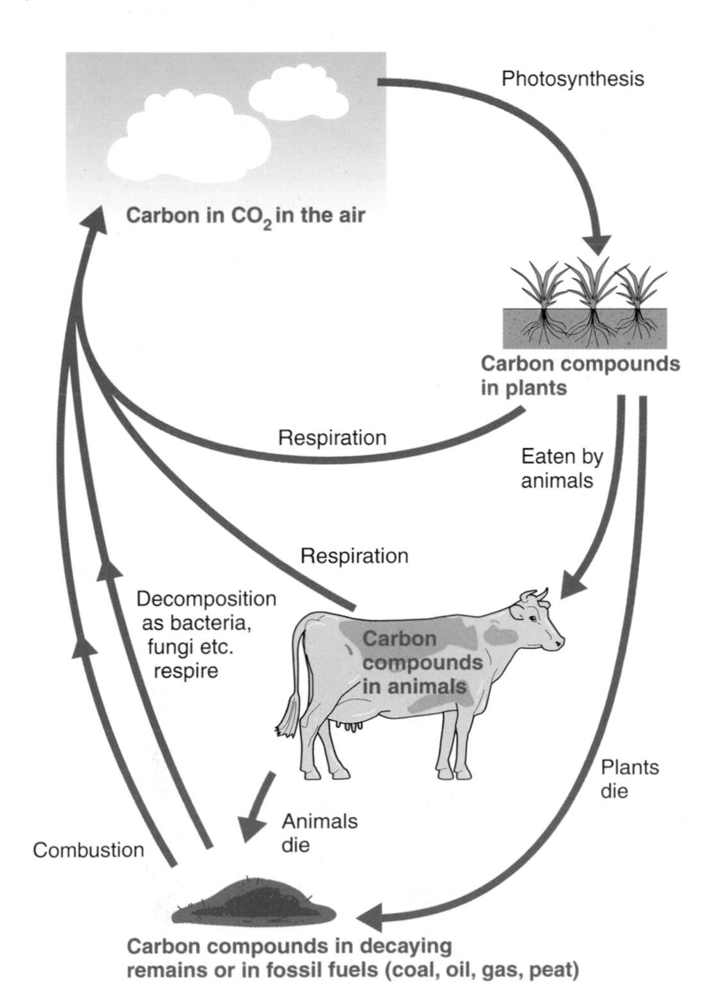

Figure 9.8 Important processes in the carbon cycle

3 When plants and animals die, their carbon compounds begin to decay in the soil or may be trapped and eventually become fossil fuels.

4 When plants and animals respire, some of their carbon compounds form carbon dioxide which is then returned to the atmosphere.

5 Bacteria, fungi and some animals are decomposers, feeding on the decaying remains of dead animals and plants. Carbon dioxide is released into the atmosphere as these organisms respire.

6 When fossil fuels or the decaying remains of animals and plants are burnt, carbon dioxide is released into the atmosphere.

9.7 The nitrogen cycle

Nitrogen, like carbon, is another element essential to plants and animals which contain nitrogen in their amino acids and protein. Neither plants nor animals can use nitrogen (N_2) directly from the air to make their amino acids and proteins, but there are some microbes which can do this. Plants can use the nitrogen compounds made available by microbes and pass these on to animals. When the plants and animals die and decay, their nitrogen is returned to the air as N_2, or it remains in the soil as nitrogen compounds for the processes to start again.

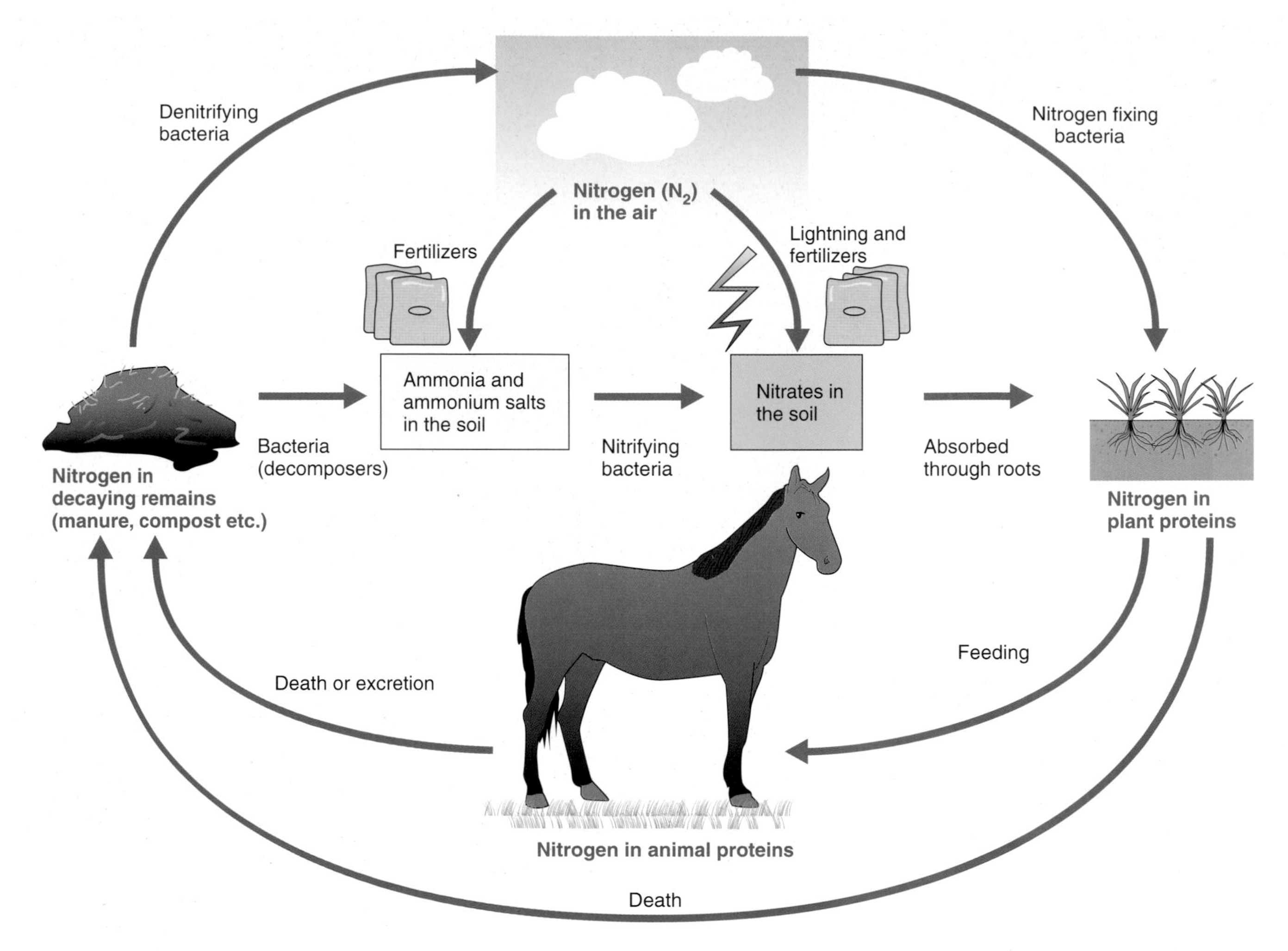

Figure 9.9 Important processes in the nitrogen cycle

These processes in which nitrogen moves from the air into the soil and through plants and animals is called the **nitrogen cycle** (Figure 9.9).

Notice the following important points from Figure 9.9:

1 Nitrogen is present as proteins in plants, animals and decaying remains, as ammonium salts (NH_4^+) and nitrates (NO_3^-) in the soil and as nitrogen gas (N_2) in the air.

2 The nitrogen in decaying remains from animals and plants is converted by decomposer bacteria in the soil into ammonia and ammonium salts.

3 Other microbes in the soil called **nitrifying bacteria** convert this ammonia and ammonium salts into nitrates.

4 Plants can absorb the ammonium salts and nitrates through their roots and use them to make amino acids and proteins.

5 Another group of microbes, called **nitrogen-fixing bacteria**, can also convert nitrogen in the air into amino acids and proteins for use by plants. Some nitrogen-fixing bacteria live freely in the soil whilst others are present in nodules on the roots of leguminous plants.

6 Lightning increases the nitrate content of the soil (Figure 9.10).

Farmers and gardeners play an important part in the nitrogen cycle by increasing the nitrogen available to plants to promote growth. They do this by:

- adding nitrogenous fertilizers to the soil in the form of ammonium salts and other fertilizers (section 21.8);
- growing leguminous crops as part of a crop rotation plan;
- ploughing or forking compost and well rotted manure back into the soil.

Figure 9.10 During a flash of lightning, the temperature is so high that nitrogen and oxygen in the air react to form nitrogen oxides. The nitrogen oxides then react with rain water to form nitric acid which gets washed into the soil to form nitrates

Leguminous plants, such as peas, beans and clover, have nodules on their roots like these. These nodules contain nitrogen-fixing bacteria which convert nitrogen from the air into amino acids and proteins which can be used by the plants

1 A supermarket chain advertises that its plastic carrier-bags are 'environmentally friendly'. These bags are made from polythene and contain added starch. A student decided to find out if the claim is true. In the investigation the student buried a piece of plastic from the bag in a pot of soil and left it for one month. When the plastic was dug up there were many small holes in it.
The drawings show the plastic, as seen through a microscope, before and after being buried.

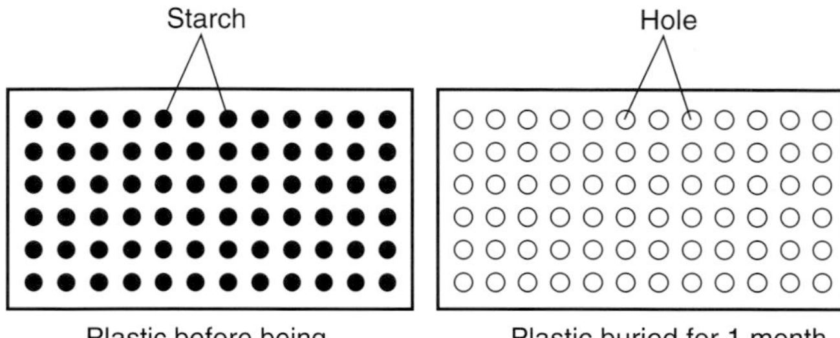

a) Explain, as fully as you can, the process that caused the holes to appear in the plastic.

b) Give **two** ways in which this process could be speeded up.

c) Explain how the carbon in the starch is eventually returned to the atmosphere. **NEAB**

2 The diagram shows a food web for some of the organisms that live in a pond.

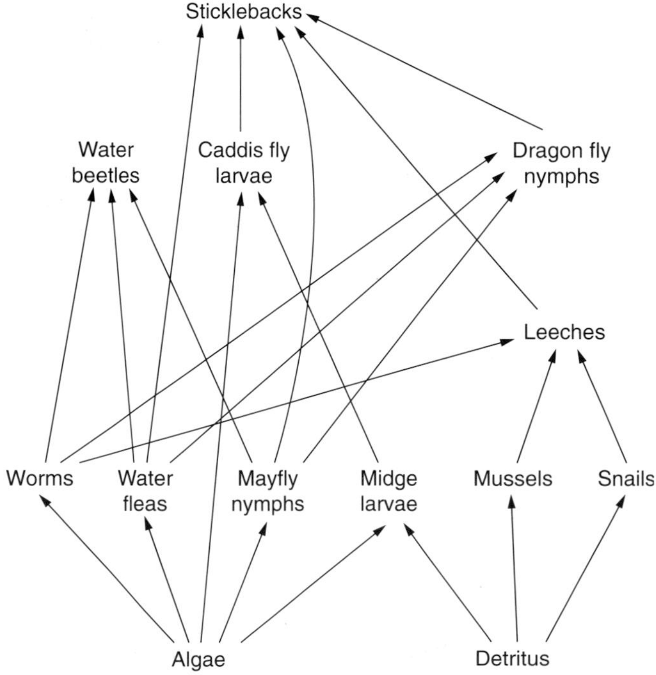

Use information from the food web to answer the following questions.

a) A disease kills large numbers of snails. Explain how this might affect:
i) the population of leeches;
ii) the population of mussels.

b) Water beetles are approximately 2 cm in length. The algae in the food web are microscopic in size. The water fleas can just be seen with the naked eye. Draw and label a pyramid of numbers for the food chain:

algae → water fleas → water beetles

c) The water beetles incorporate less than 1% of the solar energy captured by the algae.
Explain, as fully as you can, how the rest of the solar energy captured by the algae is eventually returned to the environment. **NEAB**

3 a) The diagram shows how the energy from one square metre of grassland is used by a grazing bullock.

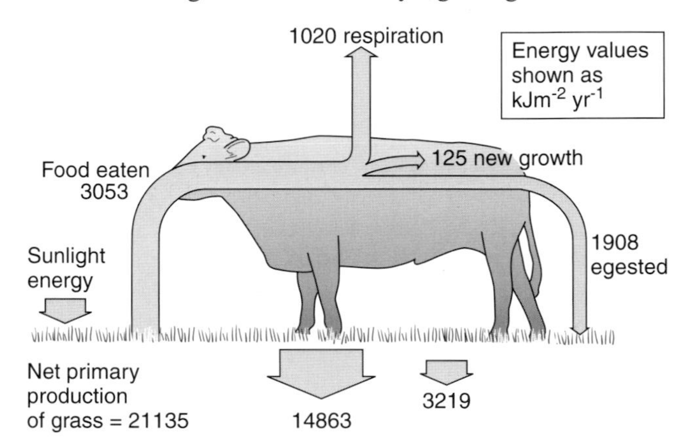

Beef cattle on grassland use 1/7th of the total energy in the grass.
i) Explain what happens to the other 6/7ths of the energy.
ii) What percentage of the energy in the grass, eaten by the bullock, was used to produce new growth? Show your working.

b) Explain why, in terms of energy used, it is more efficient to grow crops rather than graze cattle for food. Give **two** reasons. **MEG**

4 Compost is what remains of garden waste which has been partly decomposed by bacteria. A compost heap is built up from layers of garden refuse and waste food separated by layers of soil. Bonfire ash may be added.

a) i) Suggest why large pieces of wood should **not** be added to the compost heap.
ii) Suggest how the nutrients in waste wood can be recycled.

b) Some of the material in the compost heap is carbohydrate.
i) Explain how the action of bacteria on carbohydrate helps to recycle carbon into the atmosphere.
ii) Write a word equation to summarise the decomposition of the carbohydrate.

c) Some of the other material in the compost heap is protein. How is protein changed in the compost heap to make it into a useful fertilizer? **MEG**

CHAPTER

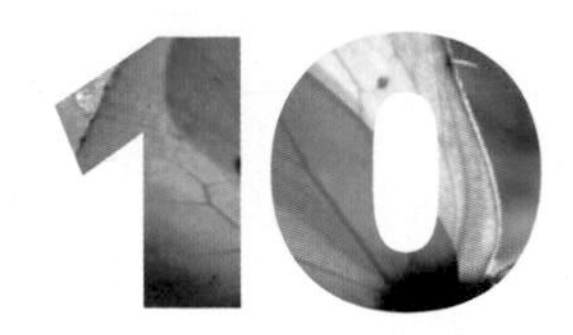

Reproduction and genetics

The human reproduction system	Cell division
Chromosomes, genes and DNA	Sexual and asexual reproduction

10.1 The human reproduction system

Living things must reproduce to ensure the continuation of the species. Humans do this by sexual reproduction.

Figure 10.1 shows a diagram of the male reproductive system and the important functions of the various parts.

Sperms are produced in the **testes** and stored in the **epididymis**. The sperms leave the male's body by passing along the sperm duct where they mix with white seminal fluid produced by the prostate gland and the seminal vesicle. This mixture of white liquid plus sperms is called **semen** (pronounced 'seemen').

During sexual intercourse, semen passes down the urethra very quickly and is ejected from the penis.

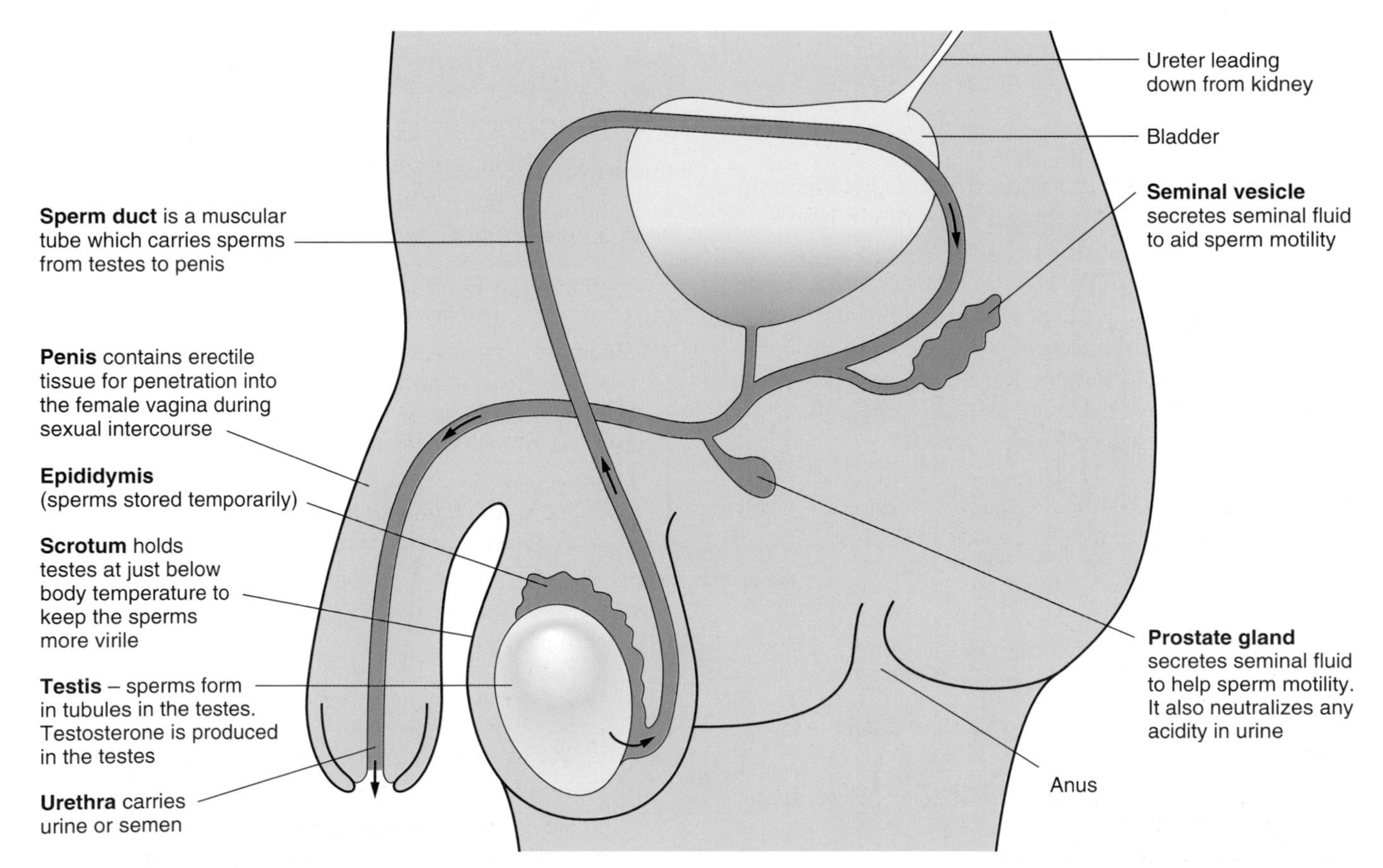

Figure 10.1 The male reproductive system showing the important functions of various parts

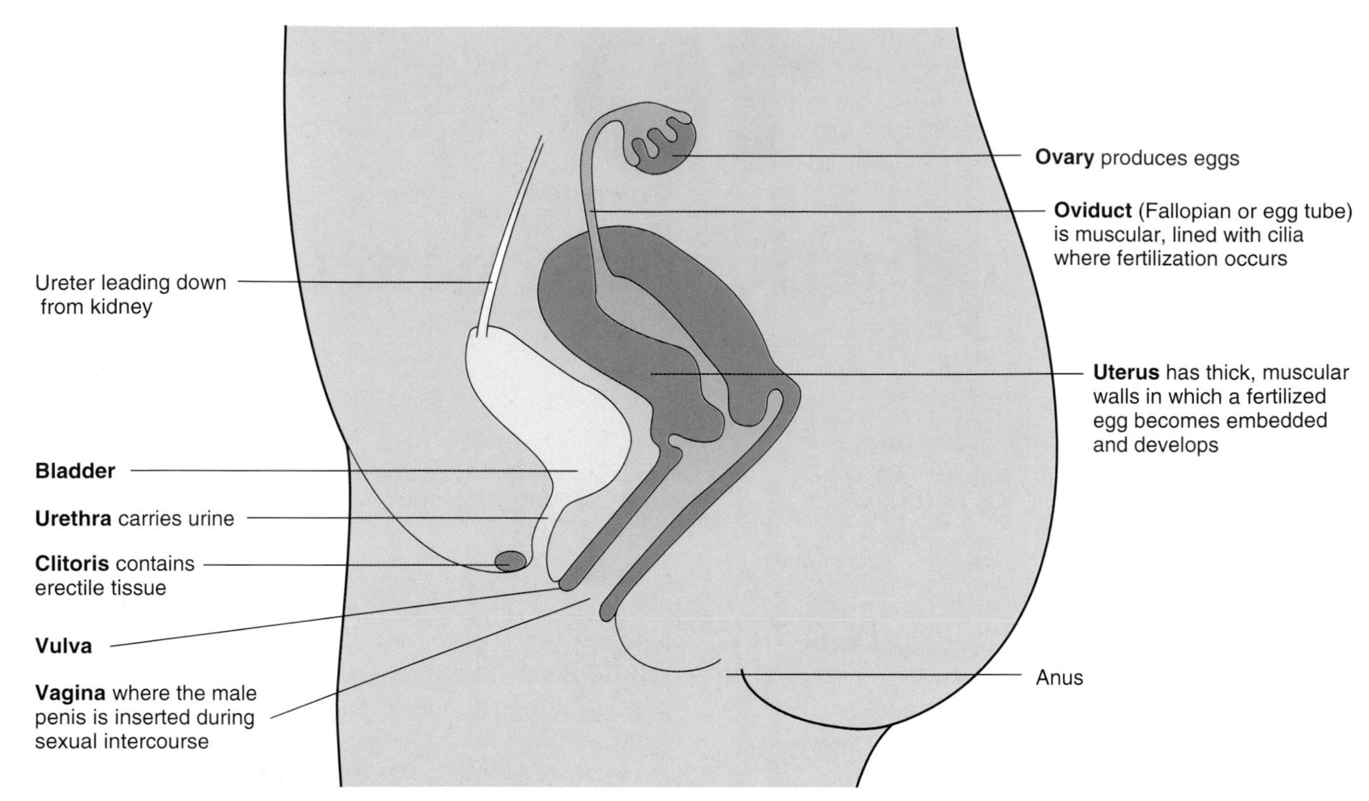

Figure 10.2 The female reproductive system showing the important functions of various parts

Figure 10.2 shows a diagram of the female reproductive system and the functions of the various parts.

Fertilization

For fertilization to occur, a **sperm cell** from the male must fuse (join) with an **egg cell** from the female. This usually occurs in an **oviduct** (Fallopian tube). At roughly monthly intervals, an egg is released from one of the ovaries into its oviduct. If the woman has sexual intercourse at this time, the sperms will swim up the oviduct. One of them may penetrate the membrane of the egg and fuse with the nucleus of the egg cell resulting in **fertilization** (Figure 10.3).

Notice in Figure 10.3 that sperm cells are much smaller than egg cells. Sperm cells have a 'tail' which allows them to move in search of an egg cell. If fertilization occurs, then the fertilized egg is called a **zygote**.

Once a sperm has penetrated the membrane of the egg cell, changes take place in the membrane to prevent the entry of other sperms.

The zygote grows by dividing into two cells, then again into four cells and so on. Eventually, the cells form an **embryo**.

Figure 10.3 Fertilization of an egg by a sperm

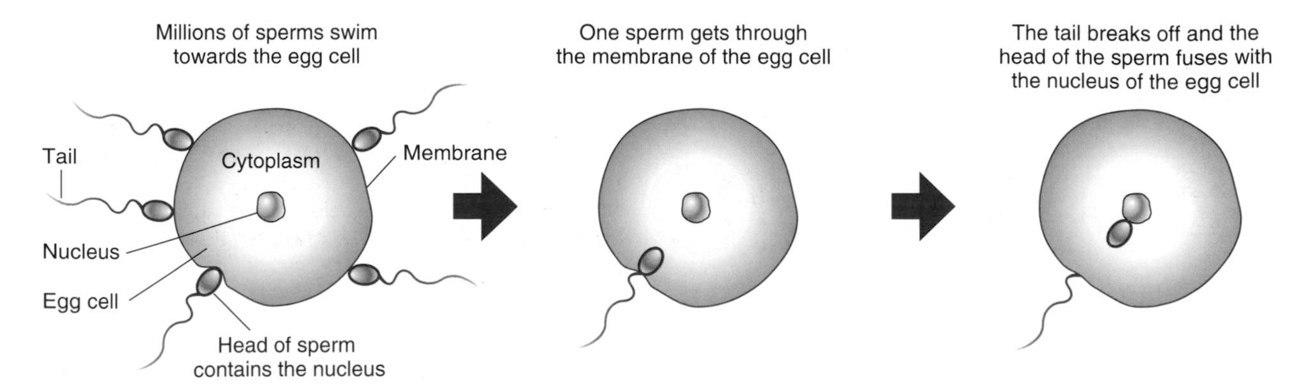

Sometimes, an embryo divides into two during its early stages of development. Each of the parts then develops normally producing two offspring. Both these offspring have developed from the same egg and the same sperm. They have **identical genes** (section 10.2) and will be **identical twins**.

On other occasions, two eggs may be released in one month and both of these may be fertilized by sperms. In this case, the embryos develop at the same time but from different eggs and different sperms. The offspring have a different combination of genes and they are **non-identical twins**.

10.2 Chromosomes, genes and DNA

The largest single structure within a cell is usually its nucleus. Although the nuclei of cells may appear to be uniform, they do in fact contain separate structures called **chromosomes**.

These chromosomes carry the information which allows cells to reproduce and which is passed on from parents to offspring during fertilization.

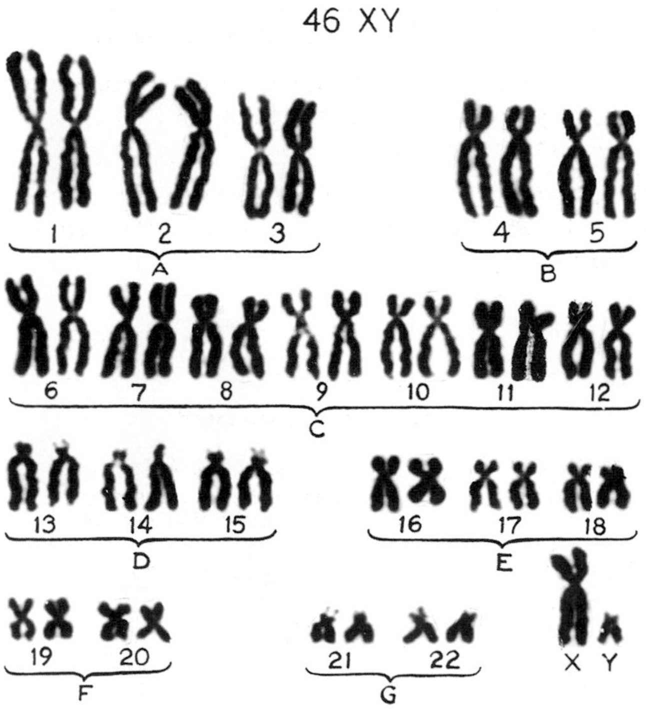

46 XY

1 2 3 4 5
A B
6 7 8 9 10 11 12
C
13 14 15 16 17 18
D E
19 20 21 22 X Y
F G

The 23 pairs of human male chromosomes just before cell division. At this stage, each chromosome has two strands joined at one point. Notice in the photograph that the last pair of chromosomes are not alike. Males have an X and a Y chromosome in this pair. In contrast, females have two X chromosomes

When the nucleus of a cell is stained with dye and viewed under a microscope, the chromosomes look like fine threadlike structures. The threadlike chromosomes have different lengths and different thicknesses. All organisms have a fixed number of chromosomes in the nuclei of their cells. This is called the **chromosome number** which differs from one organism to another. For example, human cells have 46 chromosomes, hen cells have 36 and pea cells have 14.

These chromosomes can be arranged in pairs which look alike when stained but which are not identical. The members of each pair control the same characteristics in an organism and they are called **homologous pairs**. So, humans have 23 homologous pairs (46 chromosomes in total), hen cells have 18 and pea cells have 7.

In each of these homologous pairs, one chromosome has come from the male parent and the other from the female parent.

The structure of chromosomes

Chromosomes are made of a complex polymer called **DNA** which is short for **deoxyribonucleic acid**. The long DNA molecules which make up chromosomes can be divided into sections called **genes** (Figure 10.4 over the page). Chromosomes and hence genes can be copied by cells and passed on to the next generation. This explains why parent animals and plants can pass on physical features and other characteristics to their offspring.

The chemicals in genes carry instructions which control the way cells develop and function. As Figure 10.4 on the following page shows, each gene has a precise order for the four monomers which make up DNA. The order of these monomers acts as a code for the synthesis of the proteins and enzymes which control the metabolic processes in all organisms. So, if you have blue eyes, you must have a gene which synthesizes a protein which eventually results in blue eyes rather than brown eyes.

Figure 10.4 A cell showing the relationship between its nucleus, chromosomes, genes, DNA and monomers

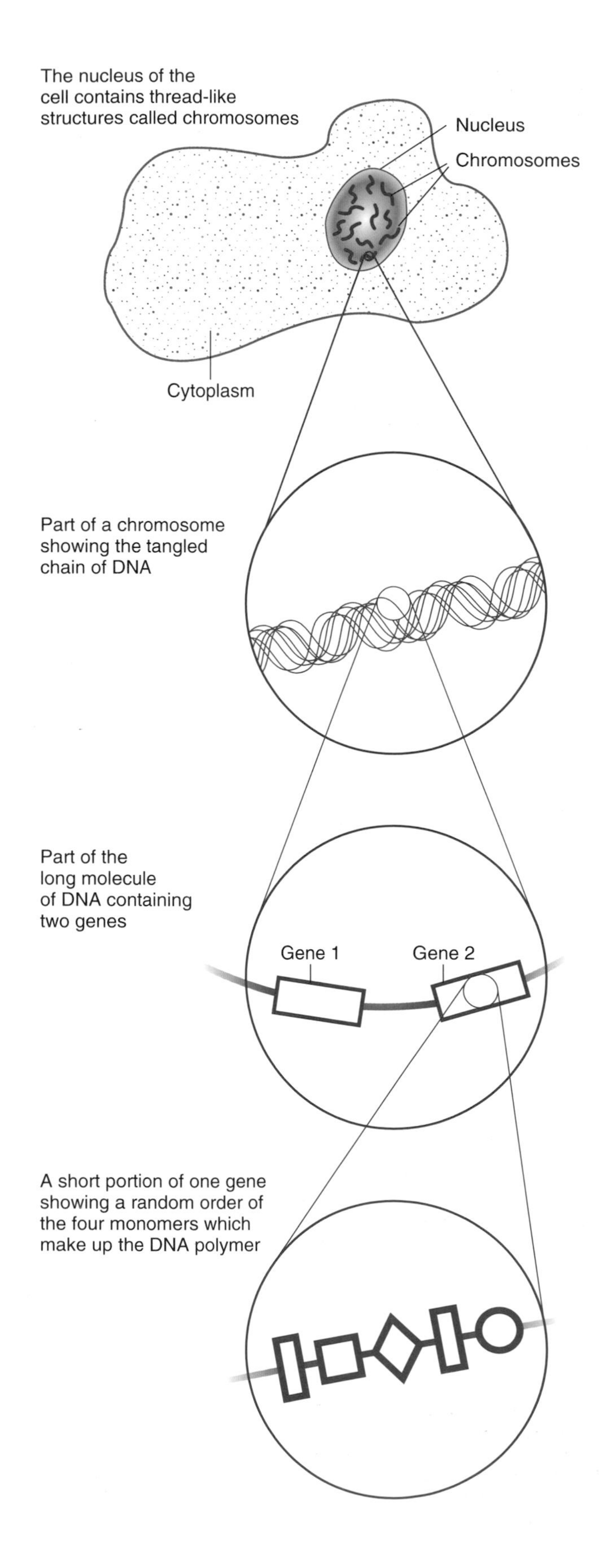

Genetic engineering

During the last twenty years, scientists have discovered ways in which they can isolate a gene from one particular organism and then insert it into the DNA of another organism. This technique is called **genetic engineering**.

It involves extracting the DNA from one particular species and breaking it into smaller fragments. Selected fragments are then isolated and attached to the DNA of a 'carrier'. The carrier organisms are usually bacteria. By culturing these carrier bacteria on a large scale, commercial quantities of proteins, and other chemicals associated with the inserted fragment of DNA, are produced. These techniques are used to manufacture antibiotics and hormones such as insulin, which is used to treat diabetes.

Scientists are now developing ways in which they can:

- replace 'faulty' genes which cause hereditary diseases like haemophilia,
- transfer genes to the cells of embryonic animals and plants so that they develop with desirable characteristics.

All the cells in your body (except sperms or eggs), contain the same 46 chromosomes. Scientists can treat the chromosomes and produce a chromatogram of the DNA which is different from one person to another. These DNA profiles are called 'genetic fingerprints' and can be used in forensic science to help identify the offender from other suspects of a crime

10.3 Cell division

Living cells can divide in two different ways – by **mitosis** or by **meiosis**.

Mitosis occurs when an organism is growing (producing more cells) and when its cells are being replaced. Mitosis takes place during the formation of all cells except male sex cells (sperms and pollen) and female sex cells (eggs).

Meiosis occurs only during reproduction with the formation of male and female sex cells.

Mitosis

Figure 10.5 shows what happens in a cell during mitosis. In order to appreciate the process more easily, the cell in Figure 10.5 has only 4 chromosomes (i.e. 2 homologous pairs).

Notice the following:

> Mitosis produces exact copies.

This is what happens when an organism grows and when an organism reproduces asexually, forming clones.

The explanation of mitosis in Figure 10.5 has been simplified by taking a cell with only 4 chromosomes. In humans, mitosis occurs in this way, but as we have 46 chromosomes, it is a very complicated process.

In healthy individuals, there is a control over this very complex process. The disruption of this control of cell division can lead to certain cancers.

Meiosis

Figure 10.6, over the page, shows what happens in a cell during meiosis. The process, like that in Figure 10.5 for mitosis, has been simplified by considering a parent cell with only 4 chromosomes.

Notice in Figure 10.6 that in stages a and b, the chromosomes form chromatids as in mitosis. In stage c the linked chromatids line up near the middle of the cell in a different way from mitosis. In this case, homologous pairs of chromosomes come together and then separate to opposite ends of the cell in stage d before cell division occurs (stage e).

Figure 10.5 Changes in chromosomes during mitosis

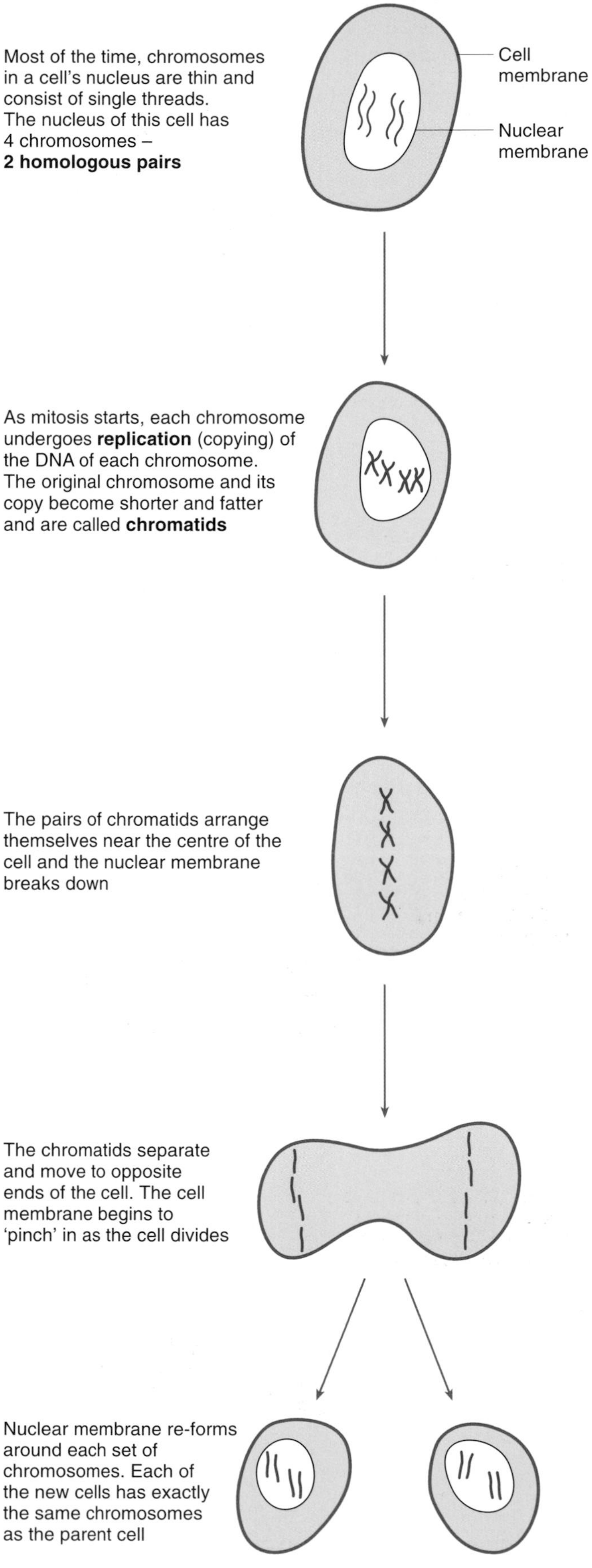

Most of the time, chromosomes in a cell's nucleus are thin and consist of single threads. The nucleus of this cell has 4 chromosomes – **2 homologous pairs**

Cell membrane

Nuclear membrane

As mitosis starts, each chromosome undergoes **replication** (copying) of the DNA of each chromosome. The original chromosome and its copy become shorter and fatter and are called **chromatids**

The pairs of chromatids arrange themselves near the centre of the cell and the nuclear membrane breaks down

The chromatids separate and move to opposite ends of the cell. The cell membrane begins to 'pinch' in as the cell divides

Nuclear membrane re-forms around each set of chromosomes. Each of the new cells has exactly the same chromosomes as the parent cell

A *second* cell division now begins (stage f) in which the chromatids separate from each other. When this second cell division is complete (stage g) the new cells have only two chromosomes, one from each pair.

So, in meiosis there are two cell divisions, one after the other forming four new cells with only half the number of chromosomes as their parent cells. These cells with only half the full complement of chromosomes are described as **haploid** whereas cells with their full number of chromosomes are described as **diploid**.

> Meiosis occurs in sexual reproduction. The testes and ovaries are where the cells divide to produce sperms or eggs. These sex cells (sperms and eggs) are called **gametes**.

When gametes (haploid cells) join at fertilization, a single new diploid cell is formed with the full complement of chromosomes. This new cell has a different set of chromosomes to either of its parents, so **meiosis results in variations**.

Once a new cell has formed by meiosis, a new individual then develops and grows as the cells divide by mitosis.

Figure 10.6 Changes in chromosomes during meiosis

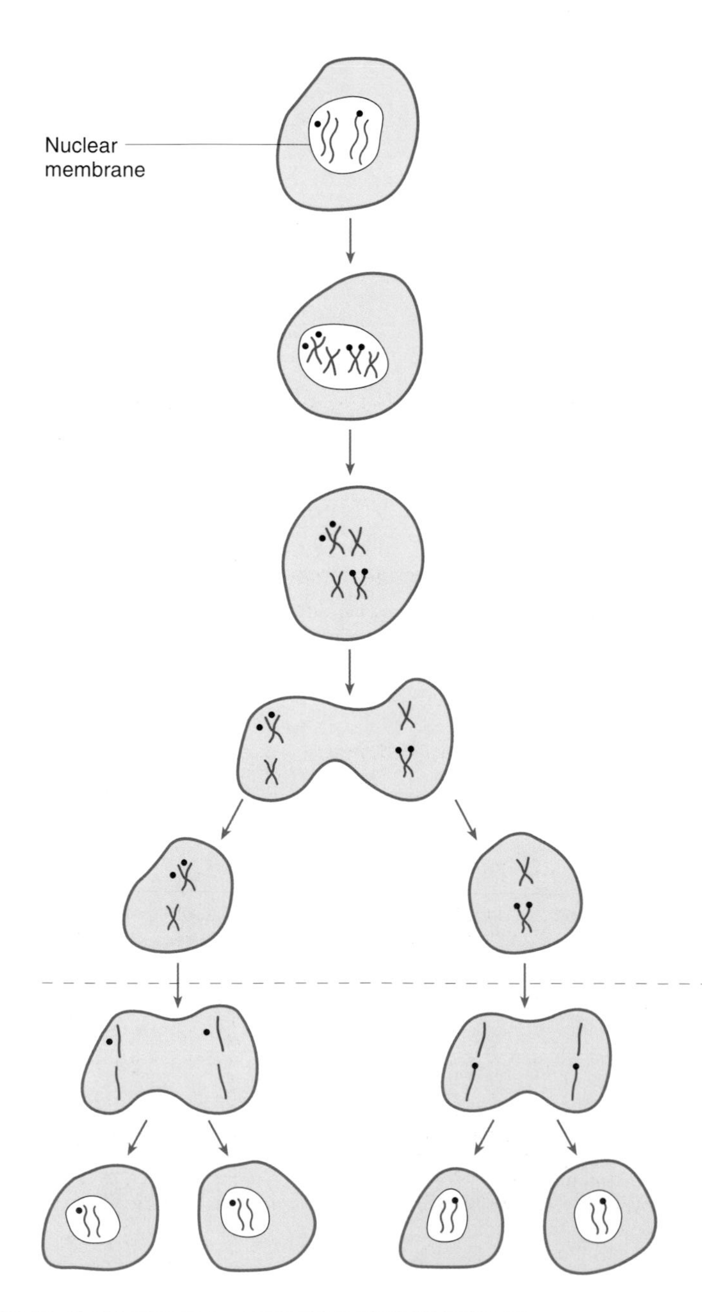

Stages in meiosis

a) Parent cell with 4 chromosomes – 2 homologous pairs

b) Chromosomes replicate forming pairs of chromatids which are shorter and fatter

c) Homologous pairs of chromosomes arrange themselves near the centre of the cell and the nuclear membrane breaks down

d) Homologous chromosomes separate and move to opposite ends of the cell. The cell membrane begins to 'pinch' in as the cell divides

e) **The first cell divison** occurs and the chromosomes move to the middle of the two new cells

f) The chromatids now separate from each other and move to the opposite ends of the cells. The cell membrane begins to'pinch' in as the cells divide

g) **The second cell division** occurs and the nuclear membranes re-form around each set of chromosomes. Each of the four new cells has only half the number of chromosomes as the original parent cell

10.4 Sexual and asexual reproduction

Living things can reproduce in two different ways – by sexual reproduction and by asexual reproduction.

Sexual reproduction

Reproduction in most animals, including humans, is sexual. It involves sexual intercourse or mating of a male and a female. Some plants also reproduce sexually.

Sexual reproduction is similar for all organisms. It involves **the formation of gametes** (sex cells) during **meiosis** and is followed by **fertilization**.

> Female gametes are the egg cells produced in the ovary of the female animal or plant. Male gametes are either sperms in animals or pollen in plants.

As both parents contribute part of themselves at fertilization, sexual reproduction results in offspring with a mixture of the genetic information of their parents. Sexual reproduction is therefore a source of **genetic variation**.

There are two occasions in sexual reproduction when variation is introduced.

- In meiosis, the homologous pairs of chromosomes separate quite randomly to produce genetic variations amongst the gametes.
- At fertilization there is a random fusion of gametes.

Asexual reproduction

Most plants and some primitive animals reproduce asexually. This does not involve the mating of male and female. It occurs when part of the parent becomes detached and then grows and develops separately. Plants which reproduce asexually usually do so by forming bulbs, tubers or runners.

Unlike sexual reproduction, asexual reproduction requires only **one parent**.

It involves **mitosis** in which each new cell and its genetic information are copies of the parent cell.

> This means that **asexual reproduction** produces identical offspring called **clones**.

Sexual and asexual reproduction are compared and summarised in Figure 10.7.

Cloning

All the offspring from one asexually reproducing parent are known as clones. These offspring have **identical genes** because they are produced by mitosis.

There are various methods of producing clones. Here are a few examples. The first one occurs naturally and the others show how we can exploit the ability of some plants to reproduce by mitosis.

Figure 10.7 Comparing and summarising sexual and asexual reproduction

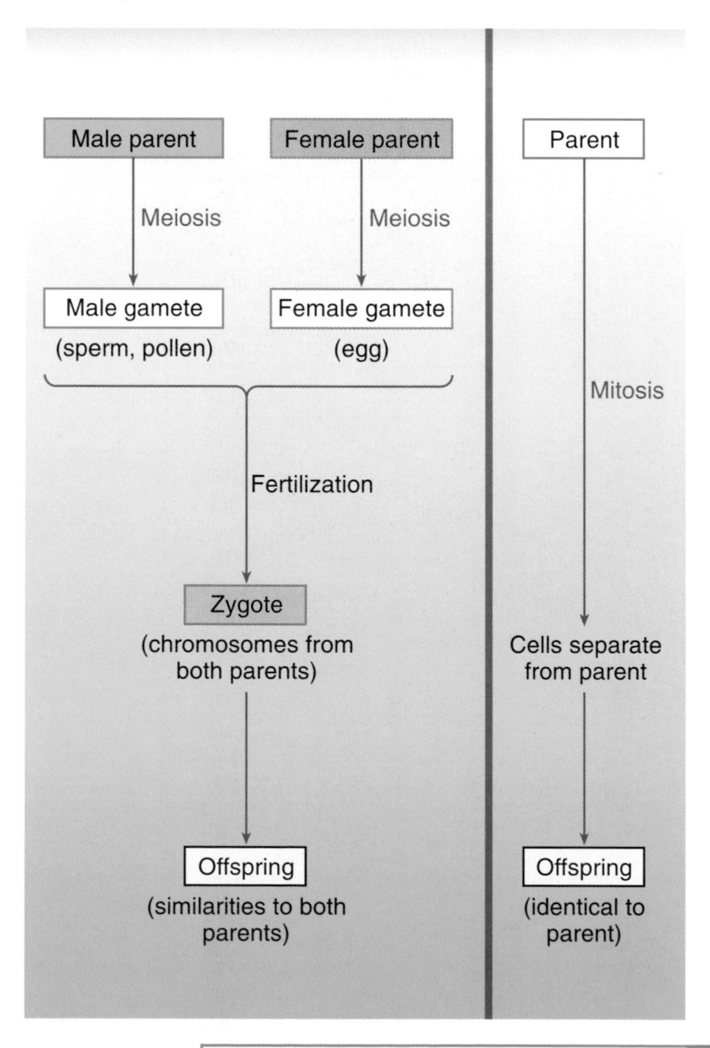

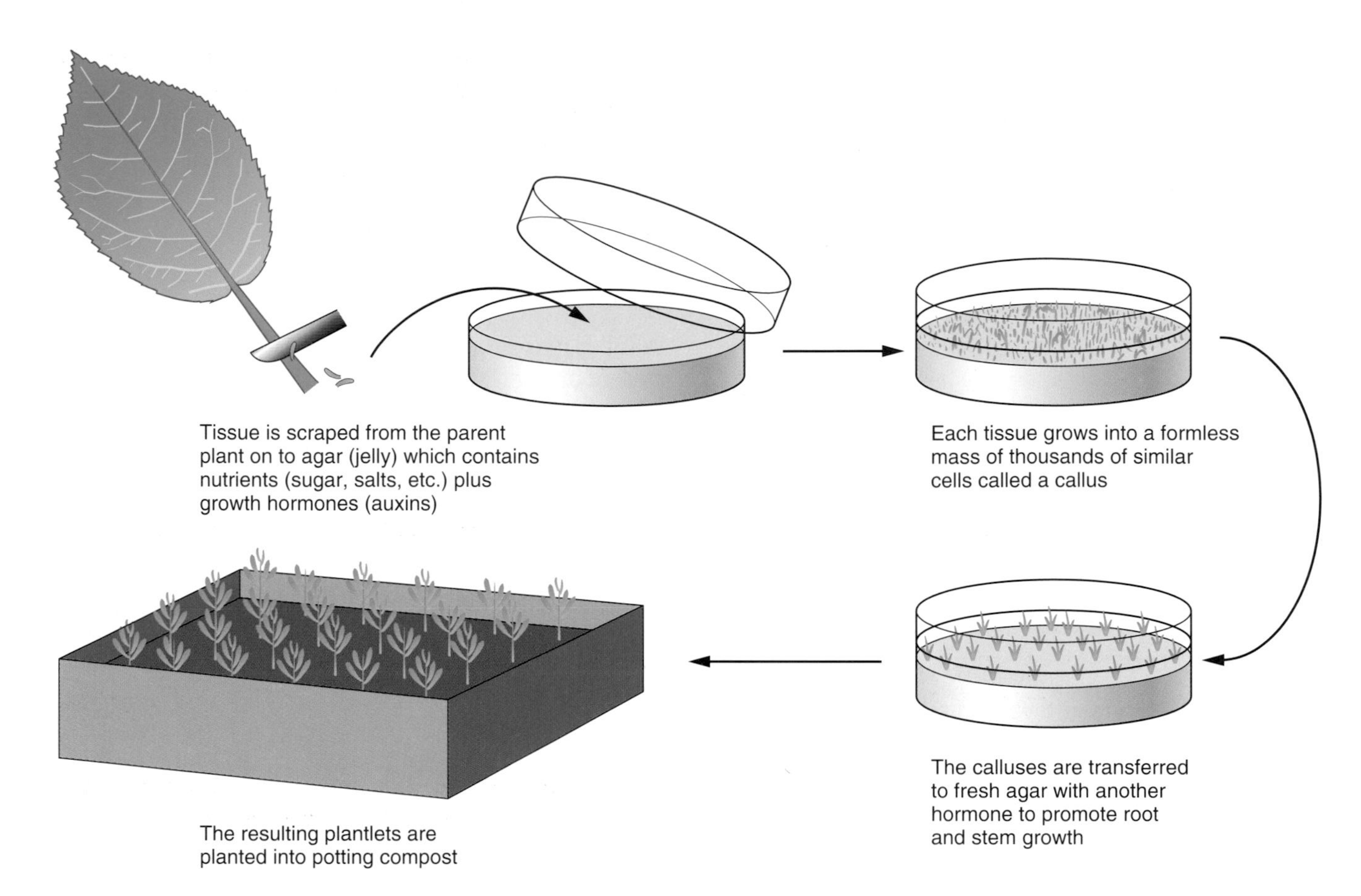

Figure 10.8 Using tissue culture for micro-propagation

i) Binary fission

This method occurs in simple organisms such as amoeba and bacteria. In this case, the 'parent' amoeba or bacterium (which are single cell organisms) divides to produce two identical 'daughters' and so on.

ii) Cuttings

A short stem with a few leaves can be cut from a parent plant such as a geranium. Hormone rooting powder may be used to encourage root growth. This is planted in damp soil, and, in ideal growing conditions a cloned geranium plant will be established.

iii) Tissue culture

Using this technique large numbers of identical plants can be grown commercially from a single parent in a relatively short time (Figure 10.8). This is sometimes called **micro-propagation**.

Seedlings of Douglas Fir trees produced by micro-propagation

1 Diabetics cannot produce their own insulin. The diagram shows some stages in the production of human insulin for diabetics, by genetic engineering.

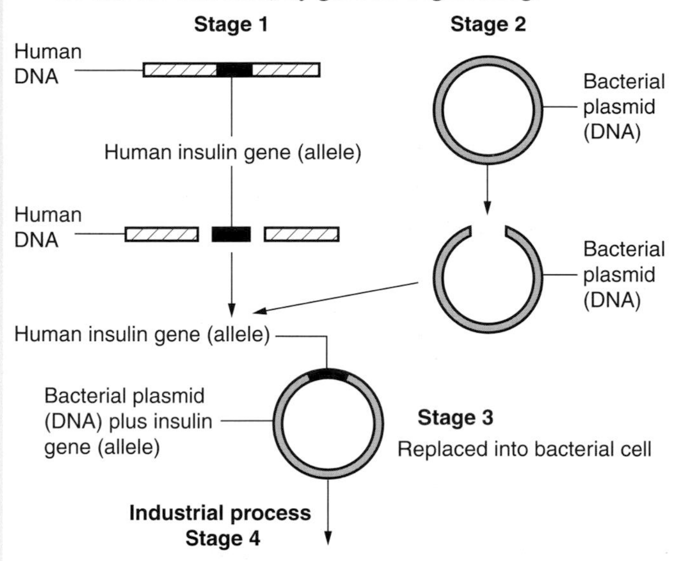

Copy and complete the table to explain what is happening at Stages 2, 3 and 4.

Stage	Explanation
2	
3	
4	

NICCEA

2 Modern farmers get much higher yields of wheat grains from each plant than was possible 800 years ago. The drawings show a modern wheat plant and a wheat plant from 800 years ago.

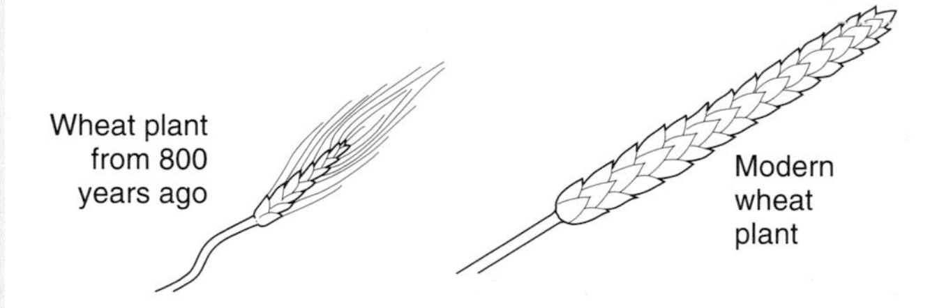

a) Explain, as fully as you can, how farmers have developed modern wheat plants from those which grew 800 years ago.

b) Farmers who grow apple trees use clones of these trees when planting new orchards.
i) What is meant by clone?
ii) Name the type of cell division involved in the production of clones.
iii) Explain **one** possible drawback to the widespread use of clones in agriculture. **NEAB**

3 The following flow chart shows, in the correct order, the principles of one technique used to improve crop plants by genetic engineering.

The table at the bottom of the page lists the stages in the genetic engineering of cotton but they are listed in the **wrong** order. Write out the correct order of these letters so that they correspond with the correct principle of genetic engineering.

Flow chart

Principles of genetic engineering
A useful gene is identified
↓
This gene is cut out of the donor DNA
↓
The gene is inserted into a vector organism
↓
The vector transfers the useful gene to a cell of a crop plant
↓
The crop plant cell is cloned to produce many transgenic plants
↓
The new transgenic crop plants undergo trials to find out if the useful gene has the desired effect

Table

Stages in the genetic engineering of cotton	Letter
Cotton plants of the new variety are tested in the field to find out if they kill insect pests.	A
Agrobacterium tumefaciens infects cotton plant cells, passing the gene which controls toxin production into their nuclei	B
Whole cotton plants are grown from the genetically changed cotton plant cell	C
Bacillus thuringiensis, a bacterium, has a gene which controls the production of an insect-killing toxin	D
The gene controlling toxin production is transferred to *Agrobacterium tumefaciens*, a bacterium	E
The useful gene is cut from the bacterial DNA using an enzyme	F

Edexcel

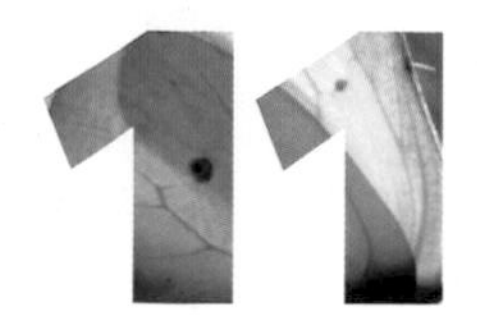

Genetics and inheritance

11.1 Genetics and sex

Genetics is the branch of science which involves the study of **genes** and the **inheritance** of characteristics.

The first characteristic that we notice when a baby is born is whether the baby is a boy or a girl. What decides the sex of this baby and why are there equal numbers of boys and girls? The explanation of this is a good example of the way in which genes dictate the physical characteristics of living things.

Every human being has 23 pairs of chromosomes. One of these pairs determines our sex and are called the **sex chromosomes**. Unlike other homologous pairs, the sex chromosomes are not necessarily identical. There are two types of sex chromosomes.

A long one, called the **X chromosome**
A short one, called the **Y chromosome**

All females have two X chromosomes, written as XX. All males have one X and one Y chromosome, written as XY.

Figure 11.1 shows what happens to the sex chromosomes when

- eggs and sperms are produced by meiosis,
- the eggs and sperms fuse to produce a fertilized egg.

All the eggs which the mother produces will contain one X chromosome, but half the sperms from the father will have one X chromosome and the other half will have one Y chromosome.

This means that there is an equal chance that an egg will be fertilized by an X sperm or by a Y sperm.

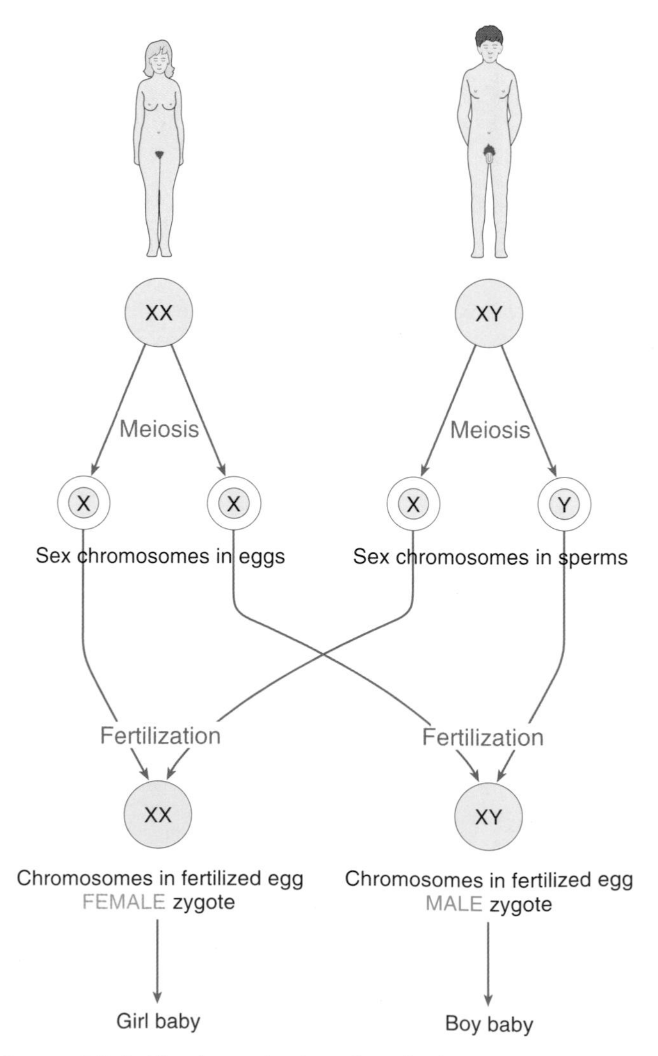

Figure 11.1 The determination of gender by sex chromosomes

If an X sperm fertilizes an egg, the zygote will contain two X chromosomes, XX. This zygote will develop into a girl. If a Y sperm fertilizes an egg, the zygote will contain an X and a Y chromosome, XY. This zygote will develop into a boy.

As there is an equal likelihood of XX pairs and XY pairs, there will be equal numbers of boys and girls. Notice in Figure 11.1 that sex is determined solely by the male. So, Henry VIII, who blamed every one of his six wives for not producing a son and heir was totally wrong. In fact, the 'fault' was entirely his own!

Chromosomes are copied during meiosis. Usually the copying process goes to plan, but sometimes the chromosomes may not be copied exactly. Sometimes bits of the chromosome may be lost, added in the wrong place or broken off. The information carried by the inaccurate chromosome will therefore be different. The changed genes which occasionally arise are known as **mutant genes** and the process is known as **genetic mutation**. Most mutations are harmful, some mutations have no effect and in rare cases a mutation can be beneficial to an organism or individual.

Down's syndrome in humans is an example of genetic mutation. In this case, the zygote receives 24 chromosomes from the egg and not the usual 23. The embryo, therefore develops with 47 rather than the normal 46 chromosomes. This leads to abnormal mental and physical development.

A mutation has caused these abnormal cancerous lung cells to invade the normal ciliated cells of the bronchus

Mutations occur naturally, but the chance of a mutation is increased by certain environmental agents such as ionising radiations and specific chemicals. For example, the chemical nicotine in tobacco smoke is known to cause mutation which leads to abnormal cell growth and cancers.

11.2 Genes and characteristics

During the 19th century, an Austrian monk called Gregor Mendel studied the way in which characteristics were passed on from one generation to the next. Mendel carried out experiments with pea plants, studying their inheritance of different characteristics such as flower colour and height.

As a result of Mendel's experiments, scientists began to make important conclusions about the inheritance of characteristics. These conclusions are sometimes called **the rules of genetics**.

- An organism gets its characteristics (e.g. flower colour, height) from its **genes**.
- Sometimes a particular characteristic is controlled by a single pair of genes. One gene in the pair comes from the father, the other from the mother. When a characteristic is determined by a single pair of genes, it is described as **monohybrid inheritance**.
- Genes may be **dominant** or **recessive**. Dominant genes always express themselves in the characteristic. Recessive genes only express themselves in the characteristic if *both* of the genes in an organism's pair are recessive.

Gregor Mendel (1822–1884) was the first person to study the inheritance of characteristics of living things

For example, humans can have free ear lobes or attached ear lobes. The gene for free ear lobes is dominant and is written as F. The gene for attached ear lobes is recessive and is written as f.

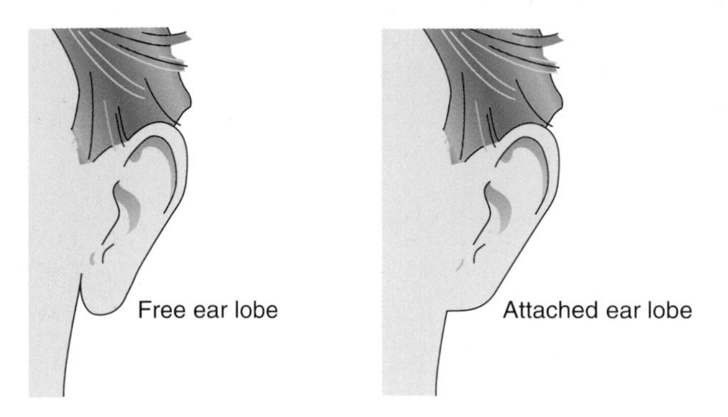

Figure 11.2 The two types of ear lobes in humans

- Dominant genes are represented by a capital letter and recessive genes by a small (lower case) letter.
- Genes which control the same characteristic but which produce different results are known as **alleles**. F and f are examples of alleles. The alleles for eye colour are B (dominant which results in brown eyes) and b (recessive – blue eyes).
- Because genes are usually paired, an individual may have three possible combinations of ear lobe genes: FF, Ff and ff. These possible pairings of genes are called **genotypes**. A person will only have attached ear lobes if their genotype is ff.

> Genotypes describe the genetic make-up of an organism.

- When an organism has a genotype with *two identical genes*, e.g. FF or ff, the organism is described as **homozygous**.
  When an organism has a genotype with *two contrasting genes*, e.g. Ff, the organism is said to be **heterozygous**.
- Sometimes it is important to describe an organism in terms of its characteristic or its appearance – red flowered or white flowered, free ear lobes or attached ear lobes. This outward appearance or characteristic is called its **phenotype**.

The important terms introduced in this section are summarised and illustrated in Figure 11.3.

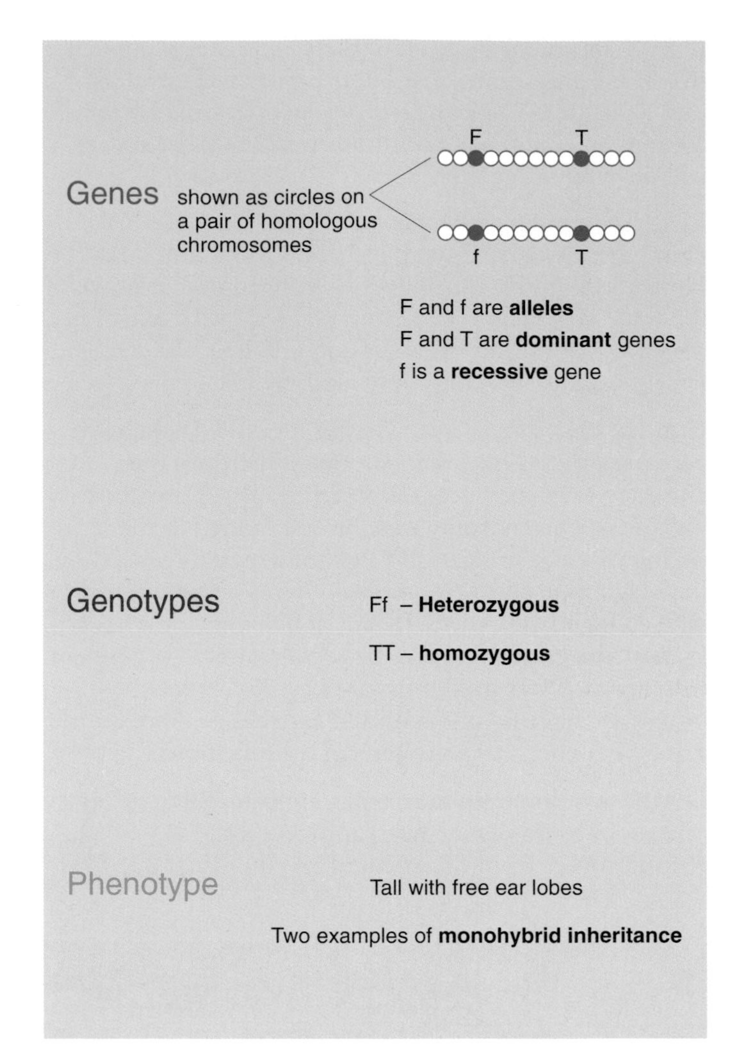

Figure 11.3 Important genetic terms

11.3 Genetic experiments

Modern genetic experiments are usually carried out with plants and fast breeding insects such as flies.

Figure 11.4 illustrates an experiment in which red and white flowers, which are strongly contrasting phenotypes of the same plant, were studied.

The parents (**P_1** or **parental generation**) were pure, homozygous plants (RR – red and rr – white). These were mated by careful cross-pollination (i.e. ovaries of RR with pollen from rr and vice versa, but never RR ovaries with pollen from RR nor rr ovaries with pollen from rr).

When careful cross-pollination was carried out and the seeds developed and then sown, all the new plants (**F_1**, or **first filial generation**) are red.

How can we explain this? Look at Figure 11.4.

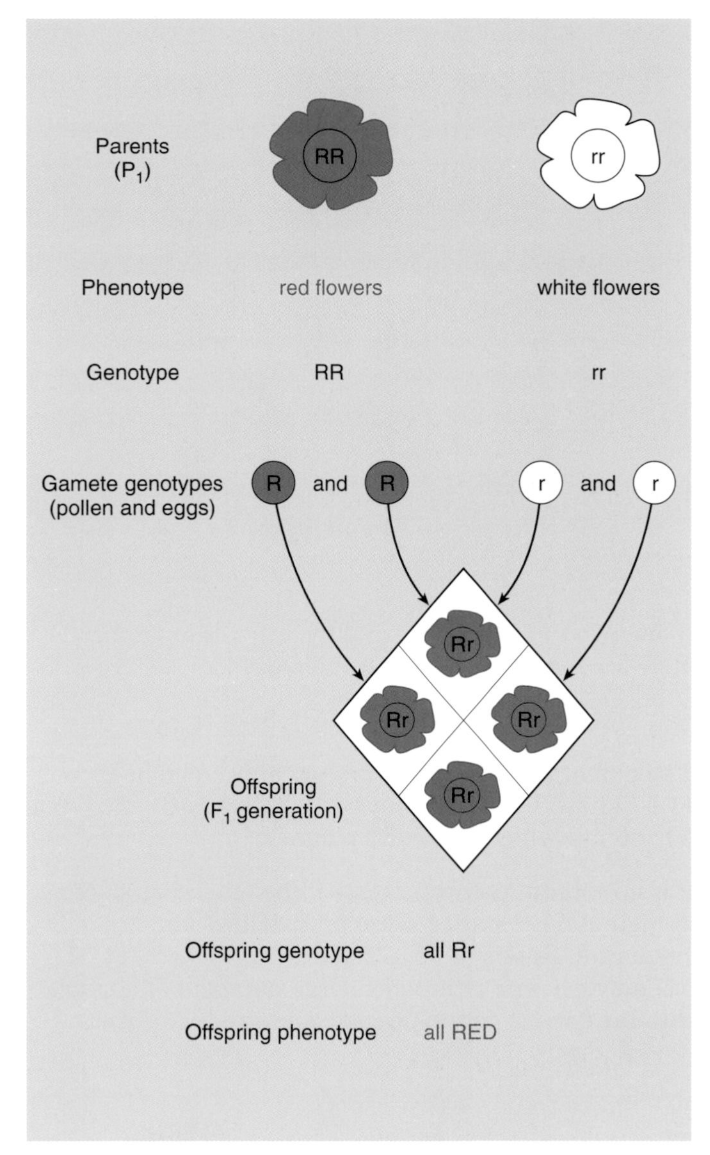

Figure 11.4 Cross-pollinating homozygous plants

Each parent (P_1) plant possesses a pair of genes controlling flower colour. The red-flowered parent contains two dominant genes, RR, which make the flowers red. The white-flowered parent contains two recessive genes, rr, which makes the flowers white.

Now, the pollen grains and egg cells (i.e. the gametes) contain only one gene: R from the red-flowered parent or r from the white-flowered parent.

When fertilization takes place, an R gene is always combined with an r gene so all the offspring (F_1) have the same genotype, Rr. The dominant R gene in the genotype means that all the first generation plants are red.

11.4 Further experiments involving monohybrid inheritance

Another characteristic controlled by a single pair of genes is the coat colour of leopards and panthers. The coat colour of these two animals is controlled by a single pair of genes. The dominant, spotted gene, S, gives rise to a yellow coat with black spots and the animal is called a leopard. The recessive gene, s, gives rise to an all black coat and the animal is called a panther.

The genotypes of these animals can be SS, Ss or ss. Animals which have the genotypes SS and Ss will have the phenotype spotted coat (leopard), whilst those with the genotype ss will have the phenotype black coat (panther).

When leopards and panthers are allowed to breed it is possible to predict the likely outcome of the genes in their cubs.

The coat colour of the leopard in the top photo is determined by a dominant gene, whereas the coat colour of the panther is determined by the corresponding recessive gene

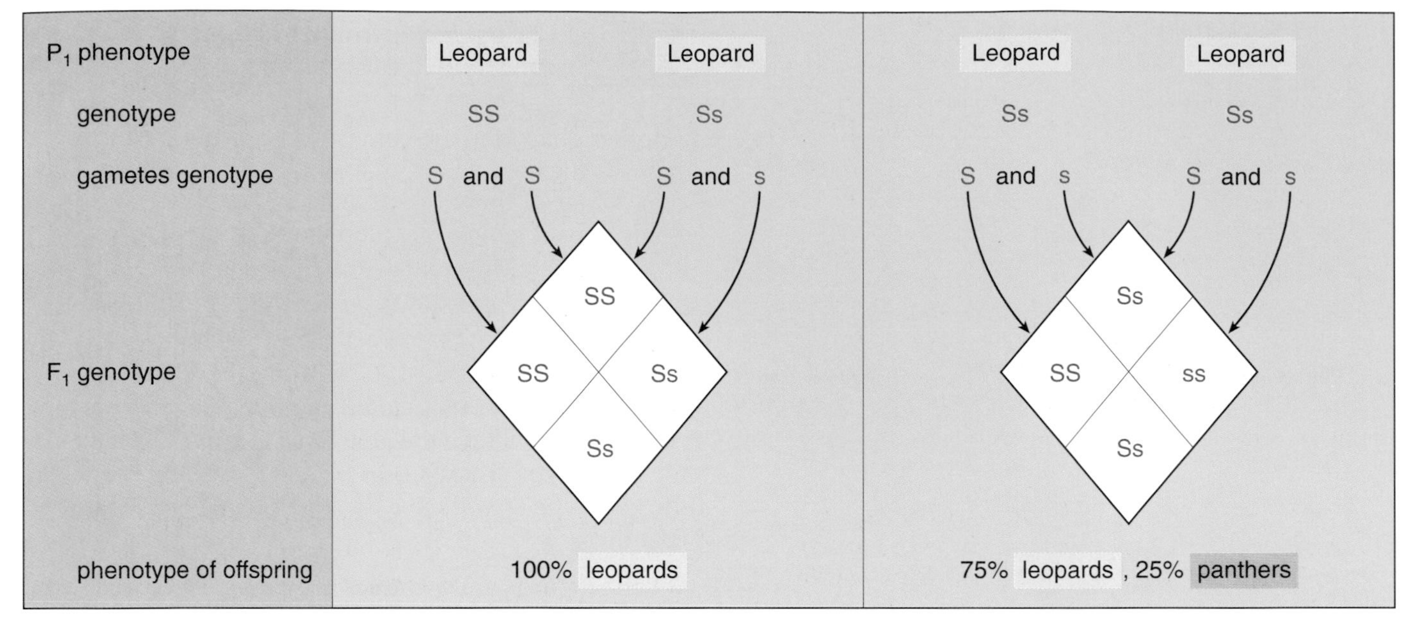

Figure 11.5 What happens when leopards breed?

If both parents are leopards, there are three possible crosses. These are:

(i) SS × SS, (ii) SS × Ss, (iii) Ss × Ss.

If both parents are homozygous leopards (SS × SS) as in cross (i), all the gametes produced by both parents will have the dominant allele S. Every fertilization will produce SS and so all the cubs (F_1 generation) will be leopards. The results for crosses (ii) and (iii) are shown as checkerboard diagrams in Figure 11.5. When cross (ii) occurs, all the cubs will be leopards. But in cross (iii), 75% of the cubs will be leopards and 25% will be panthers.

If one parent is a leopard and the other is a panther, there are two possible crosses to consider:

(i) SS × ss and (ii) Ss × ss.

The outcome of these two crosses are shown in Figure 11.6. These crosses lead to (i) 100% leopards and (ii) a 50 : 50 mix of leopards and panthers.

It is important to appreciate that the outcomes shown in Figures 11.5 and 11.6 are probabilities and not certainties. Ratios of 75 : 25 (3 : 1) and 50 : 50 (1 : 1) will only become clear when large numbers of animals with the correct genotype mate and produce cubs.

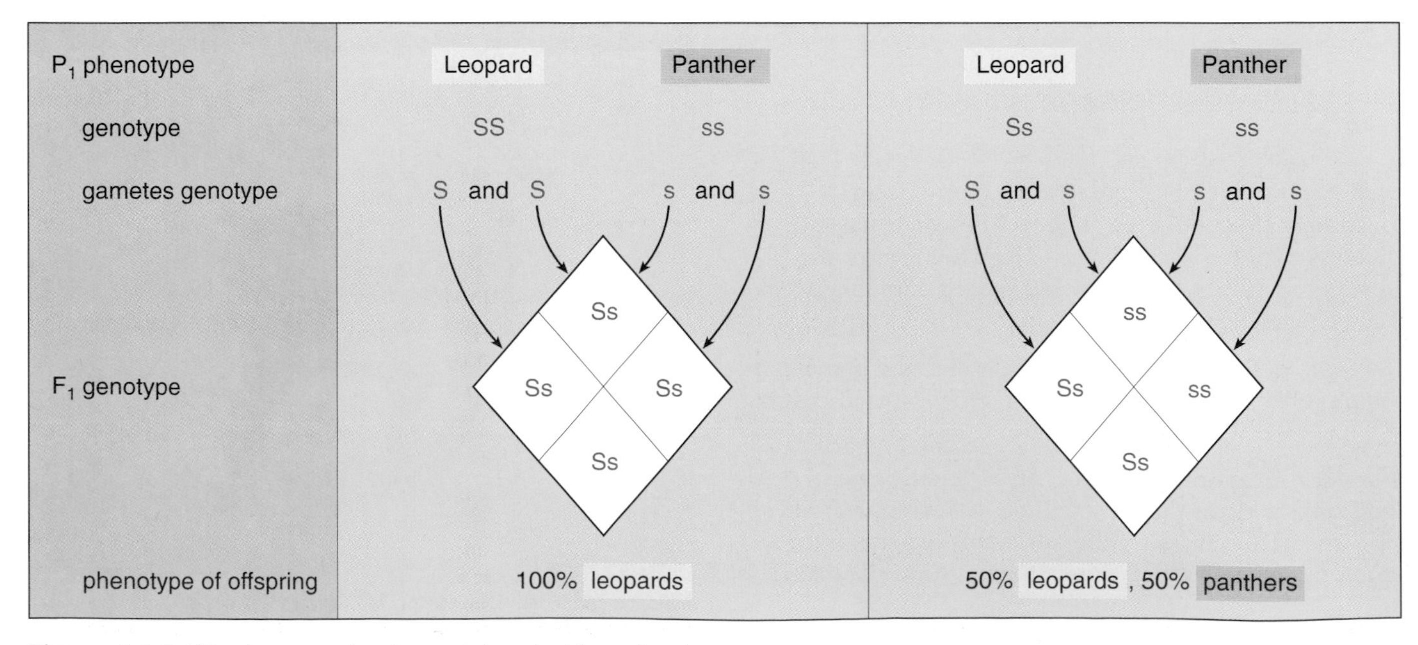

Figure 11.6 What happens when leopards breed with panthers?

11.5 Inheritance of diseases

Some diseases are genetic in origin and can be inherited from parents who do not have these diseases. These inherited diseases include cystic fibrosis, sickle-cell anaemia, muscular dystrophy, haemophilia and colour blindness.

Cystic fibrosis

This is a disease caused by abnormality in the genes. The abnormal genes cause the mucous membranes in the respiratory passages to produce excessively thick and sticky mucus. This blocks air passages in the lungs, causing chronic lung congestion and a constant risk of infection.

Cystic fibrosis is a rare disease because the gene causing it is a recessive gene. A sufferer has inherited two such genes, one from each parent. The parents themselves are usually heterozygous for this gene and therefore are healthy. They are described as **carriers**.

Sickle cell anaemia

This is another disease caused by a recessive allele like cystic fibrosis. Individuals who are homozygous for the allele (i.e. two similar recessive alleles, one from each parent) have fewer normal red blood cells and instead have abnormally shaped blood cells.

The reduced number of normal red blood cells leads to the condition of anaemia. The person has a pale complexion, may be breathless and feel generally weak and tired. The abnormal sickle shape causes blocking of capillaries, poor clotting after injury and may lead to heart failure and brain damage.

Normal red blood cells and sickle shaped cells in the blood of a person suffering from sickle cell anaemia

Sex-linked inherited diseases

Some diseases, including muscular dystrophy, haemophilia and red-green colour blindness are much more common in males than in females. This is because the diseases are caused by recessive alleles on the X chromosome. Males with an XY chromosome pair are more likely to have such diseases than females with two X chromosomes as an XX pair. This is because the female would have to have the recessive allele on *both* X chromosomes to have the disease, whereas the male only has to have it on his one X chromosome. Figure 11.7 over the page shows the possible genotypes for males and females who carry the normal (dominant) and affected (recessive) alleles associated with these diseases.

Muscular dystrophy results in muscles becoming weaker and wasting away. It is a degenerative condition and over many years the sufferer becomes totally incapacitated.

Haemophiliacs are unable to produce an essential substance for blood clotting called factor VIII. This is serious because any injury causing blood loss, either internal or external will not be halted in the normal way. Fortunately, haemophiliacs nowadays can receive factor VIII on a regular basis and can be free of constant bruising and life-threatening bleeding.

Sufferers from **colour blindness** find great difficulty in distinguishing between reds and greens. They appear to them as shades of grey.

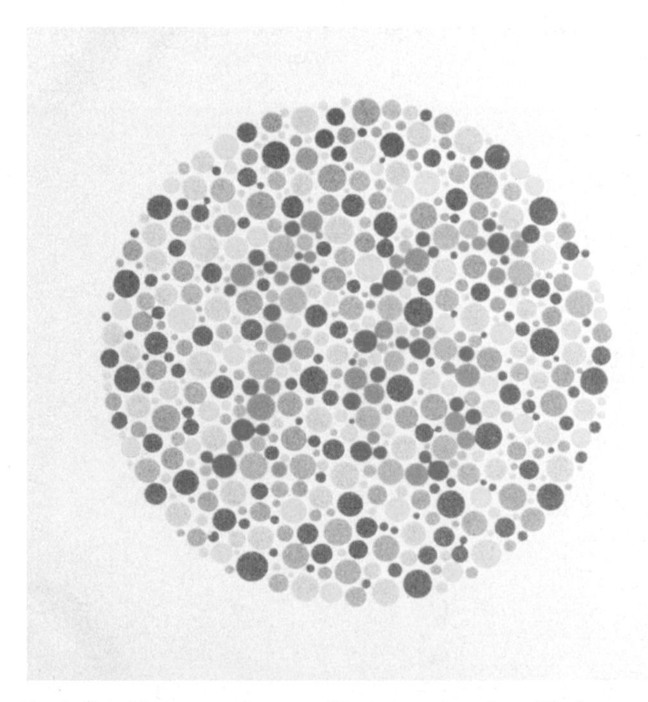

Cards like this are used to test for red-green colour blindness. A person suffering from red-green colour blindness would not be able to make out the number 16

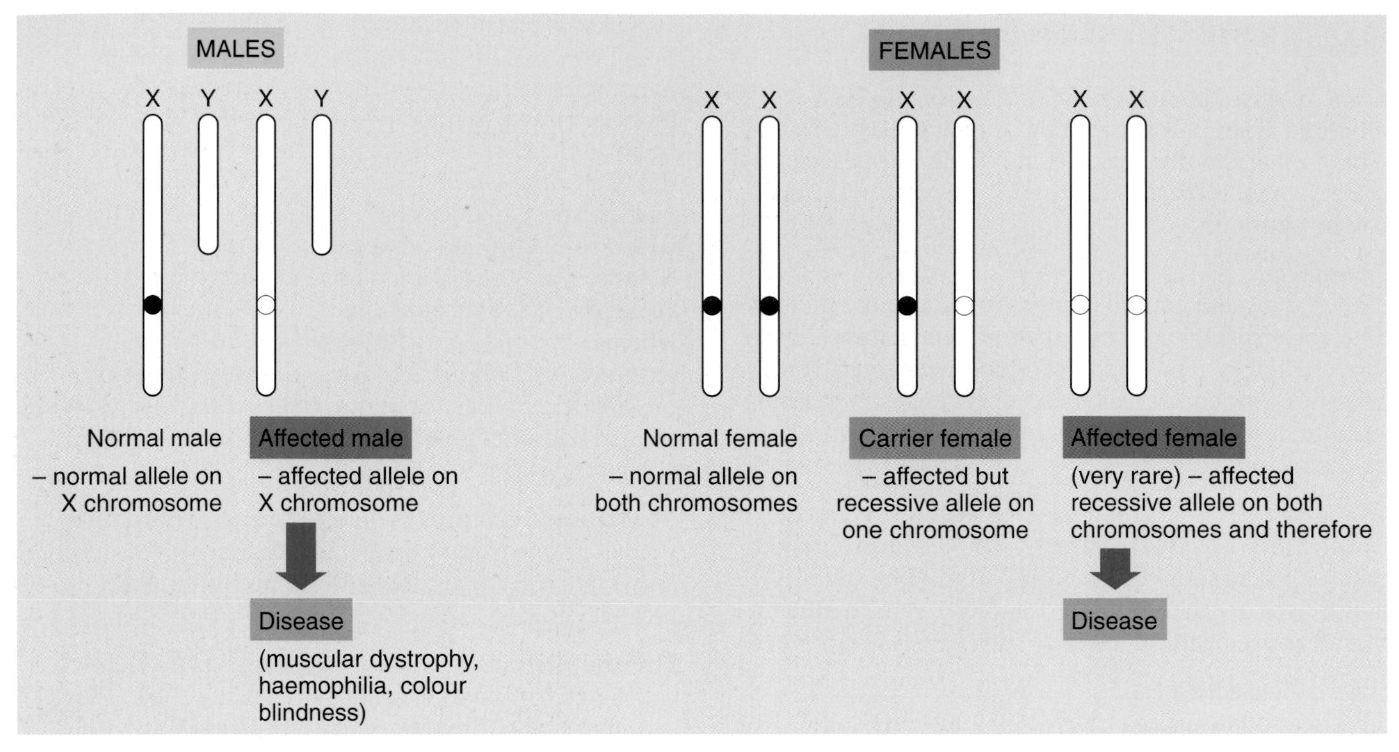

Figure 11.7 The possible genotypes for males and females carrying the normal and affected alleles associated with muscular dystrophy, haemophilia and colour blindness

11.6 Selective breeding

Farmers, animal breeders and gardeners have been applying the principles of genetics for centuries. Year by year and generation by generation, they have selected animals with characteristics such as good behaviour, high milk yield and strength. They have only saved the seeds of plants that have superior flavour, are resistant to disease or drought and will produce high crop yields. In some cases, selective breeding has led to new varieties of animals and plants.

By choosing individuals which have characteristics that are useful and then breeding from them we have produced:

- cattle with high milk yield or leaner meat,
- cereal crops resistant to plant diseases and
- fruits and vegetables with good flavour and high yields.

Nowadays, geneticists can breed varieties more systematically and advise farmers and gardeners on ways in which they can improve yields, flavour and appearance. Using these approaches, selective breeding has brought about enormous economic benefits.

This tiny, hairy Yorkshire Terrier and the massive, smooth haired Great Dane are members of the same species, *Canis familiaris*. Their differences have resulted from selective breeding over many generations

11.7 Variation within a species

Look at all the boys or all the girls in your class. Notice their differences in height, weight, skin colour, hair colour, hair texture, way of walking and intelligence. Yet these boys and girls are all of the same species – *Homo sapiens*. These differences between individuals are called **variations**.

All these children were born in the same year. What variations can you see in them from this photo?

Variations within a species can arise in two ways:

- from genetic (inherited) causes and
- from environmental causes.

Many human characteristics including birth weight, height, skin colour and hair colour are determined largely by the genes which we inherit from our parents. However, we are certain to vary from our parents because of three processes which ensure the mixing of genes.

1 **Meiosis** 'shuffles' the chromosomes. When eggs and sperms are formed during meiosis, the chromosome pairs separate in a totally random fashion (section 10.3).

2 **Fertilization** brings together an entirely new set of 46 chromosomes – 23 from the father and 23 from the mother (section 10.2).

3 **Mutations** cause a change in the chemical structure of a gene (section 11.1).

Although many human characteristics are genetically determined, they can be modified by our environment. For example, your body weight is affected by your diet, your skin colour may depend on the climate and your hair style may change with the latest fashion.

The environment can also modify the appearance of plants. Those grown in ideal conditions with plenty of sunlight, water and nutrients are very different to those which suffer from an inadequate supply of these requirements.

Two trays of pea seedlings: one grown in the dark and the other in good light. Which is which?

11.8 Natural selection

The fittest zebra with the greatest stamina will be able to escape from the lion

In the 1960s the myxoma virus killed thousands of rabbits. A few rabbits had genes which gave them immunity to the virus and they survived the myxomatosis disease. Now, the rabbit population has been restored

Ospreys compete with fishermen and other animals for the fish in a lake. Unless an osprey competes effectively, it will not survive

Alpine sunflowers grow close to the ground, achieving only ten centimetres in height. This is unlike other sunflowers which grow to a height of more than one metre. The alpine sunflowers have a very short growing season and need to survive at high altitudes in a harsh climate

Figure 11.8 Examples of the ways in which animals and plants have become better adapted to survive

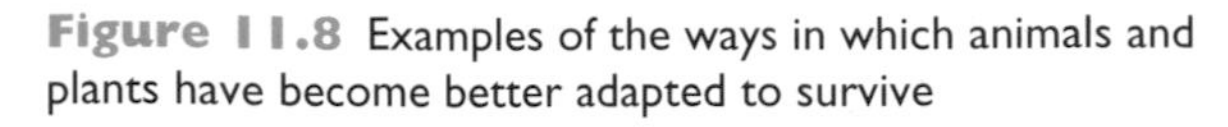

The variations in the appearance and characteristics of the individuals within a species leads to some which are taller, others which are fitter and those who can run faster, etc.

These variations will allow those organisms, who are best suited or best adapted to the environment, to survive. In many situations, animals and plants face a continual struggle to survive. For example:

- They have to avoid **predators**.
- They have to stay free from **disease**.
- They are in **competition** with others of the same species and different species for food, water and shelter.
- They have to survive the **climate**.

The animals and plants that survive are usually the fittest. They may be better at escaping their predators. They are less likely to suffer from disease. They can compete more effectively for food and shelter and they can withstand harsh weather conditions. This survival of those animals and plants best adapted to the conditions in their environment is often described as the **'survival of the fittest'**.

Figure 11.9 The dark and light forms of the peppered moth on the trunk of a tree

The photos in Figure 11.8 opposite illustrate four examples of the ways in which certain animals and plants are better adapted for survival.

Those organisms which have the characteristics which make them better suited to the conditions in their environment will survive and go on to reproduce. This will mean that those genes (alleles) which give rise to the most favourable characteristics will be passed on to their offspring and become more common. Other organisms, less well suited may die before they are able to reproduce. Over many generations, the species of animal or plant slowly changes and becomes better adapted to survive in that particular habitat.

This process by which a species slowly changes over many generations, becoming better adapted for survival is known as **natural selection**. The process of natural selection leads to the **evolution** of a species.

In some cases, the organisms within a species are unable to adapt to changing conditions or the variations which arise may lead to organisms which have become too specialised and then find it difficult to survive. This may eventually lead to the **extinction** of the species.

One of the most obvious examples of natural selection and evolution within a species concerns the peppered moth which lives in woodland areas. For much of the time, the peppered moth rests on tree trunks. During the 1840s, a dark form of the peppered moth appeared for the first time as a result of a mutation (Figure 11.9).

In unpolluted areas, the lighter form of the moth is well camouflaged on clean tree trunks. The darker form is easily seen and taken by predators like thrushes. In polluted areas, however, where tree trunks have been darkened by soot, the darker mutant moth is better camouflaged and more likely to survive. In these areas the darker mutant has, therefore, evolved as the dominant form of the species.

11.9 Darwin's theory of evolution

In 1831, Charles Darwin was appointed naturalist on a world-wide scientific and geographical expedition. The expedition visited Africa and America where Darwin studied animals and plants. As a result of his observations, Darwin became convinced that different species had evolved through a slow process of natural selection.

In 1859, Darwin published his theory of evolution by natural selection in a book called *The Origin of Species*.

The key ideas in Darwin's theory are illustrated in Figure 11.10 over the page.

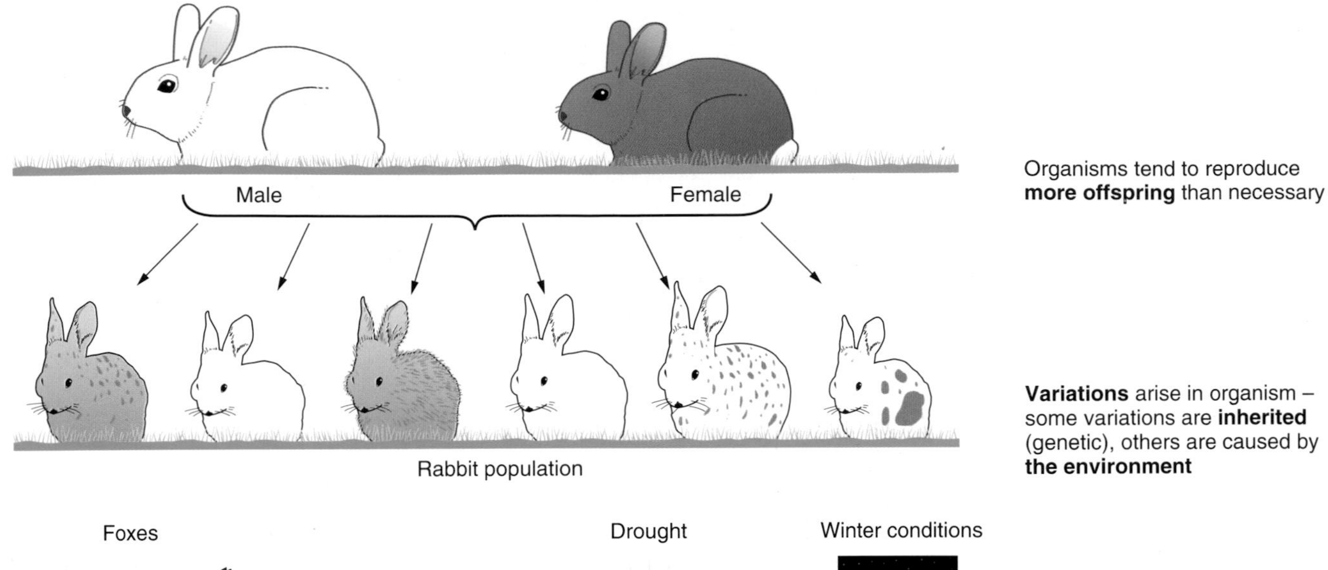

Figure 11.10 The key ideas in Darwin's theory of evolution

11.10 Evidence for evolution

Scientists now believe that the first living things to inhabit the Earth were very simple organisms. Slowly, over 3000 million years, these simple creatures and plants have evolved into thousands of different organisms by natural selection. Whilst many organisms have adapted and evolved, others like dinosaurs and dodos have become extinct or died out.

Dodos inhabited Mauritius until the end of the 17th century when the last dodo was killed by hunters

The best evidence for evolution comes from **fossils**. When a plant or an animal dies, it may be eaten or it may decay under the action of bacteria, fungi and oxygen in the air. In some cases, however, dead animals and plants have not been eaten and their decay has been slowed down and sometimes almost stopped.

In some areas, dead organisms have been covered by the sea, by sediment from rivers or by rocks from earth movements. In these places, the material has decayed in the absence of oxygen. Bacteria would have attacked it, but instead of it rotting away completely, it has been compressed leaving bones or woody tissue preserved in rock for millions of years.

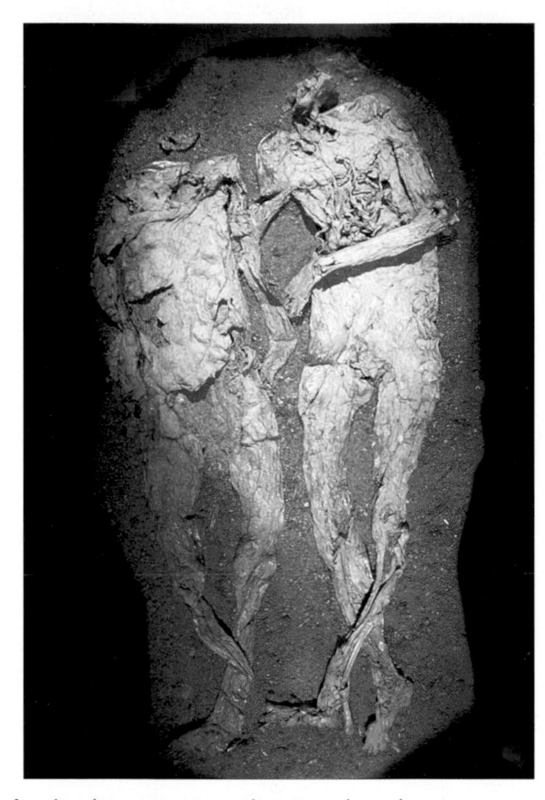

These fossilized remains were discovered in a bog in Northern Holland. Carbon dating showed that they died in about 200 BC

In other areas, dead organisms may have been preserved in ice and covered by layers of snow at such low temperatures that decay is almost halted.

As the conditions for fossil formation are rare, it is easy to see why fossils are fairly uncommon and why fossil records are incomplete. By studying fossils, scientists can deduce what the animal or plant was like when it was living. Using radioactive dating (section 29.7), the age of the fossil can be worked out.

From the fossil record, a detailed picture can now be built up showing how animals and plants have changed since life began on Earth and how they have adapted, evolved and sometimes become extinct over millions of years.

Figure 11.11 The evolution of present day horses from their prehistoric ancestors

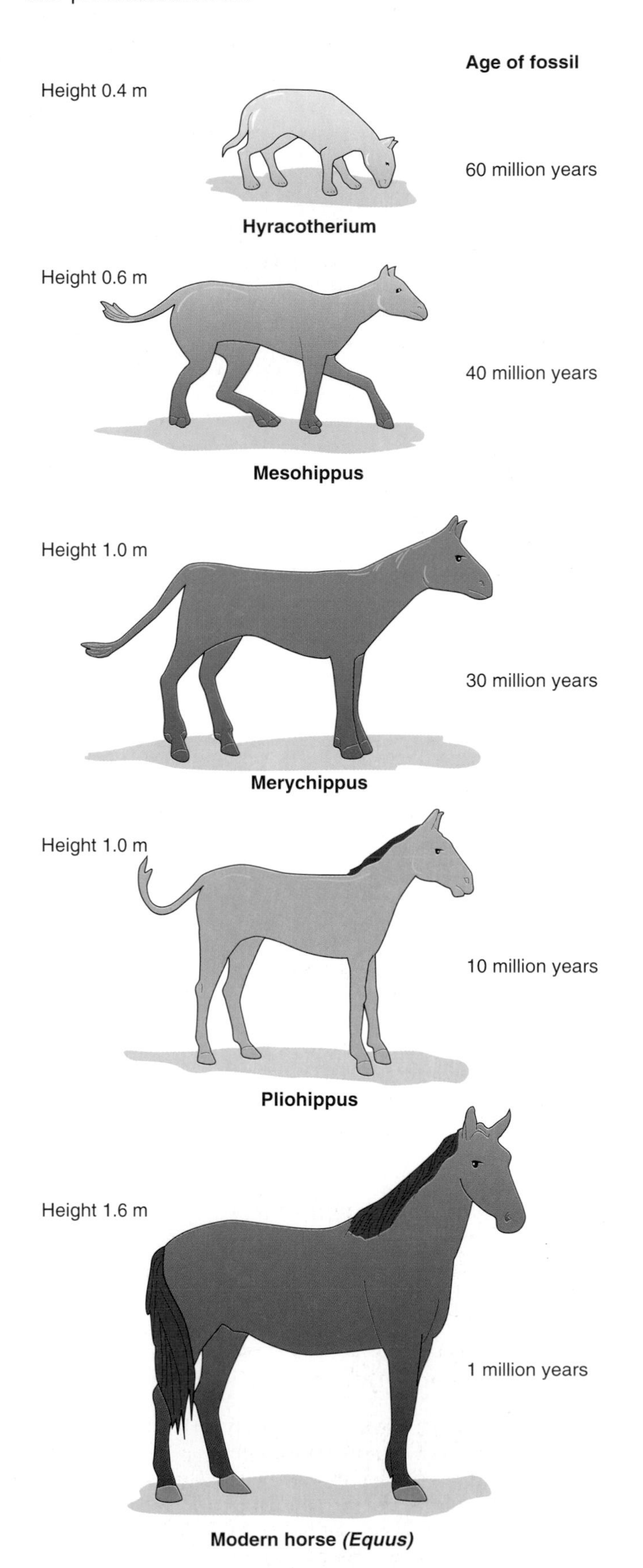

1 The histogram shows the lengths of 280 leaves picked from the same tree, in classes of 10 mm.

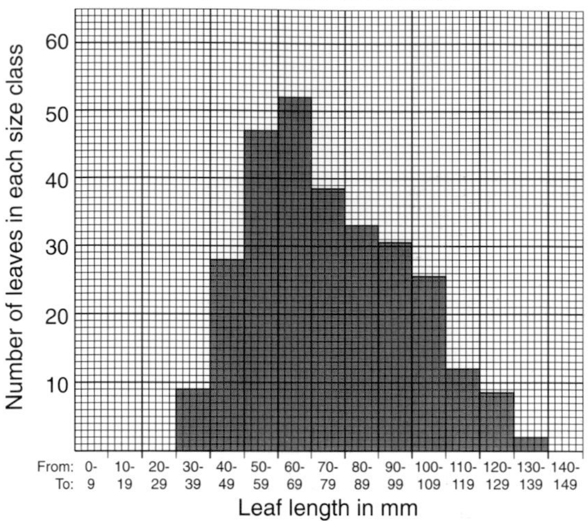

a) Which class of leaf:
   i) is the most common,
   ii) contains the longest leaves?

b) How many leaves are between 30 mm and 49 mm long?

c) Suggest a reason which could explain the difference in size of the leaves. **WJEC**

2 In 1966 a Canadian cat breeder noticed that a hairless kitten was born in a litter of kittens from normal parents. The cat breeder used some of the cats from this family to select a new type of breed which she called Sphinx. Assume that the allele for *hairless* is recessive to the allele for hairy, H.
Copy the figure below and fill in the key and parental genotypes. Then use the grids to show how the cat breeder would breed large numbers of Sphinx cats from the original hairy parent and further matings. The allele for *hairless* is not sex linked.

Key:

Parental genotypes:
(Genes found in parents)

First Cross

Second Cross

WJEC

3 The most common inherited disorder among Europeans is cystic fibrosis. The gene (allele) causing this disease is recessive and is **not** carried on the sex chromosomes.
The diagram shows a family who have a child with cystic fibrosis.

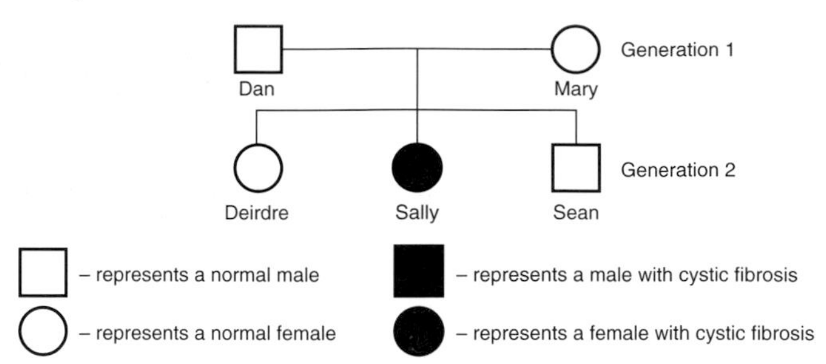

– represents a normal male
– represents a male with cystic fibrosis
– represents a normal female
– represents a female with cystic fibrosis

Use B to represent the normal gene (allele)
Use b to represent the gene (allele) for cystic fibrosis

a) What is Sally's genetic make up (genotype)?

b) Give the genetic make up (genotypes) of Dan and Mary.

c) Use a punnett square to show the possible genetic make up (genotypes) of Sally's sister, Deirdre. Give the possible genetic make up (genotypes) of Deirdre. **NICCEA**

4 a) The diagram shows the chromosomes from the nucleus of any cell in the male human body other than sex cells.

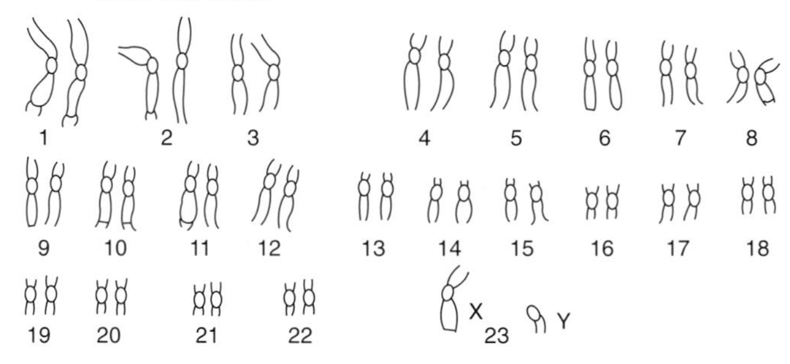

Explain how you can tell that the chromosomes came from a male rather than a female.

b) The figure below shows the chromosomes in a cell from a person suffering from a condition known as Down's syndrome.

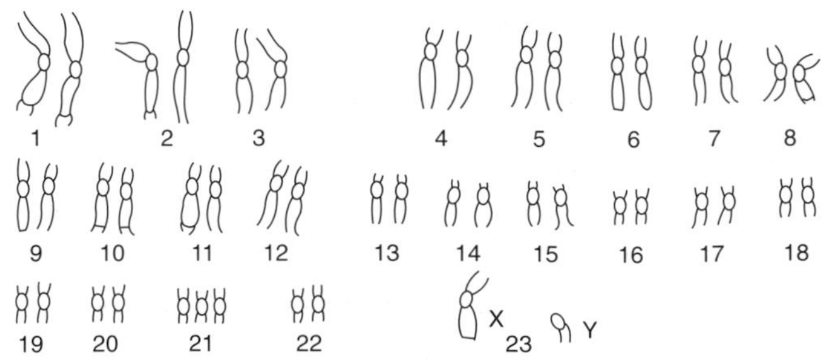

i) State how you can tell from the diagram that this person has Down's syndrome.
ii) Suggest what caused this condition to occur. **MEG**

CHAPTER

Living things in their environment

Classifying living things
Using keys to identify animals and plants
Populations and communities
Competition, adaptation and predation
The effect that we have on the environment

12.1 Classifying living things

There are an amazing number of different organisms in the world. Studying them would be impossible without sorting them into groups with similar features and similar characteristics. These groups can then be divided into smaller groups and so on. This division into groups is known as **classification**.

Living things are first divided into **kingdoms**. There are five kingdoms (Table 12.1). The best known kingdoms are those for animals and plants.

Table 12.1 The five kingdoms of living things

Kingdom	Organisms in the kingdom
Animals	Vertebrates (have backbones) e.g. mammals, reptiles, birds Invertebrates (no backbone) e.g. insects, spiders, worms
Plants	Flowering plants e.g. trees, flowers, grasses Non-flowering plants e.g. mosses, ferns, seaweeds
Fungi	Moulds and fungi living off living or dead organisms e.g. pin mould, mushrooms
Protoctists	Very small, simple organisms, usually with only one cell e.g. amoeba, algae
Bacteria	Very small, single-celled organisms with no nucleus e.g. tetanus, pneumonia bacteria

Kingdoms are then divided into **phyla** (singular, phylum) and then phyla are divided into **classes**. For example, there are two phyla in the animal kingdom – vertebrates (animals with backbones) and invertebrates (animals without backbones). There are five classes in the vertebrate phylum – mammals, birds, reptiles, amphibians and fish. The important features of the five classes of vertebrates are shown in Figure 12.1 over the page.

Further divisions of classes are still necessary. Classes are divided into **orders**, orders into **families**, families into **genera** (singular, genus) and genera into **species**. Table 12.2 shows this full classification for humans.

Table 12.2 The full classification of humans from kingdom to species

Group	Human
Kingdom	Animal
Phylum	Vertebrate
Class	Mammal
Order	Primate
Family	Hominid
Genus	*Homo*
Species	*sapiens*

Figure 12.1 The important features of vertebrates

VERTEBRATES

	Outer covering	Adaption for obtaining oxygen	Fertilization	Eggs	Parental care
Mammals 	hair	lungs	internal	develop in uterus	young cared for and fed on milk from mother
Birds 	feathers	lungs	internal	hard shell	incubate eggs and feed chicks in nest
Reptiles 	dry scales	lungs	internal	soft, leathery shell	some care of eggs
Amphibians 	moist skin	tadpoles –gills frogs –lungs	external in water	no shell	none
Fish 	scales	gills	external in water	no shell	none

The grouping which we use most commonly for organisms is probably the **species**. All the organisms in a species are very like one another and may breed together. For example, all dogs are members of the same species and may breed together. But dogs cannot breed with cats, which are a different species.

Naming living things

Each species of living things has a name. There are common names – human, cat and foxglove. There are also **systemic** or **proper** names, often derived from Latin words. Proper names for organisms give the **genus** (in italic print with a capital first letter) followed by the **species** (in italics). So, the proper name for human is *Homo sapiens*, for cat is *Panthera catus* and for foxglove is *Digitalis purpurea*.

12.2 Using keys to identify animals and plants

The animals and plants in a group of organisms can be identified using **keys**. Keys are prepared by specialists who know what all the organisms look like. A key usually has a number of paired questions. The questions focus on differences between the organisms in order to identify each one.

Use the simple key to identify the five animals in Figure 12.2.

Figure 12.2 Identifying five animals using a key

Question			
Question 1 Does it have legs?	**Yes**	go to question 2
	No	go to question 3
Question 2 Does it have fur?	**Yes**	it is a deer
	No	go to question 4
Question 3 Does it have fins?	**Yes**	it is a trout
	No	it is a cobra
Question 4 Does it have feathers?	**Yes**	it is a penguin
	No	it is a frog

12.3 Populations and communities

Living things can only survive as populations if the species is well adapted to its habitat. The population will grow if the conditions are favourable and will fall in adverse conditions. Different types of organisms will find different situations favourable at any one time.

This herd of bison in Yellowstone National Park, U.S.A. travel hundreds of miles every year to find good grazing

There are various factors which can affect organisms and the size of their populations.

The supply of food and water

In a community, the number of animals of any species (its population) is limited by the amount of available food. When food and water are abundant, animals are in prime condition and well able to reproduce and have healthy young. The effect of a plentiful supply of food is particularly evident in the offspring of animals which are predators. Ospreys will be able to rear strong fledglings if fish is in good supply. A pack of wolves will rear more cubs if there is a ready supply of deer and elk.

Plants will flourish and reproduce when there is adequate rainfall and good light.

A good food supply has ensured a successful start in life for these wolf cubs

Climate

Climate will, of course, control the supply of food and water. Drought conditions will severely test animals, especially if the plants they rely on die back because of lack of rainfall.

The temperature will also play an important part in controlling the size of both plant and animal populations. Like the availability of water, temperature will also vary from season to season, affecting the size of certain populations at different times of the year.

This male robin is strongly defending his territory

Space
Given more space, organisms can maintain a larger population. If there is too little space and overcrowding occurs, there will be competition. Plants grow weak and 'leggy' in their search for more light. Some animals such as robins and wolves will fight to maintain their territory. Female rats will not breed if they have contact with too many males.

Waste products
If an animal population increases to a high level, a regulatory system comes into being. Waste products such as faeces and urine pollute the surrounding area and disease and general malaise may ensue. This leads to premature death and poor reproduction, and so the population will fall and eventually will be regulated.

Predators
Look closely at the graph in Figure 12.3. This shows the population numbers of both the arctic lynx and snowshoe hares in Northern Canada. The lynx hunt snowshoe hares.

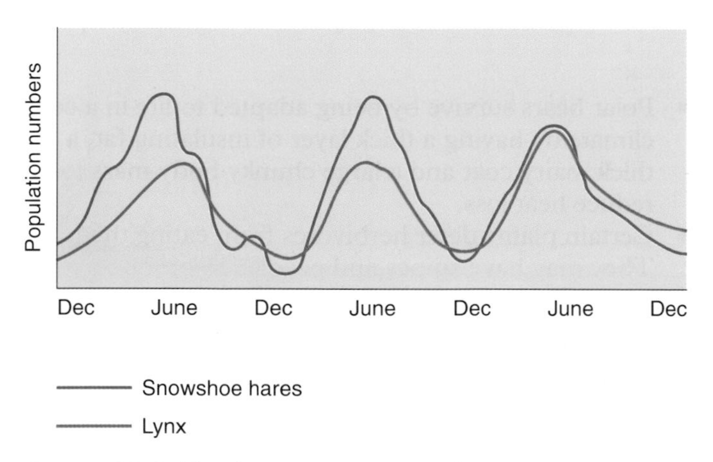

Figure 12.3 The changing populations of arctic lynx and snowshoe hares in Northern Canada

Notice that:

1 as the population of snowshoe hares (prey) increases, the population of lynx (predators) also increases,

2 as the population of lynx reaches a high value, that of the hares falls.

In general, biologists have found that:

1 if the population of prey increases, more food is available for its predators and, as you would expect, their population increases,

2 if the population of predators increases, more food is needed by these greater numbers of animals and so the population of prey decreases.

12.4 Competition, adaptation and predation

If a garden is neglected for some time, the grass grows tall, weeds appear in the flower beds and in the lawn and gradually the garden becomes a meadow.

The plants may die back in the winter months, but they will grow taller and thicker than ever the following spring. Seeds from shrubs and trees will germinate. Saplings will become trees, taking light and moisture and depriving the original garden plants. The most successful plants are weeds and they will take over. The development of plant populations in a neglected garden is a good example of competition in nature. Initially, grass on the lawn and garden flowers compete with weeds for light, space, water and nutrients in the soil. However, without a gardener to tend the garden, the flowers lose out in competition to the weeds.

In a garden, lawn grass and flowers are in competition with weeds. Provided they are tended by a gardener, the lawn and flowers can win!

Later, the weeds have to compete with the canopy of branches of leaves on trees and shrubs and will not thrive as before because of poor light and moisture.

> Life on Earth is a competition. Both plants and animals experience competition for food (nutrients), water and space.

Competition comes from many sources. Organisms may experience competition from members of their own species when there is survival of the fittest. In addition, organisms may experience competition from different species. For example, frogs compete for mates, robins compete for territory and squirrels compete with finches for pine cones.

Material being buried at a landfill quarry site. Eventually the area will be landscaped

A technician digging pellets made from refuse prior to their incineration which will generate electricity for the National Grid.

Workman inspecting crushed aluminium cans before recycling

Figure 12.5 These photos show three ways of dealing with waste in more acceptable ways.

Acid rain has affected these trees and made the lake water acidic in the Liesjarvi National Park in Finland

Air pollution

Home heating, vehicle fuels and electricity generation all rely very heavily on fossil fuels for energy. When fossil fuels are burnt in our homes, in our vehicles and in power plants, carbon dioxide is released into the atmosphere. Sulphur dioxide and nitrogen oxides may also be released when fossil fuels burn.

Sulphur dioxide and nitrogen oxides dissolve in rain water and make it acidic. Acid rain harms plants, including trees and attacks the stonework on buildings. If the acid rain enters rivers and lakes, making the water too acidic, aquatic plants and fish begin to die.

The greenhouse effect

The increasing use of fossil fuels has led to a small but relatively significant increase in the concentration of carbon dioxide in the air. This increase in the concentration of CO_2 in the air has been further exacerbated by large scale deforestation to provide timber. The huge trees of the Brazilian and Malaysian forests have been felled as have large areas of trees throughout the world. Before this deforestation, carbon dioxide was removed from the atmosphere as the trees photosynthesized.

Molecules of carbon dioxide are larger and heavier ($M_r(CO_2) = 44$) than the molecules of oxygen ($M_r(O_2) = 32$) and nitrogen ($M_r(N_2) = 28$) in clean air (M_r is the symbol given for relative molecular mass).

Because of this, radiant heat (energy waves) cannot pass through carbon dioxide as easily as it passes through clean air. So, as the concentration of carbon dioxide in the atmosphere slowly rises, less heat escapes from the Earth and the temperature slowly rises. Carbon dioxide traps heat in the Earth's atmosphere just like glass does in a greenhouse so we call it **'the greenhouse effect'**. This has led to **global warming**.

Global warming

Some people believe that global warming is already a reality. The following points are made to substantiate this belief.

- **Changes in the climate** are occurring. Some areas are experiencing higher temperatures than in the past and there are changes in the pattern of rainfall. There seem to be more violent storms. The higher average temperatures in temperate zones, including Britain, are causing birds to begin to make nests and lay eggs earlier in the spring.
- **Patterns of food production are changing.** In certain parts of the world, particularly in East Africa, rainfall has decreased so that crops are failing more frequently than in the past.
- **Sea levels are rising** and coastal flooding is causing more damage due to an overall melting of the polar ice caps.

Many world leaders are now committed to trying to find ways of halting pollution and we all have a contribution to make by not adding to the world's problems by wasting our resources.

Ornithologists in Britain have studied nest building and egg laying trends and have found that magpies are now starting this process 17 days earlier than they did in the 1940s. Is this evidence for global warming?

QUESTIONS

1 The diagram shows the design and operation of a system for farming prawns. The system is an example of a controlled environment.

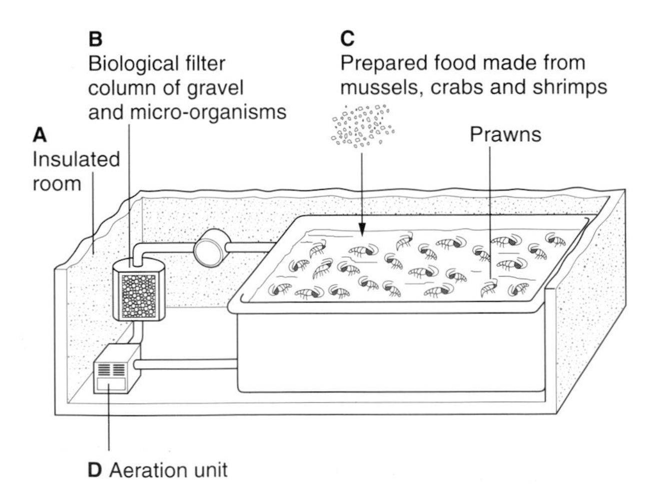

Explain why the factors A, B, C and D are carefully controlled in this environment. **MEG**

2 In a town was a large factory that produced sulphur dioxide as a waste product. Some plants can be used as indicators of air pollution and a scientist was asked to look at the sulphur levels in these plants. These are his results.

Distance from factory(km)	Amount of sulphur in plants (measured units)
1.0	3.35
1.5	3.3
2.0	2.4
2.5	2.0
3.0	1.7
3.5	1.2
4.0	0.02
4.5	0.01

a) Draw a graph to show how the sulphur content in plants changes with the distance from the factory.

b) Since 1954 the numbers of different types of flowering plants living within 1 km of the town has been counted together with the amount of sulphur dioxide produced.

Year	Amount of sulphur dioxide produced (millions of tonnes)	Numbers of types of plants
1954	5.5	542
1965	6.4	328
1970	5.8	322
1975	5.4	320
1980	5.6	317

i) How has the sulphur dioxide affected the plants?
ii) Suggest how the sulphur dioxide gas could have got into the plants.
iii) The amount of sulphur dioxide produced in 1975 was less than it was in 1954, but there were fewer plant types. Suggest a reason why. **WJEC**

3 Power stations release waste gases into the atmosphere.

a) Name **two** gases, other than carbon dioxide, which may be released into the atmosphere when fossil fuels are burned in power stations.

b) Explain how the release of these gases into the atmosphere may eventually damage the environment.

c) The drawing shows an area of forest next to another area where the forest has been cut down.

i) Carbon dioxide in the atmosphere produces the greenhouse effect. The greenhouse effect may make the mean temperature rise. Explain how deforestation could cause an increase in the greenhouse effect.

ii) Forests may be managed to provide fuel. In a managed forest, trees are grown just for use as a fuel. As soon as an area of forest has been cut down, new trees are planted. Will this system affect the greenhouse effect in the long term? Explain your answer. **NEAB**

4 Last year, half a million tonnes of detergents were flushed into rivers in Britain. Most detergents contain phosphates which prevent the formation of scum on clothes. Some detergents contain enzymes. The use of phosphates in detergents is banned in Switzerland and Holland because cadmium is a by-product of phosphate production. This could reach water supplies. There are no controls in Britain despite this hazard of the heavy metal.

A substance called zeolite can be used instead of phosphate and has no side effects. When phosphates enter water, they can be used by algae and act as a fertilizer. Rivers and lakes can become 'over-fertile' and cause a dense blanket of growth of algae on the surface of the river.

The composition of some detergents and information about their manufacture is shown in the following table:

Name	North Sea oil used in manufacture	Phosphate present	Enzymes present	Plant oils used in manufacture
Ariel	+	+	+	–
Bold	+	+	+	–
Ecover	–	–	–	+
Daz	+	+	+	–
Lux	–	–	–	–
Persil	–	+	+	–
Asda Auto	+	–	–	–
Tesco Auto	+	–	+	–

a) Name **one** detergent which would be banned in Holland.

b) i) Name **one** detergent which would be successful in boiling water.
ii) Explain your answer to i).

c) Explain why the process of producing phosphate could be harmful to the environment.

d) Explain how a dense blanket of floating algae might affect the environment where the algae live.

e) If you were a detergent manufacturer, explain how you would avoid environmental pollution and still prevent the formation of scum on clothes.

f) Which detergent is the most environmentally friendly? **WJEC**

These tiny crystals of the element silver have been photographed using an electron microscope. Elements are introduced in Chapter 13, whilst the structures of materials like silver are considered in Chapter 17. The whole of this section is about **Chemistry** – the study of different materials and their properties.

SECTION 2

Materials and their Properties

CHAPTER

Mixtures, elements and compounds

13.1 Materials

We use the word 'material' in different ways. Often it is used to describe the fabrics which are used to make clothes. In science, however, a 'material' is a form of matter which is used to make things. There are millions of different materials in the universe. Just look around at the different materials that you can see at the moment – wood, glass, paper, air, ink, water and steel. These different materials, like wood, glass, air and water which we can put a name to, are called **substances**.

Classifying materials

Materials can be classified in different ways. Some materials, like wood, sand, air and water occur naturally. These are described as **naturally-occurring materials** or **raw materials**. These naturally-occurring materials are used to make more useful materials like paper and steel. In fact, most of the materials that we use have been made or manufactured from raw materials. These materials, that we have made from natural materials, are sometimes described as **man-made materials**. The main sources of raw materials for our chemical and manufacturing industries are rocks, minerals, the sea and the atmosphere.

A second way of classifying materials is by their properties. When you classify materials in this way, there are five important groups – metals, plastics, ceramics (e.g. pottery), glasses and fibres (Figure 13.1).

Figure 13.1 What materials are these different cups, mugs and glasses made from? Can the materials be classified as naturally-occurring or man-made? Can the materials be classified as metal, plastic, glass, ceramic or wood?

Containers for food can be classified very neatly into these five groups. Just think of:

- cans made from **metals**;
- polythene wrappings and bags made of **plastics**;
- bottles and jars made from **glass**;
- paper and wooden boxes made from **fibres** and
- china jars and crockery made from **ceramics**.

Fibres are materials that have long, thin strands such as cotton, wool, cellulose in paper and polyester in clothes.

Ceramics include a whole range of materials made from clay including crockery, bricks, tiles, concrete and pottery. In fact, the word 'ceramic' comes from a Greek word meaning 'pottery'.

Table 13.1 Classifying materials as metals, plastics, fibres, glasses and ceramics

Class of material	Examples	Typical properties	Raw materials from which they are made	Constituent elements
Metals	iron, aluminium, copper, steel	• hard • strong • high density – above 5 g/cm^3 • good conductors • malleable (can be hammered and bent into different shapes) • melt on heating	rocks and ores in the Earth's crust	metals
Plastics	polythene, PVC, polystyrene	• flexible • low density (1 g/cm^3) • easily moulded and coloured • poor conductors • can be transparent • melt and may burn on heating	crude oil	mainly carbon and hydrogen with other non metals
Fibres	cotton, wool, paper, wood, polyester	• flexible • low density (1 g/cm^3) • may burn on heating • have long, stringy strands	natural fibres from plants and animals, man-made fibres from crude oil	mainly carbon and hydrogen with other non-metals
Glasses	bottle glass, crystal glass	• hard • brittle • medium density • very high melting point • very unreactive – do not burn • transparent	sand, limestone and other minerals	silicon, oxygen and various metals
Ceramics (pottery)	china, concrete, bricks, tiles	• the same as glasses – but not transparent	clay, sand and other minerals	mainly silicon, oxygen, hydrogen and aluminium

Table 13.1 shows some examples, typical properties, sources and constituents of these five major classes of materials.

Notice these two things from Table 13.1;

- Ceramics and glasses have very similar properties and constituent elements.
- Plastics and fibres have similar properties and constituent elements. In fact, plastics and some fibres belong to a larger group of materials called **polymers** (Chapter 20).

Choosing materials for different jobs

In choosing materials for different jobs we have to ask two key questions.

i) Are the properties right?
These properties include **physical properties** like hardness, strength, conductivity, density and melting point. They also include questions about the **chemical properties** of the material and the chemical reactions it may undergo. For example: does the material burn, does it corrode, does it react with water? If these properties are wrong, the material will be useless or it may be unsafe.

ii) Is the cost reasonable?
The cost of raw materials and the cost of manufacture are also important in choosing materials for different jobs. Sometimes we need a material that combines the properties of two different materials. The carbon fibre-reinforced plastic used in the shafts of golf clubs and the handles of tennis rackets is a good example of this.

Carbon-reinforced plastic has a plastic **matrix** reinforced by carbon (graphite) fibres. So, the shafts of golf clubs have the flexibility of plastics and the strength of carbon fibres – the best of both worlds.

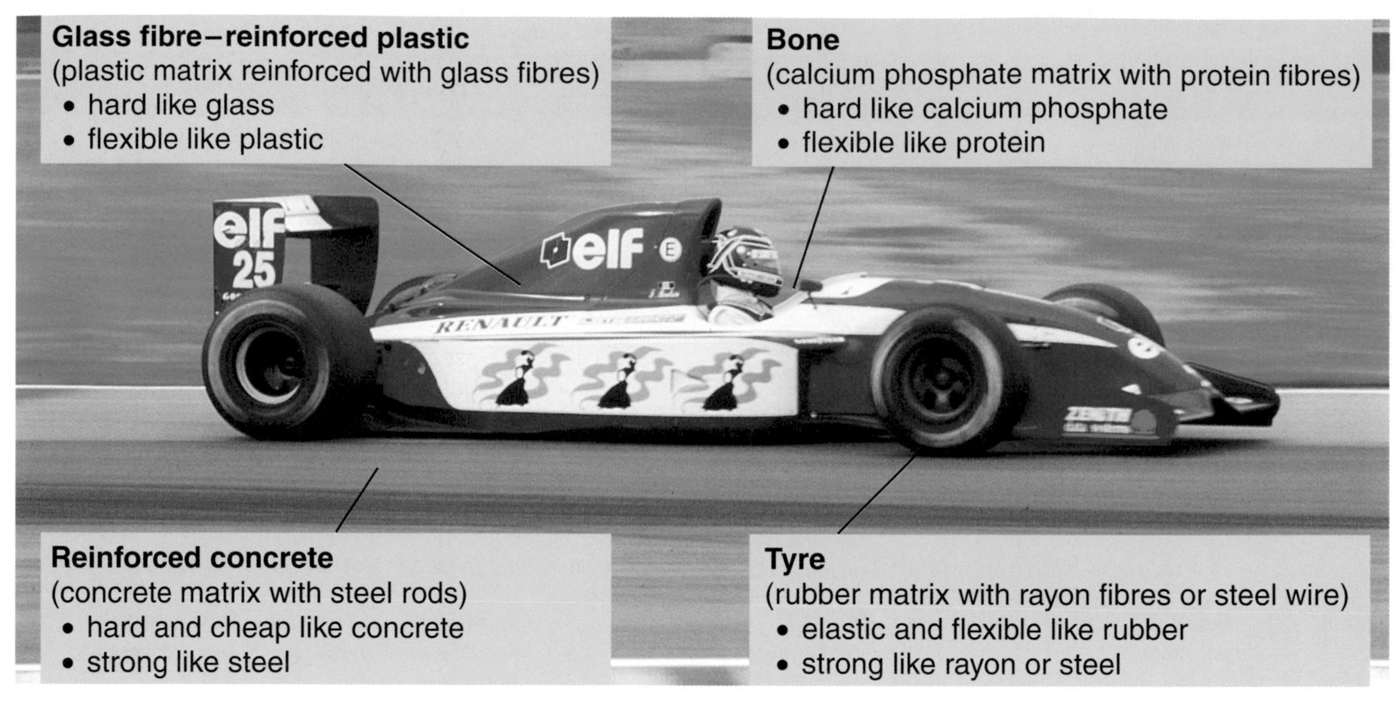

Figure 13.2 Four important composite materials

Carbon fibre-reinforced plastic is an example of a **composite material**. Four other composite materials are shown in Figure 13.2. Composite materials are made of two or more different materials which work together to give a better material for the job than any of the separate materials.

In the 1960s, tennis players used wooden-handled rackets. Nowadays, fibre-reinforced plastic rackets are used

13.2 Separating mixtures

Most naturally-occurring materials are mixtures. Very often these mixtures have to be separated before we can use the materials in them. The methods that we choose to separate mixtures will depend on the different physical properties of the substances in the mixtures. When scientists separate mixtures, they often use similar equipment to cooks. Their equipment does, of course, depend on the type of mixture being separated.

Separating an insoluble solid from a liquid

It is not difficult to separate an insoluble solid from a liquid. There are three possible methods.

i) Filtering
This method is used when the particles of solid are very small – for example when making filter coffee (Figure 13.3), in using a tea strainer and in removing fine particles from drinking water.

ii) Decanting
Decanting just means pouring the liquid off from the solid. This method is often used when the solid is in large pieces and denser than the liquid. For example, we decant boiling water from vegetables when they are cooked. A kind of decantation is used when we drink tea and the tea leaves are left in the bottom of the cup.

iii) Centrifuging
Centrifuging is used when the particles of solid are so small that they float in a liquid as a **suspension**. The mixture is poured into a tube and spun round very rapidly in a centrifuge. This forces the denser solid particles to the bottom of the tube and the liquid can be poured off (decanted) easily.

Centrifuging is used in hospitals to separate denser blood cells from blood plasma and in dairies to separate milk from cream.

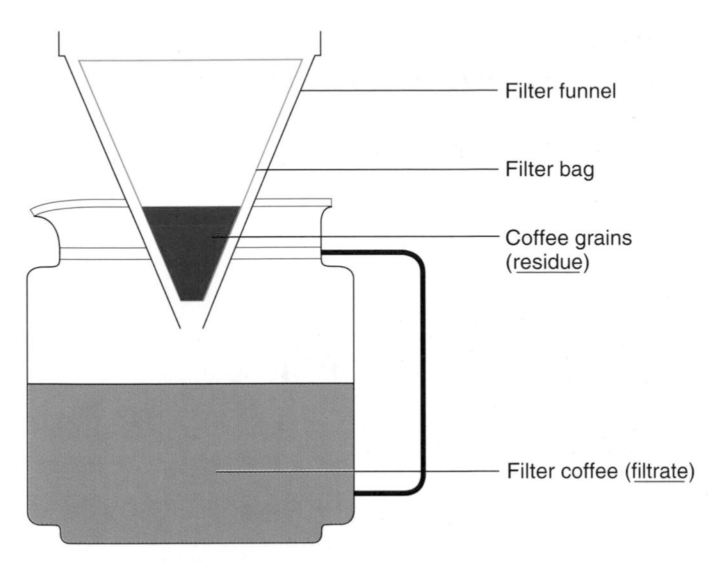

Figure 13.3 Filtering coffee. The filter bag is made of filter paper. It has tiny holes that let the liquid through, but these holes are too small for the solid particles to pass through. The liquid that runs out of the filter is called the **filtrate**. The solid left behind is called the **residue**

Evaporation is used to obtain salt from sea water in hot countries

Separating a soluble solid from a liquid

Tap water is clean but not pure. It contains dissolved gases, such as oxygen from the air, and dissolved solids from the soil and river beds over which it has flowed. Tap water is, of course, a **solution**. Sea water is another example of a solution. It contains salt (the **solute**) dissolved in water (the **solvent**). The easiest way to separate a soluble solid (such as the salt in sea water) from a liquid is by evaporation because the liquid is more volatile. When sea water is left in the sun, the water turns into a vapour and salt is left behind as a white solid. This process in which a liquid turns to a vapour is called **evaporation**.

If the solvent evaporates very slowly from a solution, the solute is often left behind as large, well-shaped crystals. This process of obtaining crystals by evaporating the solvent from a solution is called **crystallization**. Usually evaporation is carried out more rapidly by boiling the solution. In this case, the solute is left behind as small, poorly-shaped crystals.

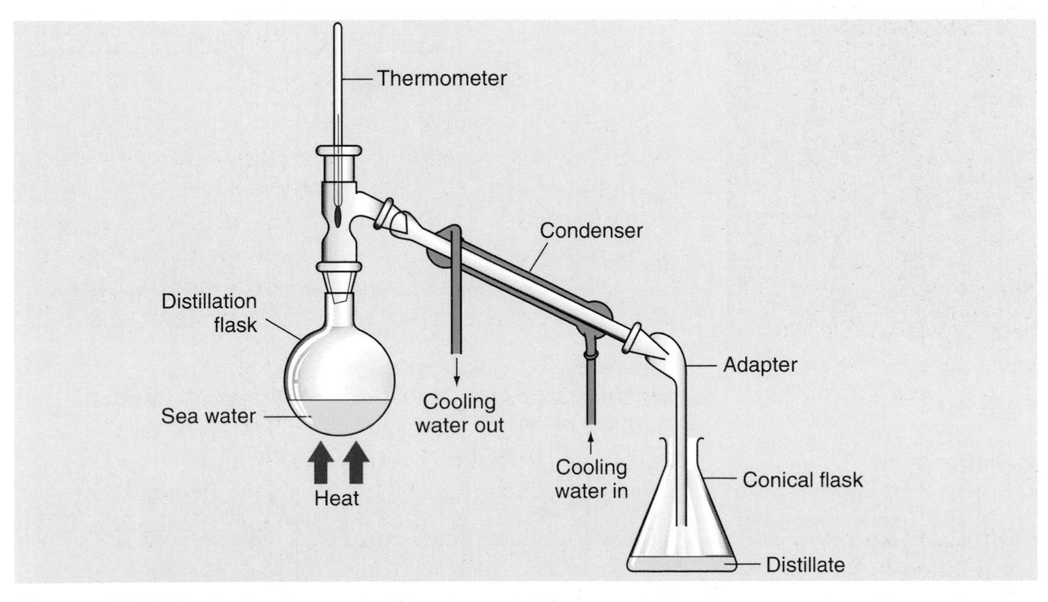

Figure 13.4 Separating pure water from sea water by distillation. The pure water (liquid) which collects after distillation is called the **distillate**

Distillation

Sometimes the part of a solution that you want is the liquid solvent and *not* the solute. In this case, distillation can be used to separate and collect the liquid part of a solution.

Distillation can be used to separate pure water from sea water (Figure 13.4 on page 111). If the sea water is boiled, water vapour comes off as steam and evaporation has taken place. The steam is then passed into a second container called a **condenser**. Here the steam is cooled and it turns back to water. This process in which a vapour changes to a liquid is called **condensation**. Notice from Figure 13.4 that:

> distillation = evaporation + condensation

Distillation is an important process in:

- making 'spirits' such as whisky, gin and vodka from weaker alcoholic liquids;
- obtaining pure drinking water from sea water in parts of the Middle East where fuel is cheap.

Separating immiscible liquids

The method used to separate liquids from one another depends on whether or not the liquids mix. When two liquids, like water and alcohol dissolve in each other to form a single layer, the liquids are **miscible**. But some liquids, like oil and water, are not like this. When you add them together, they form two quite separate layers. Liquids which do not mix and dissolve in each other are called **immiscible**.

Immiscible liquids, like cream and milk or liquid fat and gravy can be separated fairly easily by spooning off the cream or liquid fat. This is rather like decanting. In the laboratory, immiscible liquids are separated using a **separating funnel** (Figure 13.5).

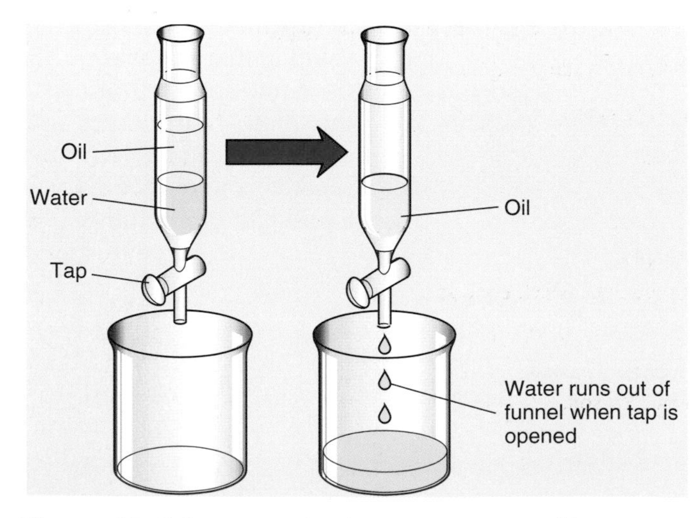

Figure 13.5 Separating oil and water – two immiscible liquids

Separating miscible liquids

If alcohol is added to water, the two liquids mix completely to form one liquid. There is only one layer and the liquids cannot be separated using a separating funnel.

Miscible liquids can, however, be separated by a special form of distillation called **fractional distillation**.

The method works because the different liquids have different boiling points. When the mixture is heated, different liquids boil off at different temperatures as each one reaches its boiling point. As the different liquids boil off, they are condensed separately.

Fractional distillation is important in:

- separating the different fractions in crude oil (section 20.3);
- separating oxygen and nitrogen from liquid air.

13.3 Chromatography – separating similar substances

Chromatography provides an important method for separating very similar substances. It is used, for example, to separate dyes in ink, different sugars in urine and drugs in the blood.

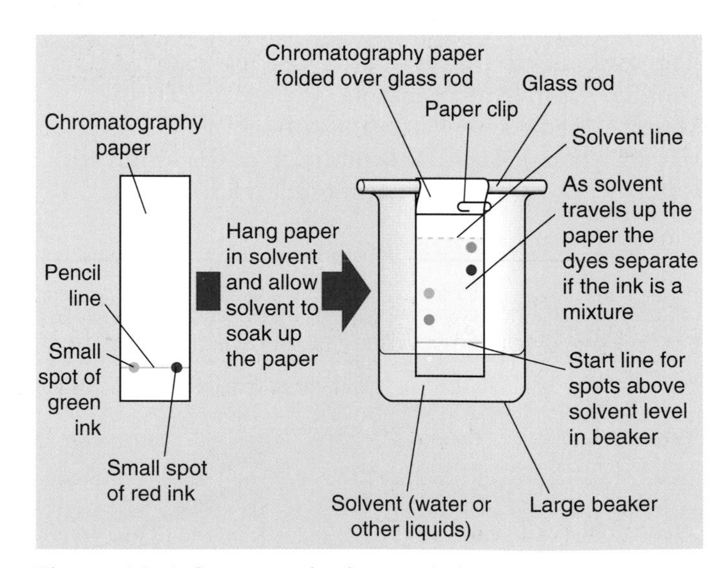

Figure 13.6 Separating the dyes in inks by chromatography. How many dyes are there in a) the green ink, b) the red ink?

Figure 13.6 shows how the dyes in inks can be separated by chromatography. As the solvent rises up the paper, the dyes separate. Some dyes stick to the paper strongly, but others tend to dissolve in the solvent. The dyes that dissolve more easily in the solvent will travel further up the paper.

This method was first used to separate mixtures of coloured substances. Hence the name **chromatography** which comes from a Greek word meaning colour. Nowadays, chromatography is also used to separate colourless substances. After the solvent has soaked up the paper, it is dried and then sprayed with a **locating agent**. The locating agent reacts with each of the colourless substances to form a coloured product.

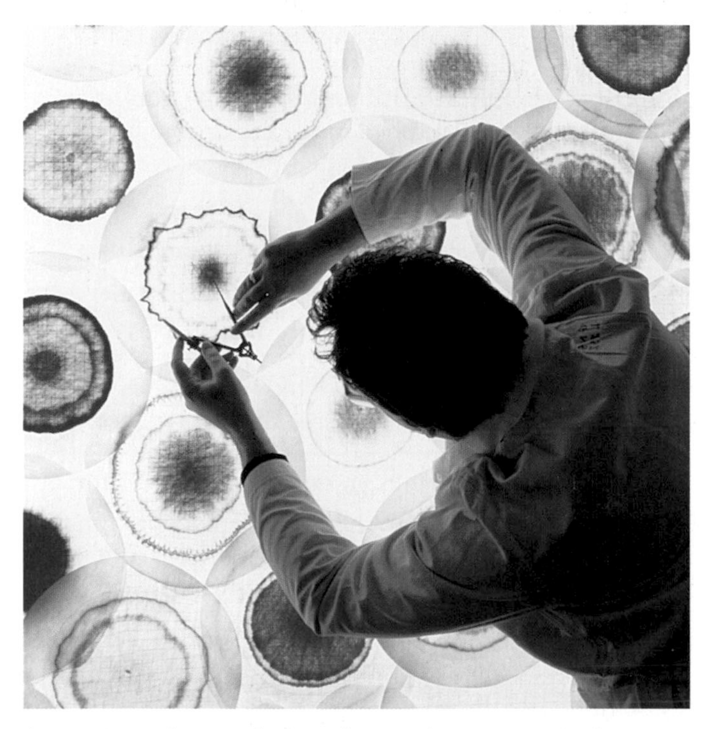

A scientist studying a selection of paper chromatograms of industrial dyes

13.4 Elements – the simplest substances

Although there are millions of different substances, we can put all of them into one of three groups – **elements**, **mixtures** or **compounds**.

> Elements are the simplest substances. They cannot be broken down into simpler substances.

So, elements are the building blocks for all other substances. They are the simplest possible materials. So far, we know of 106 elements. These include iron, copper, aluminium, gold, carbon and oxygen. Every substance in the universe is made from one or more of the 106 known elements. For example, sand is composed of silicon and oxygen, water is composed of hydrogen and oxygen and air is mainly oxygen and nitrogen.

Classifying elements

Elements can be classified into groups with similar properties. The simplest way of classifying elements is as metals and non-metals. The major differences between the properties of metals, such as iron and copper, and those of non-metals, such as oxygen and nitrogen, are summarised in Table 13.2.

The easiest way of checking whether an element is a metal or a non-metal is to see if it conducts electricity. This can be done using the apparatus in Figure 13.7 over the page.

Property	Metals (e.g. iron, copper, aluminium)	Non-metals (e.g. oxygen, nitrogen, sulphur)
State	usually solids at room temperature	mostly gases at room temperature
Appearance	shiny solids	mostly colourless gases
Density	usually high	usually low
Melting point and boiling point	usually high	usually low
Conduction of heat and electricity	good	poor (except graphite which conducts well)
Effect of hammering (malleability)	can be hammered or bent into different shapes – malleable	solids are brittle or soft

Table 13.2 Comparing the properties of metals and non-metals

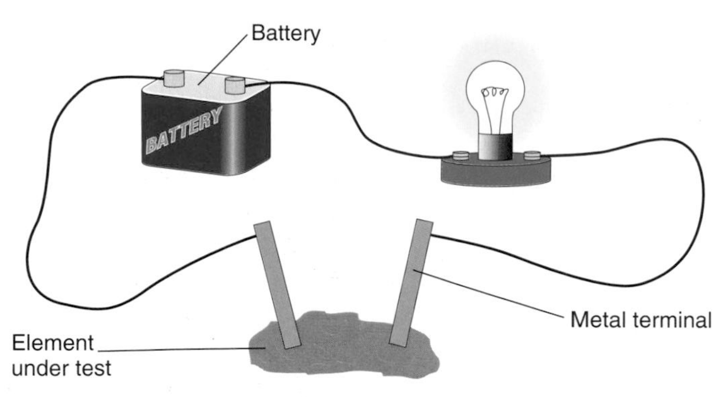

Figure 13.7 How does this apparatus show you the difference between a metal and a non-metal?

13.5 Elements and compounds

When charcoal (carbon) is heated strongly, it burns with a pale blue flame to form carbon dioxide which is a colourless gas. The carbon has **reacted** or **combined** with oxygen in the air to form carbon dioxide. The carbon dioxide is an entirely new substance. Changes like this, which result in new substances, are called **chemical reactions**. The substances which react, in this case carbon and oxygen, are called **reactants**. The new substance which forms is called the **product** of the reaction. We can represent this reaction by writing a **word equation**.

carbon + oxygen → carbon dioxide
(charcoal)

Carbon and oxygen are elements. They have combined to form a **compound** called carbon dioxide.

Charcoal (carbon) burning on a barbecue

A compound is a substance which contains two or more elements combined together.

When two elements react together to form a compound, the name of the compound ends in **-ide**. For example:

carbon	+	oxygen	→	carbon diox**ide**
iron	+	sulphur	→	iron sulph**ide**
aluminium	+	chlorine	→	aluminium chlor**ide**

When a metal reacts with a non-metal, the non-metal forms the -ide part of the name of the compound. When two non-metals react, the more reactive non-metal forms the -ide part of the name.

When elements combine to form compounds, the reaction is an example of **synthesis.**

Synthesis is the building up of more complex substances by joining together simpler substances.

Almost all substances, including those in living things, are made through chemical reactions. Photosynthesis, for example, involves a synthesis reaction between water and carbon dioxide (see Chapter 5).

Unlike elements, compounds can be split into simpler substances. For example, sodium chloride can be split into sodium and chlorine when electricity is passed through molten sodium chloride.

$$\text{sodium chloride} \xrightarrow{\text{electricity}} \text{sodium} + \text{chlorine}$$

When a compound is split into simpler substances, the reaction is an example of **decomposition**.

Decomposition is the breaking down of more complex substances into simpler substances.

Notice that decomposition is the reverse of synthesis.

$$\text{sodium} + \text{chlorine} \underset{\text{decomposition}}{\overset{\text{synthesis}}{\rightleftarrows}} \text{sodium chloride}$$

When chemical reactions occur, it is important to remember that all the mass of the reactants goes into forming the products. We say that all the mass is conserved and this is summarised in the **law of conservation of mass**.

This says:

> In any chemical or physical change, the total mass of the products equals the total mass of the reactants.

13.6 Mixtures and compounds

Most materials are mixtures. They may be *mixtures of elements*, like air which contains mainly oxygen and nitrogen, or *mixtures of compounds* like sea water which contains water and salt (sodium chloride), or *mixtures of both elements and compounds*.

Table 13.3 The differences between mixtures and compounds

Mixtures	Compounds
1 Consist of two or more substances – elements or compounds – just mixed together.	1 Consist of just one pure substance.
2 Properties are similar to the substances in them.	2 Properties are very different from the elements in them.
3 Substances in them can often be separated easily.	3 Elements in them can only be separated by a chemical reaction.
4 The percentages of substances in the mixture can vary, i.e. a mixture can have a variable composition.	4 The percentages of elements in a compound are constant, i.e. a compound has a constant composition represented by a formula.

The label on this carton says 'Pure orange juice'. Is it really pure? Does it contain only one substance?

> A mixture is two or more substances which are *not* combined together chemically.

Notice the four numbered points in Table 13.3 which shows the differences between mixtures and compounds.

1. The most important difference between a mixture and a compound is that a mixture contains more than one substance, whereas a compound is one pure substance. In a mixture there are two or more substances which are not combined together chemically.
2. The properties of a mixture are a 'kind of average' of the properties in it. So salt water is both salty like sodium chloride (salt) and wet like water.
3. Substances in a mixture can often be separated easily using the methods discussed earlier in this chapter.
4. Compounds have constant compositions which can be shown by formulas and these are discussed in the next chapter.

1 All materials can be sorted into three groups (states) called solids, liquids and gases. Water can exist in all three states, as shown in the diagrams.

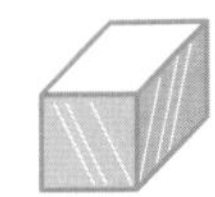
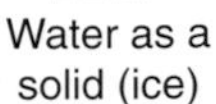
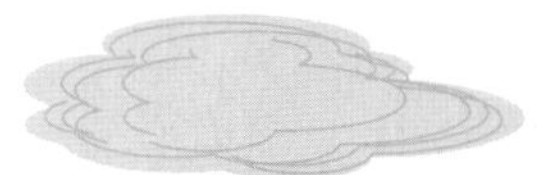

a) The table below shows simple tests to decide if a material is a solid, liquid or gas. Some spaces have been filled in. Copy and complete the table.

Test	Solid	Liquid	Gas
Stirring			can be stirred
Looking	fixed shape (cannot change)		
Pouring	cannot be poured		

b) The diagram shows the arrangement of particles in a gas.

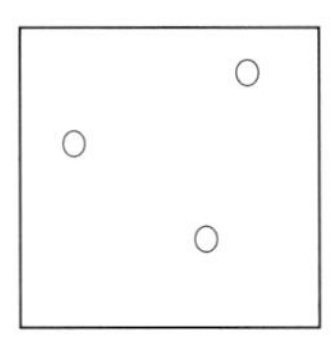

Draw similar diagrams to show the arrangement of particles in i) a solid and ii) a liquid.

Edexcel

2 A student was asked to separate salt from a mixture of sand and salt. The separation was carried out in three stages.

Stage 1: the mixture is added to hot water and stirred.

a) What is the purpose of this stage?

Stage 2: the mixture is now filtered and the residue washed with water.

b) Name the dissolved substance in the filtrate.

c) Why is the residue washed with water?

d) Describe how you would obtain a solid from the filtrate in **Stage 2**.

NICCEA

3 Substances can be put into one of these three groups:

elements – contain only **one** type of atom.

mixtures – contain the atoms of two or more elements but are **not** joined together.

compounds – contain the atoms of two or more elements joined together.

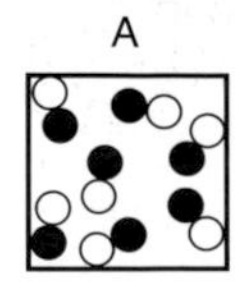

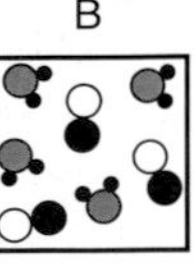

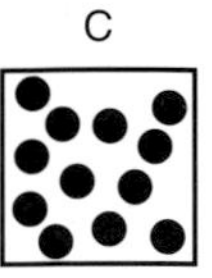

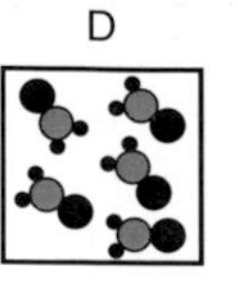

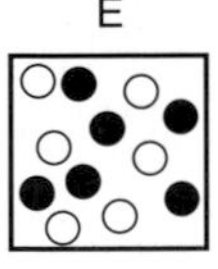

a) Write down the letter from the diagram above which **best** represents:

i) an element,

ii) a mixture of elements,

iii) two elements which have reacted together.

b) Suggest how a student can separate salt crystals from a mixture of salt and aluminium powder. You may draw labelled diagrams to show how you would do this.

MEG

4 The diagram shows a bottle of lemonade which has just been opened.

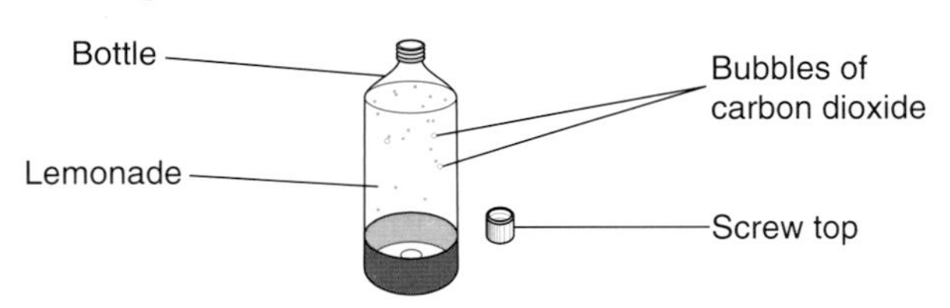

a) Copy and complete the table to show which parts are **solid**, **liquid** or **gas**. One has been done for you.

Part	Solid, liquid or gas
Screw top	solid
Lemonade	
Bottle	
Carbon dioxide	

b) Suggest **two** properties of plastic which make it more suitable than glass for making lemonade bottles.

c) Harry carried out some tests on a number of drinks containers made of aluminium, plastic and steel. The following table shows his results.

Test	Container A	Container B	Container C
Heated to 100 °C	changes shape	no effect	no effect
Conducts electricity	no	yes	yes
Attracted to a magnet	no	no	yes
Left open to air for 5 days	no effect	no effect	rusts
Conducts heat	poorly	well	well

i) Container **A** is plastic. How can you tell?

ii) Container **B** is metal. How can you tell?

iii) Container **C** is steel. How can you tell it is **not** aluminium?

MEG

CHAPTER

Particles, reactions and equations

14.1 Solids, liquids and gases

Scientists classify materials and substances in several ways. One of these classifications is as naturally-occurring materials and man-made materials. Another important method of classifying materials is as **solids**, **liquids** and **gases**. These are called the **three states of matter**. Concrete and ice are solids, petrol and water are liquids, air and steam are gases.

The important properties of solids, liquids and gases are shown in Table 14.1.

Table 14.1 The important properties of solids, liquids and gases

	Solids	Liquids	Gases
Density	have high densities, usually greater than 2 gm/cm^3	have medium densities about 1 gm/cm^3	have low densities
Shape	keep the same shape	flow easily and take the shape of their container	flow easily and take the shape of their container
Volume	keep the same volume	keep the same volume	take the volume of their container
Compressibility	cannot be compressed into a smaller volume	can be compressed very slightly	can be compressed into a much smaller volume
Example	Rocks show all the properties of a solid	Water shows all the properties of a liquid	Air inside balloons shows all the properties of a gas

14.2 Changes of state

Most substances can exist in all three states, depending on the temperature. Different materials change state at different temperatures. Water, for example, is a solid (ice) below 0°C, a liquid between 0°C and 100°C and a gas (steam) above 100°C.

The state of a substance can often be changed by heating it or cooling it. A summary of the different changes of state is shown in Figure 14.1.

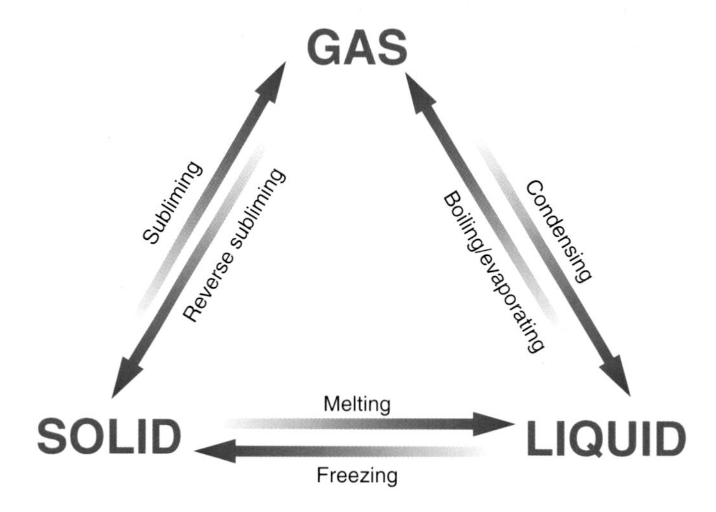

Figure 14.1 Changes of state

Notice that liquids can be turned to gases by **boiling** and by **evaporating**. Liquids evaporate at temperatures well below their boiling point. This is what happens when water evaporates from puddles.

A few solids, like iodine and solid carbon dioxide, change directly to gases on heating. This is called **subliming**. Snow sometimes sublimes straight to water vapour without forming water.

14.3 Particles in motion

How does the smell of frying bacon spread all around the house? Why can a tiny amount of curry powder flavour a large amount of sauce?

These questions can be answered using the idea that **all substances are composed of incredibly small, invisible particles**.

If you put a tiny pinch of curry powder into a soup, every spoonful tastes of curry. This means that the curry powder must contain tiny bits or particles which have spread through the whole dish.

The best evidence for moving particles comes from studies of **diffusion** and **Brownian motion**.

Diffusion

This explains how you can smell frying bacon well away from the kitchen. Particles of gas are released from the frying bacon. These particles mix with air particles and move away from the frying bacon. This moving and mixing of particles is called **diffusion**.

Gases consist of tiny particles moving at high speeds. These particles collide with each other and with the walls of their container. Sooner or later, gases like those from the frying bacon will diffuse into all the space available to them.

Diffusion also takes place in liquids, but it occurs much more slowly than in gases (Figure 14.2). This means that liquid particles move around more slowly than gas particles. Diffusion does not happen in solids.

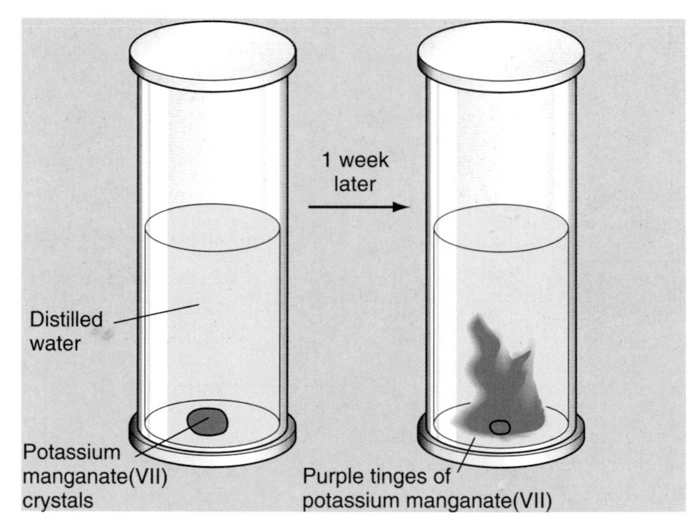

Figure 14.2 The potassium manganate(VII) dissolves in the water and then starts to diffuse away from the crystals. Notice that the solution is more purple near the bottom of the jar. What do you think will happen if the gas jar is left for several weeks?

Diffusion is important in living things. It explains how the food you eat is digested. After a meal, food passes into your stomach and then through your intestines (see section 2.5). Large particles of food are broken down into smaller particles which can diffuse through the walls of the intestines into the bloodstream.

Brownian motion

In 1827, a biologist called Robert Brown was using a microscope to look at pollen grains in water. To his surprise, the pollen grains kept moving and jittering about randomly. Similar random movements can be seen when you look at smoke particles through a microscope (Figure 14.3). This movement of tiny particles in a gas or a liquid is called **Brownian motion**.

The movement of smoke particles is caused by the random movements of air particles around them. The particles of smoke are very small, but they are much larger than air particles. Through the microscope, you can see smoke particles, but air particles are much too small to be seen. However, the air particles move very fast hitting the smoke particles at random. The smoke particles are knocked first this way, then that way so they appear to jitter about (see inset to Figure 14.3).

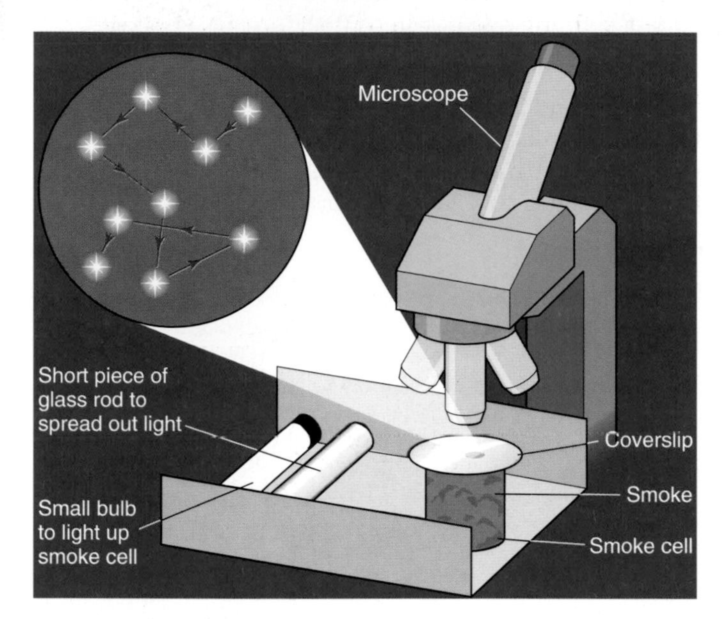

Figure 14.3 Observing the Brownian motion of smoke particles. Smoke from a smouldering piece of string is injected into the smoke cell using a teat pipette. Through the microscope, the smoke particles look like tiny points of light which jitter about

14.4 Particles in motion – the kinetic theory

The idea that all substances are composed of incredibly small moving particles is called the kinetic theory of matter.

The word 'kinetic' comes from a Greek word meaning moving.

The main points of the kinetic theory are:

- All matter is made of tiny, invisible moving particles. These particles are actually atoms, molecules and ions (section 14.6).
- The particles of different substances have different sizes. Particles of elements like iron, copper and sulphur are very small. Particles of compounds like petrol and sugar are larger, whilst particles of some complex compounds, such as polythene, PVC and proteins are much, much larger.
- Small particles move faster than larger particles at the same temperature.
- As the temperature rises, the particles have more energy and move around faster.

Figure 14.4 Particles in solids, liquids and gases

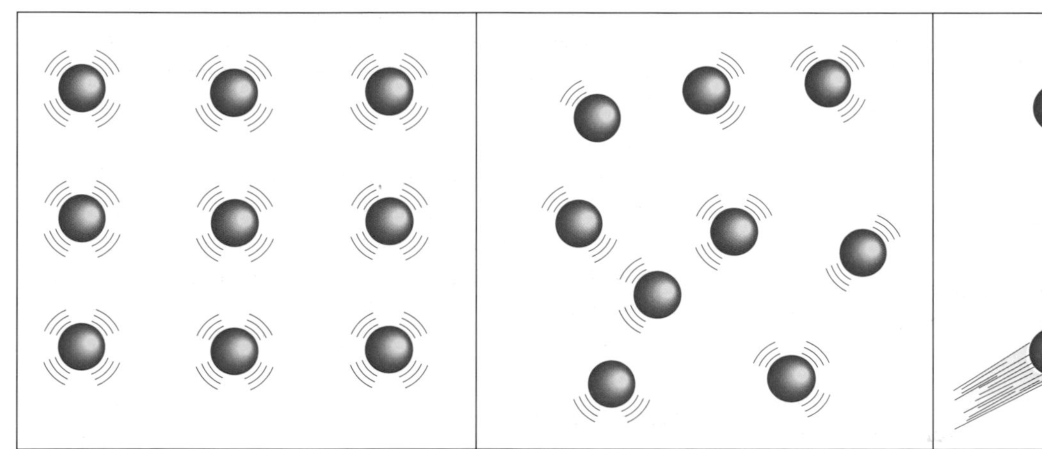

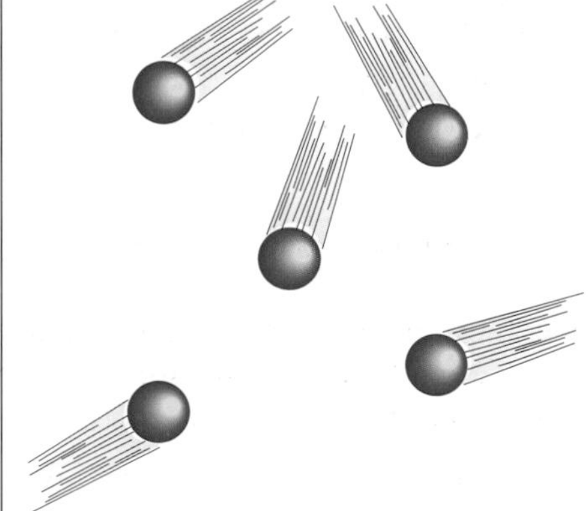

In a solid, the particles are close together with strong forces between them. This explains why solids are denser than liquids and gases and why they cannot be compressed.

Solid particles can only vibrate about fixed points. This explains why solids cannot flow and why solids have a fixed volume and a fixed shape

In a liquid, the particles are a little further apart and forces between particles are not as strong as in solids. This explains why liquids are not as dense as solids and why they can be compressed slightly.

Liquid particles can roll around each other. So liquids flow easily, they can change their shape, but keep a fixed volume.

In a gas, the particles are very far apart with virtually no forces between them. This explains why gas densities are very low and why they can be compressed so much.

Gas particles move around very fast in all the space available. So gases flow easily and fill the whole of their container.

14.5 Using the kinetic theory to explain changes of state

We can use the kinetic theory to explain changes of state. When a solid is heated, its particles gain energy. The particles vibrate faster and faster until they break away from their fixed positions. The solid has **melted** to a liquid. The temperature at which the solid melts is called the **melting point**.

In metals, the particles which hold the solid together are atoms. The forces between metal atoms in steel are so strong that thin steel cables can be used to lift heavy loads

When a liquid cools down, the particles lose energy and move around each other more slowly. Eventually, the particles are moving so slowly that they just vibrate about a fixed position. At this point, the liquid has **frozen** to form a solid.

The particles in a liquid can move around each other. Some particles near the surface may have enough energy to escape from the liquid into the air. When this happens, the liquid **evaporates** to form a gas. The gas particles have much more energy than they had in the liquid state. If the liquid is heated, its particles move faster and more of them have sufficient energy to escape from the surface. So evaporation increases as the temperature of the liquid rises.

Eventually, the liquid particles are moving so rapidly that bubbles of gas form in the liquid. The temperature at which this evaporation occurs in the bulk of the liquid is the **boiling point**. Boiling points tell us how strongly the particles are held together in liquids. Volatile liquids, like petrol, which evaporate easily and boil at low temperatures have weak forces between their particles.

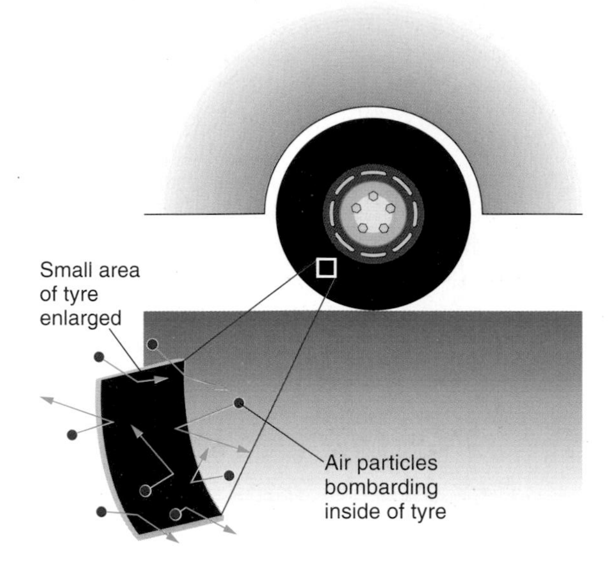

Figure 14.5 Gas pressure allows us to inflate tyres and balloons. Millions of air particles bombard the inside of the tyre every second. This causes a pressure on the inside of the tyre. The tyre is also bombarded by air particles on the outside, but the pressure inside is greater than the pressure outside. The greater pressure inside keeps the tyre inflated

When a gas is cooled, its particles lose energy and move around slower and slower. Eventually, the particles have insufficient energy to bounce off each other when they collide. The particles cling together as a liquid and **condensation** occurs.

The kinetic theory will also explain the pressure of a gas (Figure 14.5).

State symbols

Icebergs off the coast of Antarctica

This photo shows water in three different states – solid ice, liquid water and gaseous water vapour. All these are the same substance with the same formula, H_2O.

This could be confusing, so chemists use **state symbols** after a formula to show the state of a substance:

(s) means the substance is a solid;
(l) means the substance is a liquid;
(g) means the substance is a gas and
(aq) means the substance is in the aqueous state, i.e. dissolved in water.

So, ice is written as $H_2O(s)$, liquid water is $H_2O(l)$, water vapour (steam) is $H_2O(g)$ and sea water (sodium chloride solution) is $NaCl(aq)$.

14.6 Atoms, molecules and ions

All substances and materials can be classified in two ways:

1 as **solids, liquids and gases** with the properties described earlier in this chapter, or
2 as **elements, compounds and mixtures** which were discussed in Chapter 13.

- **Elements** are substances that cannot be broken down any further.
- **Compounds** contain two or more elements that have combined together. Compounds can be broken down into their constituent elements.
- **Mixtures** contain two or more different substances which are just mixed together and *not* combined.

For example, air is a mixture of mainly nitrogen and oxygen. Sea water is a mixture of mainly salt (sodium chloride) and water.

So far in this chapter, we have learnt that all substances and all materials are made of particles. There are, in fact, only three kinds of particle – atoms, molecules and ions.

What is an atom?

An atom is the smallest particle of an element.

The word 'atom' comes from a Greek word meaning 'indivisible' or 'unsplittable'. At one time, scientists thought that atoms could not be split. We now know that atoms can, in fact, be split. But, if an atom of one element is split, it is no longer the same element.

Elements are made up of only one kind of atom.

So, copper contains only copper atoms, sodium contains only sodium atoms and carbon contains only carbon atoms.

In 1807, John Dalton put forward his **atomic theory of matter**. In this theory, Dalton was the first scientist to use the word **atom** for the smallest particle of an element

What is a molecule?

A molecule is a particle in which two or more atoms are chemically joined together.

The atoms in molecules are not just mixed together, they are held together by **chemical bonds**. For example, a molecule of water contains two atoms of hydrogen combined with one atom of oxygen, whilst a molecule of carbon dioxide contains one atom of carbon combined with two atoms of oxygen (Figure 14.6).

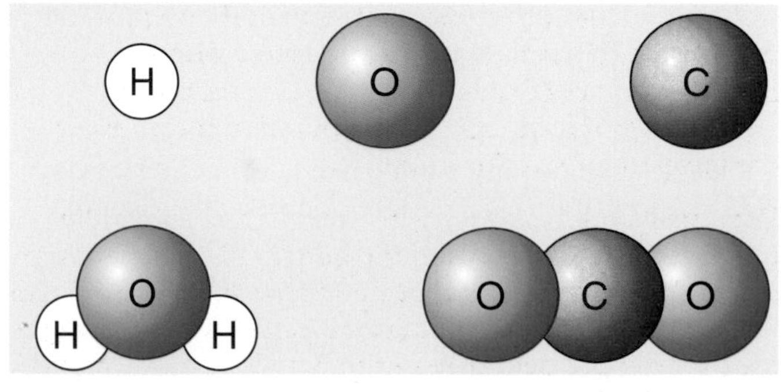

Figure 14.6 Atoms of hydrogen, oxygen and carbon above molecules of water and carbon dioxide

Element	Symbol	Element	Symbol	Element	Symbol
Aluminium	Al	Helium	He	Oxygen	O
Argon	Ar	Hydrogen	H	Phosphorus	P
Barium	Ba	Iodine	I	Platinum	Pt
Bromine	Br	Iron	Fe	Potassium	K
Calcium	Ca	Krypton	Kr	Silicon	Si
Carbon	C	Lead	Pb	Silver	Ag
Chlorine	Cl	Magnesium	Mg	Sodium	Na
Chromium	Cr	Mercury	Hg	Sulphur	S
Cobalt	Co	Neon	Ne	Tin	Sn
Copper	Cu	Nickel	Ni	Uranium	U
Gold	Au	Nitrogen	N	Zinc	Zn

Table 14.2 The symbols for some of the most common elements in alphabetical order

When the atoms of different elements join together, they form **compounds**. The chemical symbols in Table 14.2 can also be used to represent compounds. For example, water is represented as H_2O – two hydrogen atoms (H) and one oxygen atom (O). Carbon dioxide is written as CO_2 – one carbon atom (C) and two oxygen atoms (O).

'H_2O' and 'CO_2' are called **chemical formulas**. A formula shows the numbers of atoms of the different elements in a molecule. Notice that the numbers of the different atoms are written after their symbols as subscripts. Some other pictures of molecules and their formulas are shown in Figure 14.7.

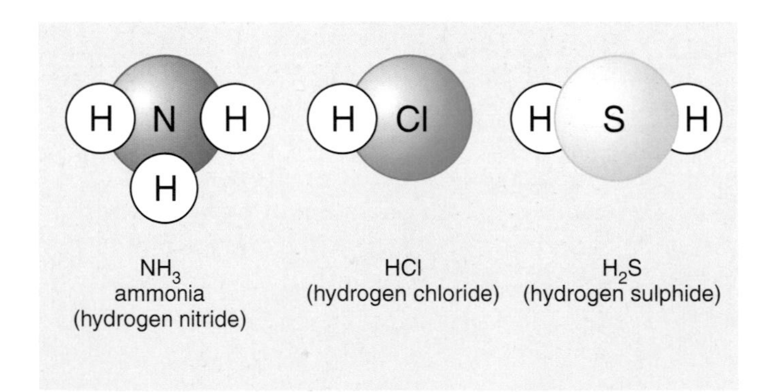

Figure 14.7 Pictures of three molecules and their formulas

Molecules can be very simple like hydrogen chloride (HCl) and water (H_2O), more complex like alcohol (C_2H_6O) and octane (C_8H_{18}) or very complicated indeed like chlorophyll ($C_{51}H_{72}O_4N_4Mg$).

Atoms and molecules of elements

Most elements, like iron (Fe), aluminium (Al) and copper (Cu), can be represented by their symbols because they contain single atoms. But this is not the case with oxygen, hydrogen, nitrogen or the halogens. Under most conditions, these elements exist as molecules containing two atoms combined together. So, oxygen is best represented as O_2 and not O, hydrogen as H_2 not H, nitrogen as N_2 and chlorine as Cl_2. These molecules containing two atoms are described as **diatomic molecules**.

What is an ion?

Compounds containing only non-metals, like water (H_2O), carbon dioxide (CO_2), ammonia (NH_3) and tetrachloromethane (CCl_4), consist of molecules.

Compounds containing metals and non-metals consist of ions, not molecules.

> An ion is a charged particle formed from an atom by the loss or gain of one or more electrons.

(There is more about electrons in sections 15.8, 15.9, 16.4, 16.8 and 17.8)

The atoms of all elements are neutral (uncharged). But, when metal atoms react with non-metal atoms, the compound formed contains ions (charged particles). In these metal/non-metal compounds;

> metal ions are always positive and non-metal ions (except hydrogen) are always negative.

When sodium reacts with chlorine, sodium chloride is formed. This contains sodium ions (Na^+) and chloride ions (Cl^-) and the formula of sodium chloride is Na^+Cl^-, or just NaCl if we leave out the charges.

Table 14.3 shows a list of some common ions with their charges.

Table 14.3 Some common ions and their charges

Positive ions (cations)		Negative ions (anions)	
Aluminium	Al^{3+}	Bromide	Br^-
Calcium	Ca^{2+}	Carbonate	CO_3^{2-}
Chromium	Cr^{3+}	Chloride	Cl^-
Copper(I)	Cu^+	Hydroxide	OH^-
Copper(II)	Cu^{2+}	Iodide	I^-
Hydrogen	H^+	Nitrate	NO_3^-
Iron(II)	Fe^{2+}	Nitride	N^{3-}
Iron(III)	Fe^{3+}	Oxide	O^{2-}
Lead	Pb^{2+}	Sulphate	SO_4^{2-}
Magnesium	Mg^{2+}	Sulphide	S^{2-}
Nickel	Ni^{2+}	Sulphite	SO_3^{2-}
Potassium	K^+		
Silver	Ag^+		
Sodium	Na^+		
Zinc	Zn^{2+}		

Notice the following points from Table 14.3.

- *Many ions have more than one unit of charge.*
- *Copper can form two ions, Cu^+ and Cu^{2+}.* We show this in the name of its compounds by using the names copper(I) (pronounced 'copper-one') and copper(II) (pronounced 'copper-two'). Thus, copper forms two oxides, two chlorides, two sulphates, etc. The correct names for its oxides are copper(I) oxide which is red and copper(II) oxide which is black (Figure 14.8). Most of the common copper compounds are copper(II) compounds.

Figure 14.8 The oxides and chlorides of copper. Pure copper(I) chloride is white, but samples of it are usually pale green due to contamination from copper(II) chloride

- *Iron can also form two different ions, Fe^{2+} and Fe^{3+}* and we use the names iron(II) and iron(III) for their respective compounds.
- *Most metal ions have a charge of 2+.* All the common metal ions without a charge of 2+ are shown in Table 14.3.
  - *the only common metal ions with a charge of 1+ are Ag^+, Na^+ and K^+* (say 'AgNaK' to remember this);
  - *the only common metal ions with a charge of 3+ are Cr^{3+}, Al^{3+} and Fe^{3+}* (say 'CrAlFe' to remember this).
- *Some negative ions contain more than one kind of atom.* For example, hydroxide, OH^-, contains one oxygen and one hydrogen atom, whilst nitrate, NO_3^-, contains one nitrogen and three oxygen atoms.

Bonds between ions

Compounds containing only non-metals are composed of molecules. They are called **molecular compounds**. However, compounds containing both metals *and* non-metals contain ions. They are called **ionic compounds**. Ionic compounds include common salt (sodium chloride), limestone (calcium carbonate) and iron ore (iron(III) oxide).

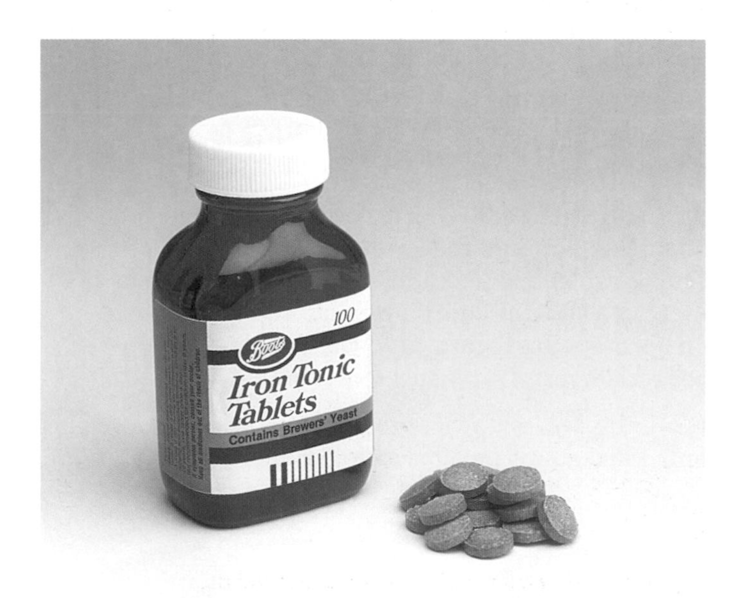

Iron tablets are really iron(II) sulphate. Why are pregnant women advised to take iron tablets? What is their formula?

In ionic compounds, the charges on the positive ions must balance the charges on the negative ions. So, as the sodium ion is Na^+ and the chloride ion is Cl^-, we can predict that the formula of sodium chloride will be Na^+Cl^-. In this case, the one positive charge on Na^+ is balanced by one negative charge on Cl^-.

Farmers often put lime on the soil. Lime is an ionic compound. Its chemical name is calcium oxide. Why is it added to the soil? What is its formula?

By balancing charges we can predict the formula of calcium chloride will be $Ca^{2+}(Cl^-)_2$, or simply $CaCl_2$ with two positive charges on one Ca^{2+} balancing two negative charges on two Cl^- ions. The formula of aluminium oxide is $(Al^{3+})_2(O^{2-})_3$ or simply Al_2O_3. In this case, the six positive charges on two Al^{3+} ions are balanced by six negative charges on three O^{2-} ions.

You can find out more about ionic structures in section 17.5 and ionic bonds in section 17.8.

14.7 Comparing the masses of atoms

Atoms are incredibly small. Atoms of chlorine are about one hundred millionth (1/100 000 000) of a centimetre in diameter. So, if you put 100 million of them in a straight line, they would still only measure one centimetre. Figure 14.9 will also help you to appreciate just how small atoms are.

From this you will realise that it is impossible to weigh individual atoms on a balance. Scientists have, however, developed a method of finding the masses of different atoms relative to one another. The method uses an instrument called a **mass spectrometer** in which atoms are deflected using a magnetic field. The heavier an atom, the less it is deflected. So, by comparing the deflections of different atoms, it is possible to compare their masses and make a list of their relative masses. The relative masses which scientists use for different atoms are called **relative atomic masses**.

The element carbon has been chosen as the standard for relative atomic masses. Carbon atoms are given a relative mass of exactly 12.

Relative atomic mass of carbon = 12.0

The relative masses of other atoms are obtained by comparison with carbon. For example, carbon atoms are 12 times as heavy as hydrogen atoms, so the relative atomic mass of hydrogen is 1.0. Magnesium atoms are twice as heavy as carbon atoms, so the relative atomic mass of magnesium is 24.0. A few relative atomic masses are listed in Table 14.4.

Table 14.4 Some relative atomic masses

Element	Symbol	Relative atomic mass
Carbon	C	12.0
Hydrogen	H	1.0
Magnesium	Mg	24.0
Oxygen	O	16.0
Aluminium	Al	27.0
Iron	Fe	55.8
Copper	Cu	63.5

The symbol for relative atomic mass is A_r. So, we write $A_r(C) = 12.0$, $A_r(Fe) = 55.8$, etc. or simply C = 12.0, Fe = 55.8, etc.

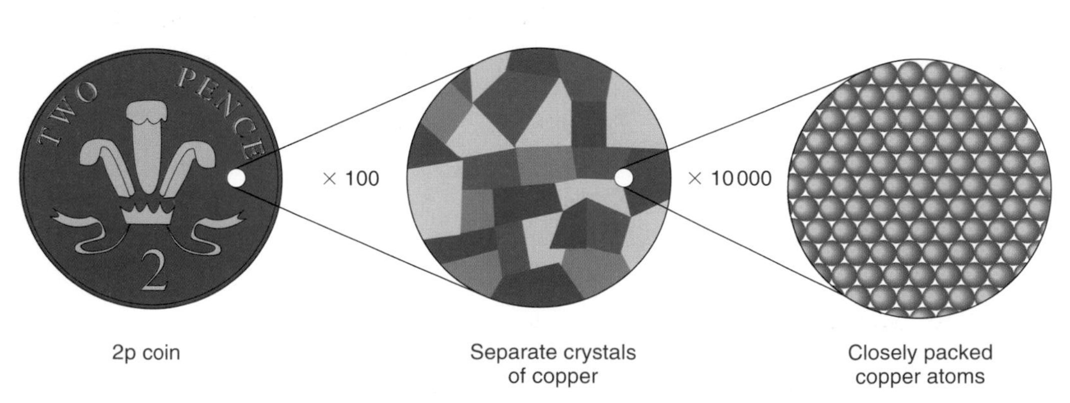

Figure 14.9 If the surface of a 2p coin is magnified one hundred times using an ordinary microscope, it is possible to see separate crystals of copper. If these crystals are now magnified 10 000 times, it would be possible to see individual copper atoms. Altogether, the coin would have been magnified first 100 times, then 10 000 times, i.e. one million times in total (100 × 10 000)

Relative formula masses

Relative atomic masses can be used to compare the masses of atoms in different elements. Relative atomic masses can also be used to compare the masses of molecules in different compounds. These relative masses of molecules are called **relative formula masses**.

The relative formula mass of a compound is obtained by just adding up the relative atomic masses of all the atoms in the formula. For example:

the relative formula mass of water, H_2O
= 2 × relative atomic mass of hydrogen + relative atomic mass of oxygen
= $(2 \times 1) + 16 = 18$

the relative atomic mass of iron(III) oxide, Fe_2O_3
= $2 \times A_r(Fe) + 3 \times A_r(O)$
= $(2 \times 55.8) + (3 \times 16) = 159.6$

14.8 Counting atoms

Relative atomic masses show that one atom of carbon is twelve times as heavy as one atom of hydrogen. Therefore, 12 g of carbon will contain the same number of atoms as 1 g of hydrogen. An atom of oxygen is 16 times as heavy as an atom of hydrogen, so 16 g of oxygen will also contain the same number of atoms as 1 g of hydrogen.

In fact, the relative atomic mass in grams of every element will contain the same number of atoms.

This number is called **Avogadro's constant** in honour of the Italian scientist Amedeo Avogadro. Experiments show that:

Avogadro's constant
= 600 000 000 000 000 000 000 000
= 6×10^{23}

The amount of a substance that contains 6×10^{23} particles is known as one **mole**.

So,

> one mole of an element is equal to its relative atomic mass in grams.

∴ 12 g of carbon contains 6×10^{23} atoms and is 1 mole
24 g of magnesium contains 6×10^{23} atoms and is 1 mole
1 g of carbon contains $\frac{1}{12} \times 6 \times 10^{23}$ atoms and is $\frac{1}{12}$ mole
10 g of carbon contains $\frac{10}{12} \times 6 \times 10^{23}$ atoms and is $\frac{10}{12}$ mole

Notice that:

$$\text{number of moles} = \frac{\text{mass in grams}}{\text{mass of 1 mole}}$$

and

$$\text{number of atoms} = \text{number of moles} \times 6 \times 10^{23}$$

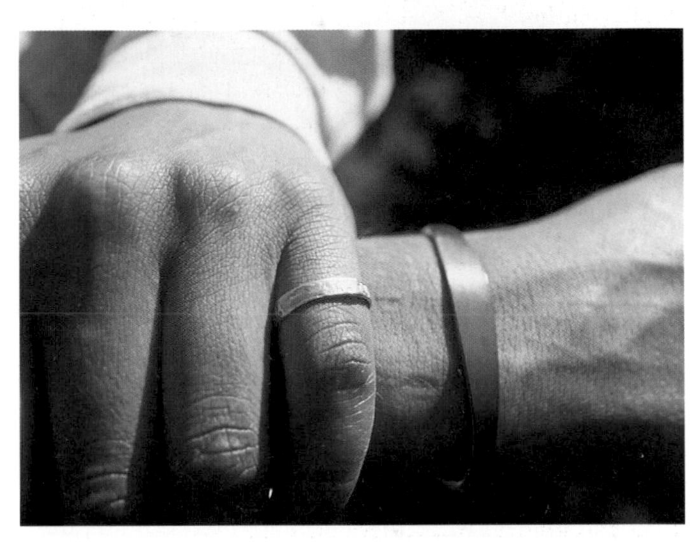

This gold ring has a mass of 19.7 g. This is 19.7/197 moles of gold which contains $19.7/197 \times 6 \times 10^{23}$ atoms. The copper bracelet has a mass of 31.75 g. How many moles of copper is this? How many copper atoms does it contain?
($A_r(Au) = 197$, $A_r(Cu) = 63.5$)

Chemists often need to count atoms. In industry, nitrogen is reacted with hydrogen to make ammonia, NH_3, which is then used to manufacture fertilizers. In a molecule of ammonia, there is one nitrogen atom and three hydrogen atoms. In order to make ammonia, chemists must therefore react:

1 mole of nitrogen with 3 moles of hydrogen
i.e. 14 g of nitrogen (as $A_r(N) = 14$) with
3×1 g = 3 g of hydrogen (as $A_r(H) = 1$)
and *not* 1 g of nitrogen and 3 g of hydrogen.

To get the right amounts, chemists must measure in moles. Thus, the mole is the chemist's counting unit.

Chemists are not the only people who 'count by weighing'. Ironmongers like the one in the photo count nails by weighing them. For example, if one hundred nails weigh 356 g, it is easier to take one hundred by weighing out 356 g of them rather than counting

14.9 Finding formulas

We have used some formulas already, but how are they obtained? How do we know that the formula of water is H_2O and that of carbon dioxide is CO_2?

All formulas are obtained by experiment. To find a formula we need to:

1. Find the masses of the different elements in a sample of the compound.
2. Calculate the number of moles of the different elements in the sample.
3. Calculate the whole number ratio of moles and hence the formula.

Example

A sample of iron ore was purified. 4.8 g of the purified ore contained 3.36 g of iron and 1.44 g of oxygen. What is its formula?

Solution

$A_r(Fe) = 56, A_r(O) = 16$

Masses of iron and oxygen in the sample = 3.36:1.44

Moles of iron and oxygen in the sample $= \frac{3.36}{56}:\frac{1.44}{16}$

$= 0.06:0.09$

Whole number ratio of moles of iron and oxygen = 2:3

So, the ratio of atoms of iron and oxygen = 2:3

∴ the formula of iron ore = Fe_2O_3

14.10 Balancing equations

A chemical equation is a summary of the starting substances (reactants) and the products in a chemical reaction.

The sparks from a sparkler are caused by tiny bits of burning magnesium. What is the equation for this reaction?

When magnesium burns in air, it reacts with oxygen to form magnesium oxide. We can summarise this in a **word equation** as:

magnesium + oxygen → magnesium oxide

In this reaction, magnesium and oxygen are the **reactants**. Magnesium oxide is the **product**.

Chemists usually write symbols and formulas rather than names in equations. So for the word equation above, we should write Mg for the element magnesium, O_2 for oxygen and $Mg^{2+}O^{2-}$ (or simply MgO) for magnesium oxide.

$$Mg + O_2 \rightarrow MgO$$

But notice that this equation does not balance. There are two oxygen atoms in O_2 on the left and only one oxygen atom in MgO on the right. So MgO must be doubled to give:

$$Mg + O_2 \rightarrow 2MgO$$

Unfortunately, the equation still doesn't balance. There are now two Mg atoms on the right in 2MgO, but only one on the left in Mg. This is easily corrected by writing 2Mg on the left to give:

$$2Mg + O_2 \rightarrow 2MgO$$

The numbers of different atoms are the same on both sides of the arrow and this is a **balanced chemical equation**.

Writing equations

The only way to be sure of the balanced equation for a reaction is to do an experiment. First you need to determine the reactants and products. Then you need to find the number of moles of each substance involved.

In practice, it just isn't possible to do an experiment every time we want to write an equation. If we know the reactants and products, we can write in their formulas and then predict a balanced equation.

The following example shows the three steps involved.

Step 1 Write a word equation
e.g. hydrogen + oxygen → water

Step 2 Write symbols and formulas for reactants and products
e.g. $H_2 + O_2 \rightarrow H_2O$

Remember that oxygen, nitrogen, hydrogen and chlorine are diatomic and written as O_2, N_2, H_2 and Cl_2. All other elements are shown as single atoms i.e. Zn for zinc, C for carbon, etc.

Step 3 Balance the equation by making the number of atoms of each element the same on both sides.
e.g. $2H_2 + O_2 \rightarrow 2H_2O$

Remember that you must *never change a formula* to make an equation balance. The formula of water is always H_2O and never HO_2 or H_2O_2. You can only balance an equation by putting numbers in front of symbols or in front of a whole formula, i.e. $2H_2$ and $2H_2O$.

Balanced chemical equations are more useful than word equations because they show:

- *the symbols and formulas of the reactants and products* and
- *the relative numbers of atoms and molecules* of the reactants and products.

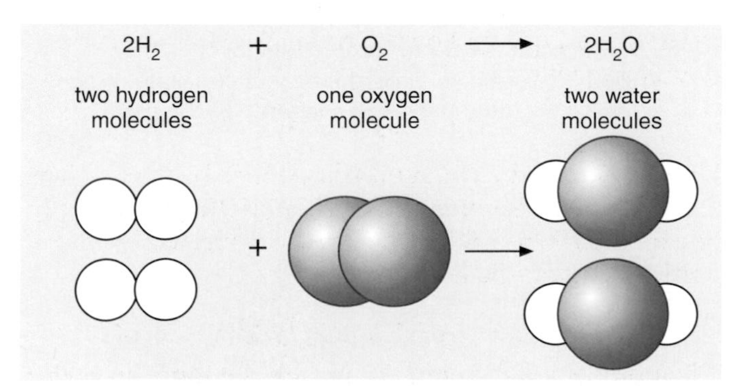

Figure 14.10 Using models to represent the reaction between hydrogen and oxygen to form water

Balanced equations also help us to see how the atoms are re-arranged in a chemical reaction. You can see this even better using models to represent an equation as in Figure 14.10.

14.11 Using equations

In industry, it is important to know the amounts of reactants that are needed for a chemical process and the amount of product that can be obtained. Very often, industrial chemists need to know how much product they can obtain from a given amount of starting material. In order to do this, they use equations and relative atomic masses. As an example, let us calculate how much lime (calcium oxide) we could obtain by heating 1 kg of pure limestone (calcium carbonate).

An old lime kiln used in Kent in the 19th century

A modern lime kiln

The equation for the reaction is

calcium carbonate → calcium oxide + carbon dioxide
$CaCO_3 \rightarrow CaO + CO_2$
∴ 1 mole of $CaCO_3$ → 1 mole of CaO
i.e.
$(40 + 12 + (3 \times 16))$ g $CaCO_3$ → $(40 + 16)$ g CaO
⇒ 100 g $CaCO_3$ → 56 g CaO
∴ 1000 g $CaCO_3$ → 560 g CaO

So 1 kg of pure limestone could produce 560 g of lime

This, of course, assumes that *all* the limestone is converted to lime and the yield of lime is 100%. In practice, this may not be the case. Some limestone may remain unchanged or it may not be pure.

Using equations, it is also possible to calculate the volumes of gases which react. At room temperature (20°C) and atmospheric pressure, 1 mole (relative formula mass in grams) of *any* gas occupies 24 dm^3. At standard temperature and pressure, s.t.p., (i.e. 273K and 1 atm. pressure) 1 mole of a gas occupies 22.4 dm^3.

For example, ammonia (NH_3) is manufactured from nitrogen (N_2) and hydrogen (H_2).

nitrogen + hydrogen → ammonia
$N_2(g) + 3H_2(g) \rightarrow 2\,NH_3(g)$
∴ 1 mole N_2 + 3 moles H_2 → 2 moles NH_3
i.e. 24 dm^3 N_2 + 3 × 24 dm^3 H_2 → 2 × 24 dm^3 NH_3
So
1 dm^3 nitrogen + 3 dm^3 hydrogen → 2 dm^3 ammonia

Ammonia is important in producing fertilizers such as ammonium nitrate, known commercially as Nitram.

Assuming the yield is 100%, what mass of ammonia is needed to react with nitric acid to make 1 tonne of ammonium nitrate?

ammonia + nitric acid → ammonium nitrate
$NH_3 + HNO_3 \rightarrow NH_4NO_3$
∴ 1 mole of NH_3 → 1 mole of NH_4NO_3
$(14 + (3 \times 1))$ g NH_3 → $(14 + (4 \times 1) + 14 + (3 \times 16))$ g NH_4NO_3
17 g NH_3 → 80 g NH_4NO_3
⇒ 80 g of NH_4NO_3 is obtained from 17 g of NH_3
∴ 1 g of NH_4NO_3 is obtained from $\frac{17}{80}$ g (= 0.21 g) of NH_3

So 1 tonne of ammonium nitrate is obtained from 0.21 tonne of ammonia.

QUESTIONS

1 Motor car bodies, and other things made from iron or steel, can be damaged by rusting.

a) Why is rusting of iron described as a chemical change?

b) A scientist collected information about the rate at which iron rusts.
Information was taken in places which have different annual rainfalls.

Test site	Annual rainfall in mm	Corrosion rate in units
A	425	80
B	570	100
C	600	110
D	900	150
E	1000	180

Copy the graph on the right and plot the values of corrosion rate against annual rainfall. Draw a suitable line to complete the graph. The first point has been plotted for you.

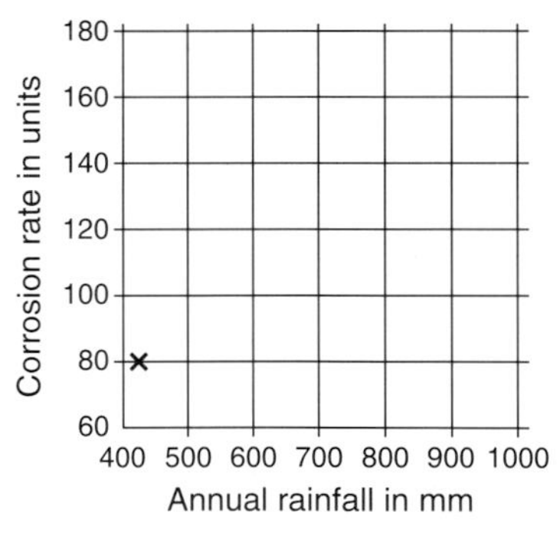

c) Use the graph to describe how corrosion is linked to annual rainfall.

d) i) Suggest **one** other factor which may affect the rate of corrosion.

ii) Explain how this factor would affect the rate of corrosion.

e) A new paint is said to 'protect against rusting'. Briefly describe an experiment you could do to test this, given some nails and a sample of the paint.

MEG

2 Ammonium dihydrogen phosphate, a common fertilizer, may be prepared by neutralizing ammonia solution with phosphoric acid according to the following equation:

$$NH_4OH + H_3PO_4 \rightarrow (NH_4)H_2PO_4 + 3H_2O$$

(Relative atomic masses: N = 14, H = 1, P = 31, O = 16)

a) Find the relative molecular masses of:
i) phosphoric acid,
ii) ammonium dihydrogen phosphate.

b) Calculate the minimum amount of phosphoric acid needed to manufacture 11.5 tonnes of ammonium dihydrogen phosphate fertilizer. Show clearly how you obtain your answer.

c) Suggest a reason why ammonium dihydrogen phosphate is sold in pellet form rather than powder form. **NICCEA**

3 a) 11.6 g of an oxide of iron was found to contain 8.4 g of iron. Calculate the formula of the oxide using the following procedure.

(A_r (O) = 16; A_r (Fe) = 56.)

i) Calculate the mass of oxygen in 11.6 g of the oxide.
ii) Using the masses and relative atomic masses of elements, calculate the ratio of the number of atoms of iron to the number of atoms of oxygen.
iii) Calculate the formula of the oxide.

b) Iron reacts with a solution of copper(II) sulphate in accordance with the following equation.

$$Fe(s) + CuSO_4(aq) \rightarrow FeSO_4(aq) + Cu(s)$$

Copper used to be extracted from Parys Mountain in Anglesey. One method was to throw scrap iron into shallow lakes containing copper(II) sulphate solution.
Calculate the mass of copper that would be obtained from 14 tonnes of iron.

(A_r (Cu) = 63.5; A_r (Fe) = 56.)

c) A packet found in a garden shed contained a pale blue powder which was thought to contain Bordeaux mixture, a fungicide used to treat diseases in plants. Bordeaux mixture contains copper(II) sulphate. Describe a chemical test to prove that the powder contains a copper(II) compound and what you would expect to observe. **WJEC**

4 a) When a crystal of potassium manganate(VII) is placed in cold water in a beaker, the crystal sinks to the bottom. The crystal dissolves and a pink colour slowly spreads throughout the liquid.

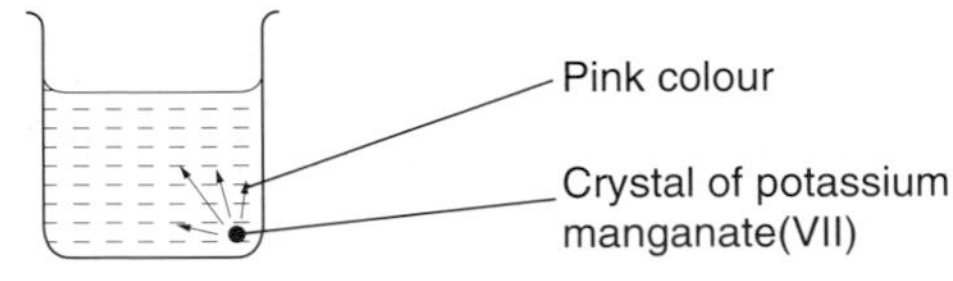

Give **two** reasons why this process takes place faster in hot water than in cold.

b) When ammonia gas and hydrogen chloride gas mix, they form solid ammonium chloride.
A pad of cotton wool soaked in ammonia solution and another soaked in hydrogen chloride solution were put into opposite ends of a dry glass tube at the same time.

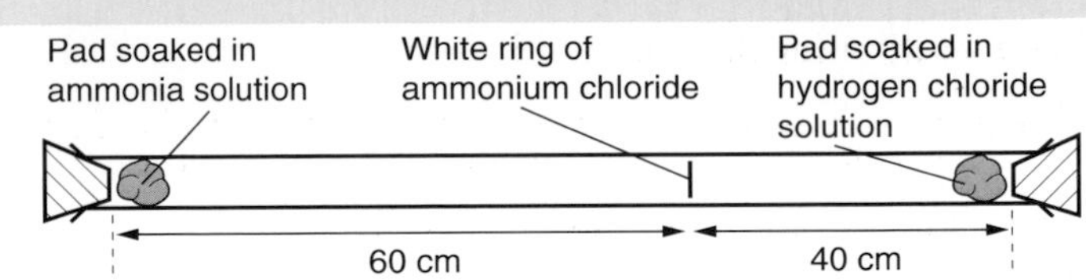

After five minutes a ring of solid ammonium chloride formed. This is shown in the diagram above.
i) What name is given to the spreading of these gases?
ii) What does the position of the ring suggest about the speeds of movement of particles of ammonia gas and hydrogen chloride gas?
iii) The experiment was repeated at a higher temperature. How would you expect this to affect the time taken for the ring to form? Explain your answer.
iv) Particles of gases move at speeds of between 400 m/s and 1000 m/s at room temperature. Give **two** reasons why the ring took so long to form. **MEG**

5 A manufacturer is making a container for a new perfume product.

a) i) Write down the most suitable material for producing this container from the following list. The customer must be able to see the liquid in the container.

china glass metal stone

ii) Suggest **one** disadvantage of the material you chose in i).

The diagram shows the container with a sprayer to produce a fine mist. In the container is a solution of perfume dissolved in ethanol.

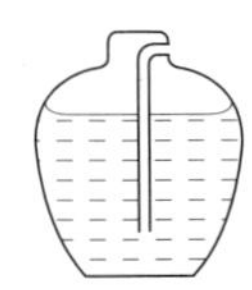

b) Why does your hand feel cold when perfume solution is sprayed onto it?

c) Copy and complete the following figure to show the particles in
i) liquid pure perfume before it is dissolved in ethanol,
ii) a solution of the perfume in ethanol,
iii) the perfume as a gas in a room.

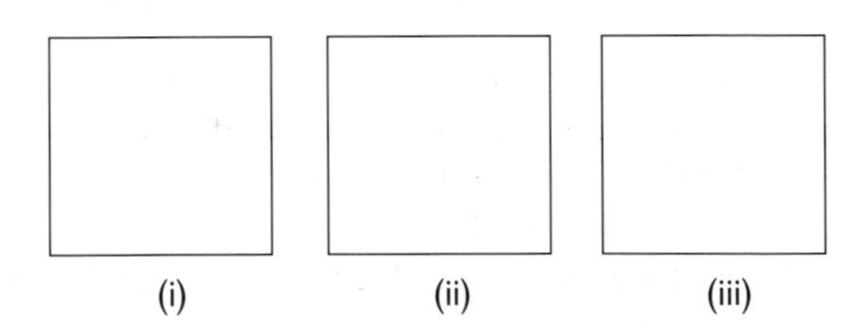

d) Suggest a reason why perfume is dissolved in ethanol. **MEG**

CHAPTER

Metals and the reactivity series

The properties of metals
Reactions of metals
Exothermic and endothermic reactions
Displacing metals
Summarising the reactions of metals
Extracting metals
Extracting iron from iron ore
Extracting aluminium by electrolysis
Purifying copper by electrolysis

15.1 The properties of metals

Metals are among the most important and useful materials. Just look around and you will notice the uses of different metals – cutlery, ornaments, handles and locks for doors, pipes, radiators, cars, girders and bridges.

The photographs in Figure 15.1 show three important uses of different metals. What properties make the metals ideal for these uses?

15.2 Reactions of metals

1 How do metals react with air (oxygen)?

As soon as sodium is exposed to the air, its shiny surface starts to go dull. This is because the sodium reacts with oxygen in the air forming white sodium oxide.

$$\text{sodium} + \text{oxygen} \rightarrow \text{sodium oxide}$$
$$4Na + O_2 \rightarrow 2Na_2O$$

Why is tungsten used for the filament of this light bulb?

Why is copper used for these hot water pipes?

Why is aluminium foil used to wrap food?

Figure 15.1

Other metals, like aluminium, react more slowly. A shiny aluminium pencil sharpener may take several weeks before it goes dull with the formation of white aluminium oxide.

$$\text{aluminium} + \text{oxygen} \rightarrow \text{aluminium oxide}$$
$$4Al + 3O_2 \rightarrow 2Al_2O_3$$

Unreactive metals, like copper, take months or even years before they become darker with the formation of a thin layer of black copper oxide.

$$\text{copper} + \text{oxygen} \rightarrow \text{copper oxide}$$
$$2Cu + O_2 \rightarrow 2CuO$$

We can summarise these reactions of metals with oxygen as:

metal + oxygen → metal oxide

Very unreactive metals, like gold, have no reaction with oxygen and they remain permanently shiny.

Gold is so unreactive that it stays shiny for centuries, like Tutankamun's face mask which dates from about 1340 BC

From these observations and from experiments in which metals are heated in air or pure oxygen, it is possible to put metals in an order of reactivity. This order of reactivity is usually called the **reactivity series** (Figure 15.2). Metals at the top of the series, such as sodium and calcium are the most reactive. Metals at the bottom of the series, like silver and gold, are the least reactive.

The reactivity series can be used to predict and summarise other reactions of metals as well as their reaction with oxygen. As expected, reactive metals like sodium also react vigorously with other non-metals such as chlorine and sulphur to form their compounds. Unreactive metals, like gold, have little desire to react with non-metals. So, unreactive metals tend to remain uncombined.

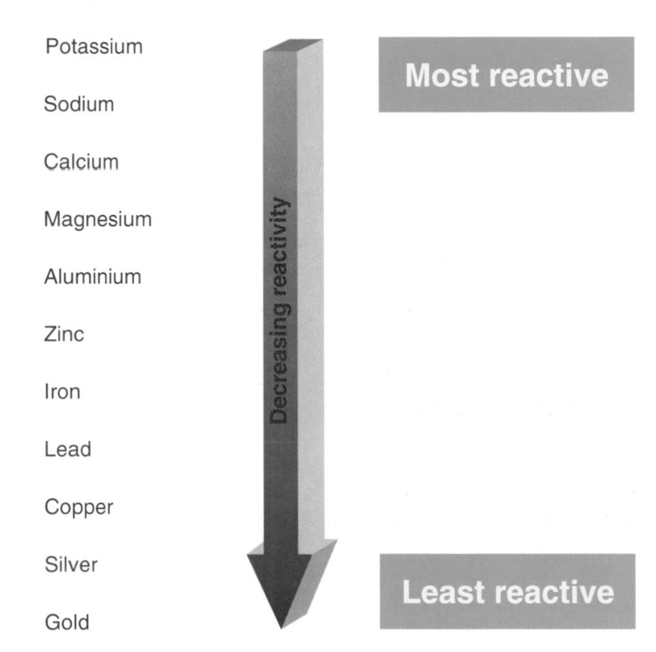

Figure 15.2 The reactivity series

2 How do metals react with water?

Those metals which are keen to react with oxygen and form their oxides will also react with water. The metals take the oxygen from water (H_2O), leaving hydrogen. For example:

$$\text{sodium} + \text{water} \rightarrow \text{sodium oxide} + \text{hydrogen}$$
$$2Na + H_2O \rightarrow Na_2O + H_2$$

The oxides of reactive metals then react further with water forming the metal hydroxide.

$$\text{sodium oxide} + \text{water} \rightarrow \text{sodium hydroxide}$$
$$Na_2O + H_2O \rightarrow 2NaOH$$

Table 15.1 over the page summarises the reactions of six metals with water.

Table 15.1 The reactions of six metals with water

Metal	Reaction with water
Calcium	Sinks in the water. Steady stream of hydrogen produced. The solution becomes alkaline and cloudy due to the formation of calcium hydroxide.
Copper	No reaction
Iron	No reaction. (Iron will only react with cold water if air (oxygen) is also present for rusting.)
Magnesium	Tiny bubbles of hydrogen appear on the surface of the magnesium after a few minutes. The solution slowly becomes alkaline.
Potassium	A violent reaction occurs. A globule of molten potassium skates over the water surface, hissing and burning with a lilac flame. Hydrogen and potassium hydroxide form.
Sodium	A vigorous reaction occurs. A globule of molten sodium skates about the water surface. Hydrogen and sodium hydroxide form.

Look at the results in Table 15.1 and work out the order of reactivity of the metals with water. Notice that this order of reactivity is the same as the order of reactivity with air or oxygen in Figure 15.2.

Only the most reactive metals, from potassium down to magnesium in Figure 15.2, react with *cold* water. However, metals below magnesium, such as zinc and iron will react with *hot* water or steam to form hydrogen. For example:

$$\text{zinc} + \text{water} \rightarrow \text{zinc oxide} + \text{hydrogen}$$
$$Zn + H_2O \rightarrow ZnO + H_2$$

These reactions of metals with water can be summarised as:

> metal + water → metal oxide + hydrogen

Why is this hot water tank made of copper rather than steel or plastic?

The boilers in steam trains are made from iron (steel). What is the disadvantage in using iron (steel)? Why is copper not used?

3 *How do metals react with acids?*

Various foods, including vinegar, oranges, lemons and rhubarb, contain acids. Have you noticed that these acids can attack metal cutlery? Figure 15.3 shows what happened when five different metals were added to cold dilute hydrochloric acid at 20°C. This acid is more reactive than the acids in food, but it shows how different metals are attacked. If you try this experiment, wear safety spectacles and remember that hydrogen is very flammable.

Notice in Figure 15.3 that aluminium does not react at first. This is because its surface is protected by a layer of aluminium oxide. The oxide reacts slowly with the acid. When the oxide has reacted away, the aluminium is exposed which reacts more vigorously. The metals used most commonly for pans and cutlery are aluminium, copper and iron (in the form of steel).

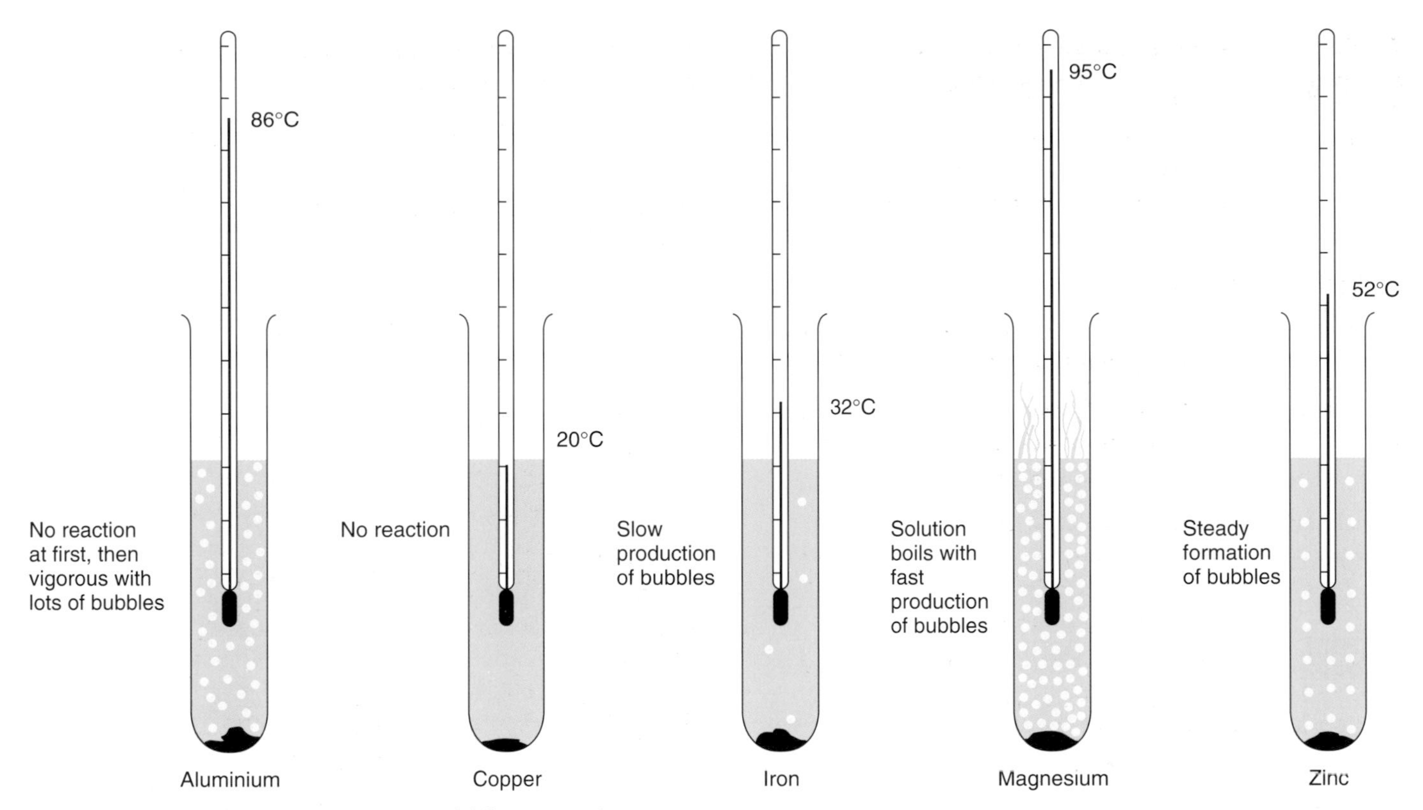

Figure 15.3 Investigating the reaction of different metals with dilute hydrochloric acid

Copper is the only one which does not react with acids in food, but it is an expensive metal. The thin oxide coating on aluminium protects it from any weak acids in food, but oxalic acid in rhubarb and acetic (ethanoic) acid in vinegar react with the oxide and 'clean' the pan during cooking. Iron (steel) which is used for pans and cutlery is also attacked by acids in food.

Foods which contain acids are best stored in unreactive glass or plastic containers. Tin cans may contain acidic foods like pineapple pieces. They are made of steel coated on both sides with tin and then lacquered on the inside. The lacquer forms an unreactive layer between the tin and its contents.

All the metals in Figure 15.3, except copper, react with the dilute hydrochloric acid. The metals take chlorine from the hydrochloric acid (HCl) leaving hydrogen which forms the bubbles. For example:

$$\text{metal} + \text{hydrochloric acid} \rightarrow \text{metal chloride} + \text{hydrogen}$$

$$\text{magnesium} + \text{hydrochloric acid} \rightarrow \text{magnesium chloride} + \text{hydrogen}$$

$$Mg(s) + 2HCl(aq) \rightarrow MgCl_2(aq) + H_2(g)$$

These reactions are similar to the reactions of metals with water. With water, the metals are taking oxygen from H_2O. With hydrochloric acid, the metals are taking chlorine from HCl. A similar reaction occurs between metals and dilute sulphuric acid. In this case, the products are a metal sulphate and hydrogen. The reactions of metals with acids can be summarised as:

$$\text{metal} + \text{acid} \rightarrow \underset{\text{(salt)}}{\text{metal compound}} + \text{hydrogen}$$

Metal compounds such as metal chlorides and metal sulphates are usually called **salts**. The best known salt is sodium chloride – common salt.

15.3 Exothermic and endothermic reactions

When metals react with acids, heat is given out and the temperature of the mixture rises (Figure 15.3). Changes in temperature often occur when chemical reactions happen. The reactions are accompanied by energy changes which can be detected as a temperature change for reactions in solution.

Reactions like those involving metals with acid which *give out heat* are called **exothermic reactions**.

Perhaps the most obvious exothermic reactions are those in which fuels, like natural gas, burn and give out heat.

fuel + oxygen → carbon dioxide + water

$$CH_4(g) + 2O_2(g) \rightarrow CO_2(g) + 2H_2O(g)$$

A few reactions, such as the decomposition of calcium carbonate (limestone) to form calcium oxide (lime) and carbon dioxide, take in heat.

calcium carbonate (limestone) →(heat) calcium oxide (lime) + carbon dioxide

$$CaCO_3(s) \xrightarrow{heat} CaO(s) + CO_2(g)$$

These reactions which *take in heat* are called **endothermic reactions**.

15.4 Displacing metals

In section 15.2, we studied the reactions of metals with acids. Metals above copper in the reactivity series react with acids to form a salt and hydrogen. For example:

zinc + sulphuric acid → zinc sulphate + hydrogen

$$Zn + H_2SO_4 \rightarrow ZnSO_4 + H_2$$

In this case, zinc has *displaced* hydrogen (H_2) from H_2SO_4. Now if zinc can displace H_2 from H_2SO_4, it may be able to displace Cu from $CuSO_4$ and other metals from their compounds.

Figure 15.4 shows what happens when strips of zinc are placed in solutions of various metal ions. Notice that zinc displaces lead from lead nitrate solution, copper from copper sulphate solution and silver from silver nitrate solution. But, zinc does *not* displace magnesium from magnesium nitrate solution.

When zinc is placed in copper sulphate solution, the zinc becomes coated with red-brown copper. At the same time, the blue colour of the solution fades.

zinc + copper sulphate → zinc sulphate + copper

$$Zn(s) + CuSO_4(aq) \rightarrow ZnSO_4(aq) + Cu(s)$$

In the reaction, copper ions have been displaced from the solution as copper atoms. The copper atoms have been deposited on the zinc and zinc ions from the zinc metal have replaced the copper ions in solution. The equation for the reaction is:

$$Zn(s) + Cu^{2+}(aq) + SO_4^{2-}(aq) \rightarrow Zn^{2+}(aq) + SO_4^{2-}(aq) + Cu(s)$$

Table 15.2 summarises the results of the experiment in Figure 15.4 and four other experiments in which magnesium, iron, lead and copper were used in place of zinc.

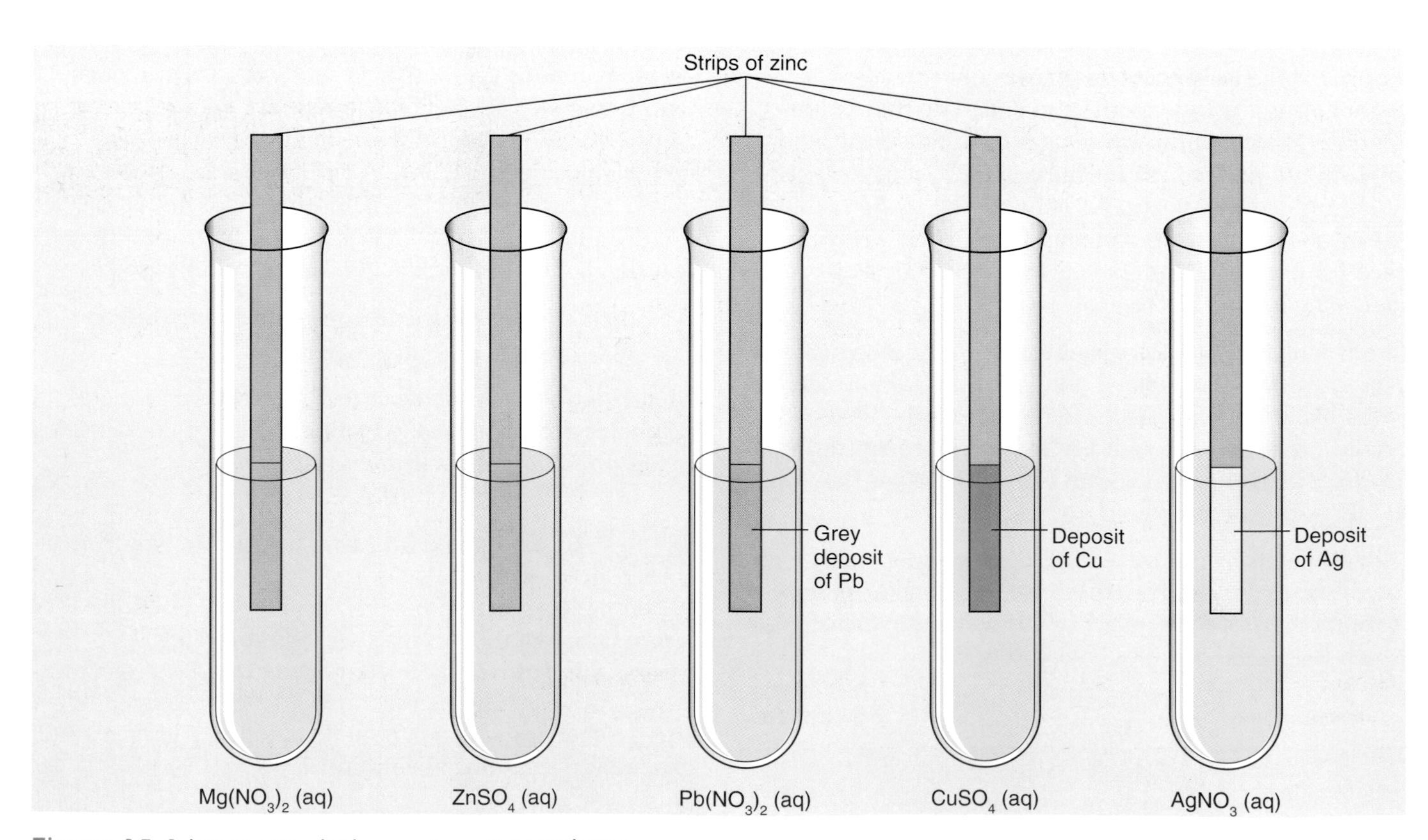

Figure 15.4 Investigating displacement reactions with zinc

Table 15.2 The results of some displacement reactions with metals

Metal used	Results with an aqueous solution of: $Mg(NO_3)_2$	$ZnSO_4$	$Pb(NO_3)_2$	$CuSO_4$	$AgNO_3$
Mg		dark grey deposit of zinc	grey deposit of lead	red-brown deposit of copper	black deposit of silver
Zn			grey deposit of lead	red-brown deposit of copper	black deposit of silver
Fe	no apparent reaction		grey deposit of lead	red-brown deposit of copper	black deposit of silver
Pb				red-brown deposits of Cu forms slowly	grey-black deposit of silver
Cu					black deposit of silver

Notice two things in Table 15.2

1 The deposits form because the metal added displaces the second metal from a solution of its ions.

2 The metals and their solutions are written in the order of the reactivity series. **Each metal only displaces metals which are below it in the reactivity series, from a solution of their ions.** So, zinc can displace lead, copper and silver, but not magnesium. Lead can displace copper and silver, but not magnesium and zinc.

These experiments confirm that magnesium is more reactive than the other metals in the table. Magnesium reacts and forms its ions as it displaces the other metals from solution.

$$Mg(s) + Zn^{2+}(aq) \rightarrow Mg^{2+}(aq) + Zn(s)$$

Thus, magnesium is higher in the reactivity series than the other metals in Table 15.2. In the same way, zinc is more reactive than lead, copper and silver, but less reactive than magnesium. So, zinc is above lead, copper and silver in the reactivity series, but below magnesium. These results provide further evidence for the order of metals in the reactivity series. They show that metals become less reactive (i.e. less likely to form their ions) going down the reactivity series.

15.5 Summarising the reaction of metals

Table 15.3 A summary of the reactions of metals based on the reactivity series

Reactivity series	Reaction with oxygen when heated in air	Reaction with water	Reaction with dilute acids	Reactions with solutions of metal salts	Symbols
Potassium Sodium Calcium Magnesium	burn with decreasing reactivity down the series to form their oxide	react with water with decreasing reactivity down the series forming hydrogen	react with dilute HCl and dilute H_2SO_4, less and less vigorously down the series, producing hydrogen	each metal will displace metals below it in the reactivity series from solutions of their salts	K Na Ca Mg
Aluminium Zinc Iron		react with steam, but not water forming hydrogen			Al Zn Fe
Lead Copper	only form a layer of oxide	do not react with water or steam	do not react with dilute acid		Pb Cu
Silver Gold	do not react				Ag Au
General equation	$2M + O_2 \rightarrow 2MO$	$M + H_2O \rightarrow MO + H_2$	$M + 2HCl \rightarrow MCl_2 + H_2$ $M + H_2SO_4 \rightarrow MSO_4 + H_2$	$M_h + M_l^{2+} \rightarrow M_h^{2+} + M_l$	

Table 15.3 on page 135 summarises the reactions of metals with oxygen (air), water, dilute acids and solutions of metal salts. Notice that the order of reactivity is the same in all four columns. In fact, the reactivity series provides a very helpful summary for all the reactions of metals.

Table 15.3 also shows a general equation for each of the reactions concerned. In the general equations, the metal is given the symbol M and assumed to form ions with a charge of 2+.

15.6 Extracting metals

Most metals are too reactive to exist on their own in the earth. They are usually combined with non-metals and found in rocks and minerals as impure substances called **metal ores**.

A few metals, like gold and silver, are so unreactive that they occur in the ground as uncombined metal.

A few metals, like gold, occur in the ground as the uncombined metal. This picture shows a prospector panning for gold in California in the late 19th century

Extracting metals from their ores usually involves three stages:

1 mining and concentrating the ore,
2 reducing the ore to the metal,
3 purifying the metal.

We should now look at these three stages in more detail.

1 Mining and concentrating the ore

Natural processes in the Earth have formed rich deposits of metal ores in some areas. But, even the richest deposits are impure. The ore is mixed with soil, rocks and other impurities from which it must be separated. This may involve crushing the rocks and then separating the ore from the impurities by the process of **flotation**.

There are important economic, social and environmental issues involved in the mining of minerals such as metal ores. These issues can create both advantages and disadvantages (Table 15.4).

Table 15.4 Advantages and disadvantages of mining for metal ores

Advantages	Disadvantages
metals are produced and used to manufacture **thousands of useful articles**	mining **damages the environment** with spoil heaps, quarries and mines
mining ores and the manufacture of metal articles **creates jobs**	mining often **destroys wildlife habitats**
the sale of ores, metals and metal products can **increase the wealth of a community**	mining can cause **subsidence**, affecting buildings

2 Reducing the ore to the metal

Table 15.5 shows the ores from which some important metals are obtained.

Table 15.5 The ores from which some important metals are obtained

Metal	Name of the ore	Name and formula of metal compound in the ore
Sodium	rock salt	sodium chloride, $NaCl$
Aluminium	bauxite	aluminium oxide, Al_2O_3
Zinc	zinc blende	zinc sulphide, ZnS
Iron	iron ore (haematite)	iron(III) oxide, Fe_2O_3
Copper	copper pyrites	copper sulphide and iron sulphide $CuS + FeS$ ($CuFeS_2$)

Notice in Table 15.5 that metal ores are often oxides, sulphides and chlorides. This is because oxygen, sulphur and chlorine are reactive non-metals.

The method used to convert a particular ore to the metal depends on two factors:

- the position of the metal in the reactivity series,
- the cost of the process.

i) Heating the ore in air

This is the cheapest way to extract a metal. However, it only works with the compounds of metals at the bottom of the reactivity series which can be reduced to the metal fairly easily. For example, copper is extracted from copper sulphide by first heating the ore in air to obtain copper oxide.

$$\text{copper sulphide} + \text{oxygen} \rightarrow \text{copper oxide} + \text{sulphur dioxide}$$
$$2CuS + 3O_2 \rightarrow 2CuO + 2SO_2$$

The supply of air (oxygen) is then cut off and more copper sulphide is added to convert the copper oxide to copper.

$$\text{copper oxide} + \text{copper sulphide} \rightarrow \text{copper} + \text{sulphur dioxide}$$
$$2CuO + CuS \rightarrow 3Cu + SO_2$$

ii) Reduction with carbon and carbon monoxide

The metals in the middle of the reactivity series, such as zinc, iron and lead, cannot be obtained simply by heating their ores in air. Instead, they are obtained by reducing their oxides with carbon (coke) or carbon monoxide.

For example,

$$\text{zinc oxide} + \underset{\text{(coke)}}{\text{carbon}} \rightarrow \text{zinc} + \text{carbon monoxide}$$
$$ZnO + C \rightarrow Zn + CO$$

Sometimes, these metals exist as sulphide ores. These ores must be converted to oxides before reduction. This is done by heating the sulphides in air.

$$\text{zinc sulphide} + \text{oxygen} \rightarrow \text{zinc oxide} + \text{sulphur dioxide}$$
$$2ZnS + 3O_2 \rightarrow 2ZnO + 2SO_2$$

In the next section, we will study the extraction of iron from iron ore in more detail.

iii) Electrolysis of molten compounds

Metals at the top of the reactivity series, like sodium, magnesium and aluminium, cannot be obtained by reduction of their oxides with carbon or carbon monoxide. This is because the temperature needed to reduce their oxides is too high.

Table 15.6 A summary of the methods used to extract different metals

Reactivity series of metal	Method of extraction
Potassium Sodium Calcium Magnesium Aluminium	electrolysis of molten compound
Zinc Iron Tin Lead	reduction of ore using carbon or carbon monoxide
Copper Mercury	heating ore in air
Silver Gold	metals occur uncombined in the ground

Copper is obtained from copper pyrites (chalcopyrites). This vast copper mine is near Salt Lake City, USA

Also, these metals cannot be obtained by electrolysis of their *aqueous* solution because hydrogen from the water is produced at the cathode instead of the metal.

The only way to extract these reactive metals is by electrolysing their *molten* compounds. Potassium, sodium, calcium and magnesium are obtained by electrolysis of their molten chlorides. Aluminium is obtained by electrolysis of its molten oxide and this is described in section 15.8.

Table 15.6 on page 137 summarises the methods used to extract different metals. Notice that the method used depends on the metal's position in the reactivity series.

3 Purifying the metal

The copper which comes straight from the furnace in the process described on page 137 contains about 3% impurities. Sheets of this copper are purified by electrolysis with copper sulphate solution (see section 15.9).

Similarly, iron obtained directly from the furnace contains about 7% impurities. The purification of this pig-iron is described in the next section.

15.7 Extracting iron from iron ore

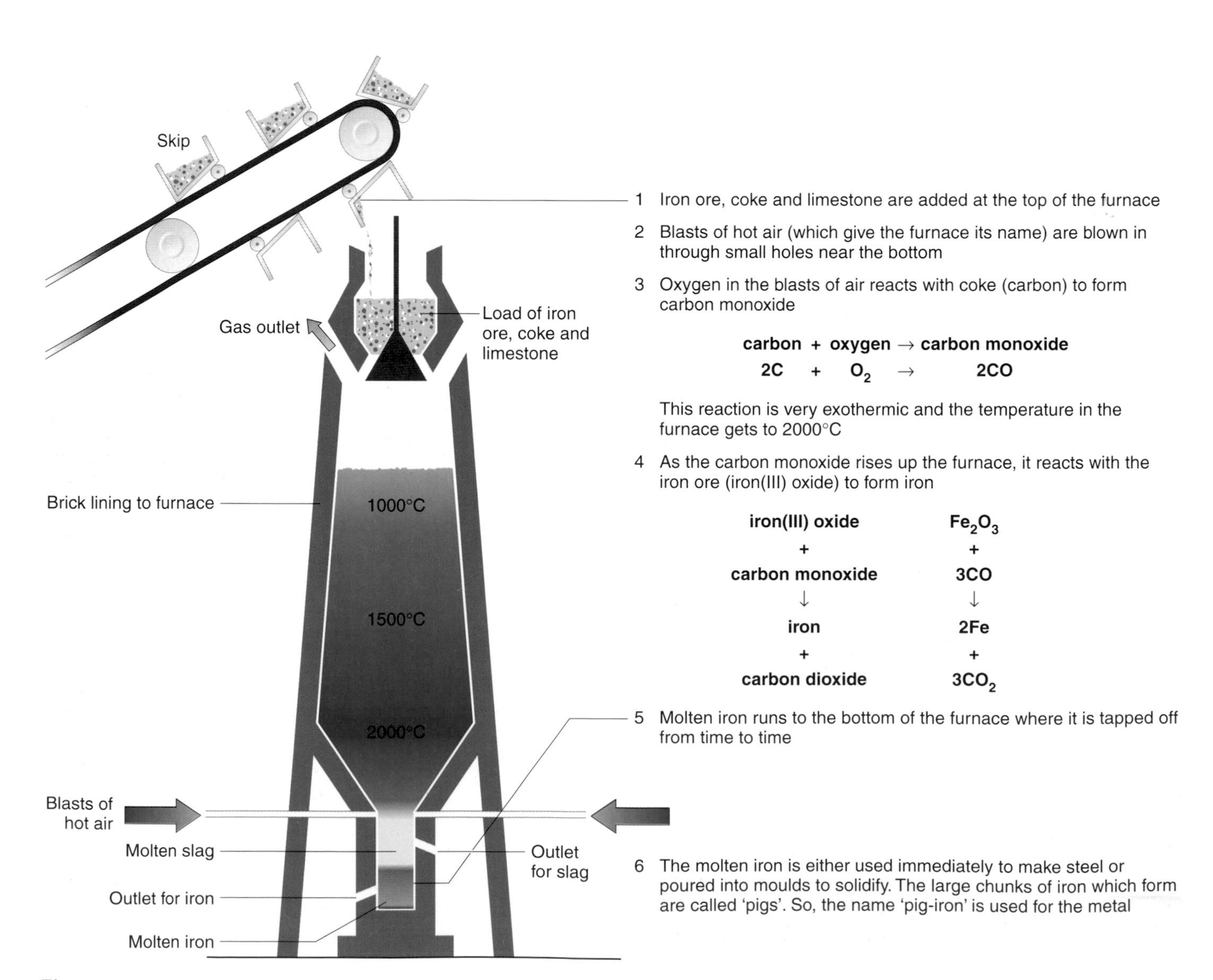

Figure 15.5 Extracting iron from iron ore in a blast furnace

The main raw material for making iron is iron ore (haematite). This ore contains iron(III) oxide. The largest deposits of iron ore in the UK are in Northamptonshire and Lincolnshire, but most of this is poor quality. The best quality ores are found in Scandinavia, America, Australia, North Africa and Russia.

Iron ore is converted to iron in a special furnace called a **blast furnace**. This furnace is a tapered cylindrical tower about twenty metres tall. Figure 15.5 shows a diagram of a blast furnace with a summary of the reactions to produce iron.

Why is limestone used in the furnace?

Iron ore usually contains impurities like earth and sand (silicon dioxide, SiO_2). Limestone helps to remove these impurities. The limestone decomposes at the high temperatures inside the furnace to form calcium oxide and carbon dioxide.

calcium carbonate → calcium oxide + carbon dioxide

$$CaCO_3 \rightarrow CaO + CO_2$$

The calcium oxide then reacts with sand (silicon dioxide) and other substances in the impurities to form slag (calcium silicate).

calcium oxide + silicon dioxide → calcium silicate

$$CaO + SiO_2 \rightarrow CaSiO_3$$

The molten slag falls to the bottom of the furnace and floats on the molten iron. The molten slag can be tapped off separately and used in building materials and in cement manufacture.

Large blocks of red-hot steel emerging from their production from pig-iron at a modern steel-making plant

From pig-iron to steel

Pig-iron still contains about 7% of impurities. The main impurity is carbon which makes pig-iron very hard and brittle compared to iron and steel.

In order to make pig-iron into steel, the percentage of carbon must be reduced to about 0.15%. This is done by blowing oxygen onto the hot, molten pig-iron. The oxygen converts carbon to carbon dioxide which escapes as a gas.

15.8 Extracting aluminium by electrolysis

Aluminium is manufactured by the electrolysis of molten aluminium oxide. This is obtained from bauxite. **Electrolysis** involves the decomposition of compounds by electrical energy. The compound which is decomposed is called an **electrolyte**. The terminals through which the electric current enters and leaves the electrolyte are called **electrodes**. The electrode connected to the positive terminal of the battery is called the **anode** and the electrode connected to the negative terminal of the battery is called the **cathode**.

Aluminium is obtained from bauxite, impure aluminium oxide. The photo shows bauxite being mined

Pure aluminium oxide cannot be used as the electrolyte in extracting aluminium because it does not melt until 2045°C. This makes its electrolysis uneconomical. The aluminium oxide is therefore dissolved in molten cryolite (Na_3AlF_6) which melts below 1000°C. Figure 15.6 over the page shows a diagram of the electrolytic method used.

During electrolysis, positive aluminium ions in the electrolyte are attracted to the negative carbon cathode lining the cell. Here, they combine with electrons to form neutral aluminium atoms.

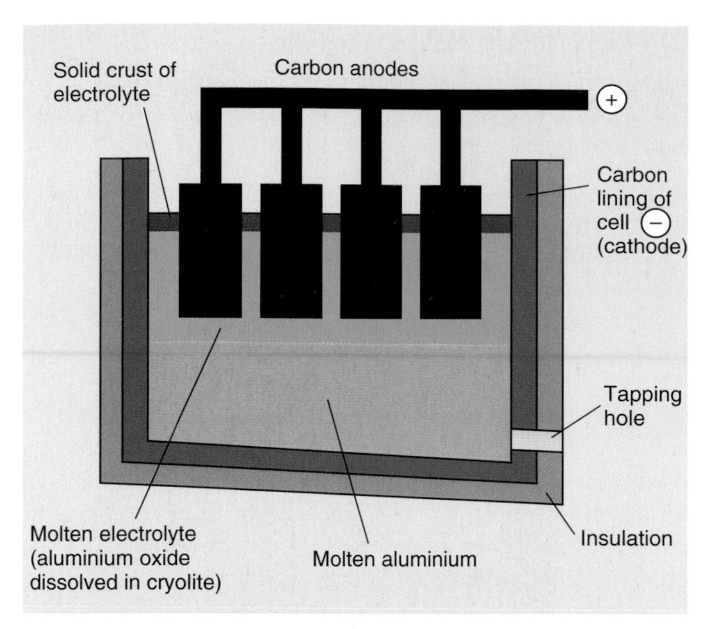

Figure 15.6 The electrolytic cell for aluminium manufacture

Cathode (−)

$$Al^{3+} + 3e^- \rightarrow Al$$

Notice that each aluminium ion, Al^{3+}, has three positive charges so it needs three electrons, each with one negative charge to form an aluminium atom.

Molten aluminium collects at the bottom of the cell and is tapped off at intervals. It takes about 16 kilowatt-hours of electricity to produce 1 kg of aluminium. The extraction plants are therefore sited near sources of cheap electricity, such as hydro-electric power stations.

Oxide ions, O^{2-}, in the electrolyte are attracted to the positive carbon anodes. Here, they give up their electrons to the anode forming neutral oxygen atoms.

Anode (+)

$$O^{2-} \rightarrow O + 2e^-$$

In this case, each oxide ion has two negative charges and so it loses two electrons in this process. The oxygen atoms join in pairs to form oxygen gas, O_2.

15.9 Purifying copper by electrolysis

When *aqueous* copper sulphate solution is electrolysed with copper electrodes, copper is deposited on the cathode and the copper anode loses weight (Figure 15.7).

The aqueous copper sulphate contains copper ions (Cu^{2+}) and sulphate ions (SO_4^{2-}). During electrolysis, positive Cu^{2+} ions are attracted to the negative cathode where they gain electrons and form neutral copper atoms. This copper is deposited on the cathode.

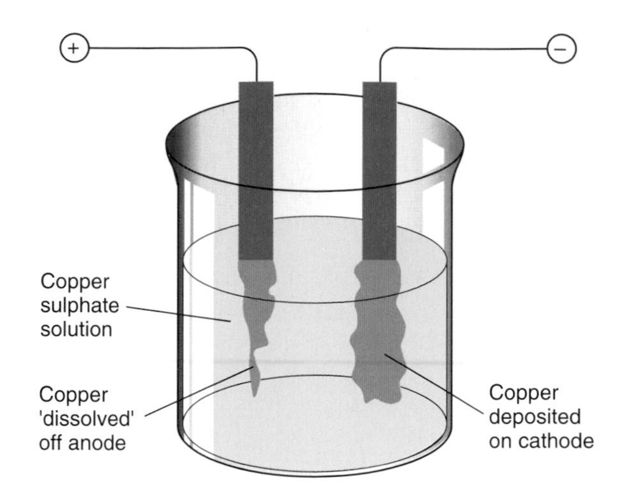

Figure 15.7 Purifying copper by electrolysis

Cathode (−)

$$Cu^{2+} + 2e^- \rightarrow Cu$$

During electrolysis, sulphate ions (SO_4^{2-}) are attracted to the positive anode but they are so stable they are *not* discharged. Instead, copper atoms, which make up the anode, each give up two electrons to form Cu^{2+} ions.

Anode (+)

$$Cu \rightarrow Cu^{2+} + 2e^-$$

The Cu^{2+} ions then go into solution. The overall result of this electrolysis is that the anode loses weight and the cathode gains weight. Copper metal goes into solution at the anode and copper metal is deposited on the cathode.

This method is used industrially to purify impure copper. The impure copper is the anode of the cell. The cathode is a thin sheet of pure copper. The electrolyte is copper sulphate solution. The impure copper anode 'dissolves' away and pure copper deposits on the cathode.

Impure copper anodes being transferred to an electrolysis tank for purification

1 Iron is manufactured by reduction of iron oxide in a blast furnace.

a) Name the **two** substances which are put into the **top** of the blast furnace along with the iron ore.

b) Name the reducing agent in the extraction of iron from iron oxide.

c) Name the main impurity present in the iron produced in a blast furnace.

d) Why is iron **not** manufactured by electrolysis of iron oxide? **NICCEA**

2 Aluminium oxide is an ionic compound. It is found in the Earth's crust in the ore, bauxite. Aluminium oxide is used to make aluminium.

a) i) What holds the aluminium ions (Al^{3+}) and oxide ions (O^{2-}) together in aluminium oxide?
ii) Why has aluminium oxide a very high melting point?

b) Work out the chemical formula of aluminium oxide.

c) To make aluminium, the aluminium oxide is melted and then electricity is passed through it. Why must aluminium oxide be melted before electricity will pass through it?

d) The diagram shows an aluminium cell.
i) Copy the diagram and show where aluminium is formed.

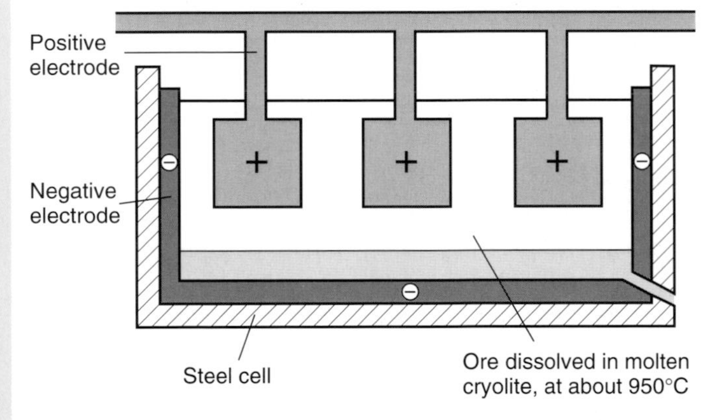

ii) Why do the positive electrodes in this cell need to be replaced at regular intervals?

e) Copy and complete the equation to show how aluminium is formed in this cell.

Al^{3+} .... $e^{-} \rightarrow Al$ **NEAB**

3 A class studied five different metals, labelled A, B, C, D and E. Some of the experiments were carried out by the teacher and some by the pupils. The results of the experiments are shown in the table at the bottom of the page.

a) Use the results of the experiments to place the five metals, A, B, C, D and E, in order of reactivity, starting with the most reactive.

b) Why were the reactions of metal C with steam or with acid **not** carried out?

c) At the end of the lesson the teacher told the class that the five metals were calcium, copper, gold, iron and magnesium. Identify each of the metals A, B, C, D and E.

d) Explain, in terms of their structure, **why** metals are good conductors of heat and electricity. **NEAB**

4 Antimony (Sb) is obtained from ore which contains 15% of antimony sulphide (Sb_2S_3).
The sulphide ore is roasted in air, and antimony oxide is formed. The antimony oxide (Sb_2O_3) is converted to antimony by heating with carbon.

a) How does the information above indicate that antimony is an element and that antimony sulphide is a compound?

b) i) Write a chemical equation, using symbols, for the reaction between antimony oxide and carbon.
ii) What name do we give to reactions of this type?
iii) Calculate the mass of antimony (to the nearest tonne) that could be made from 1000 tonnes of an antimony ore which contains 15% by mass of antimony sulphide. (Relative atomic masses: S = 32, Sb = 122)

c) Antimony is often alloyed with lead to improve its useful properties.
i) What is an alloy?
ii) Suggest **two** ways in which the mechanical properties of the alloy are likely to differ from those of antimony. **MEG**

Table for question 3

Metal	Result of heating in air	Reaction with cold water	Reaction with steam	Reaction with dilute acid
A	Did not burn, oxide formed on surface	No reaction	Slow reaction	Slow reaction
B	No reaction	No reaction	No reaction	No reaction
C	It burned violently, oxide formed.	Bubbles of hydrogen were quickly produced	Not attempted	Not attempted
D	Did not burn, oxide formed on surface	No reaction	No reaction	No reaction
E	It burned quickly, oxide formed.	A few bubbles on the surface of the metal	Vigorous reaction, hydrogen and oxide produced	Dissolved quickly, hydrogen produced

CHAPTER

Atomic structure and the periodic table

16.1 Patterns of elements

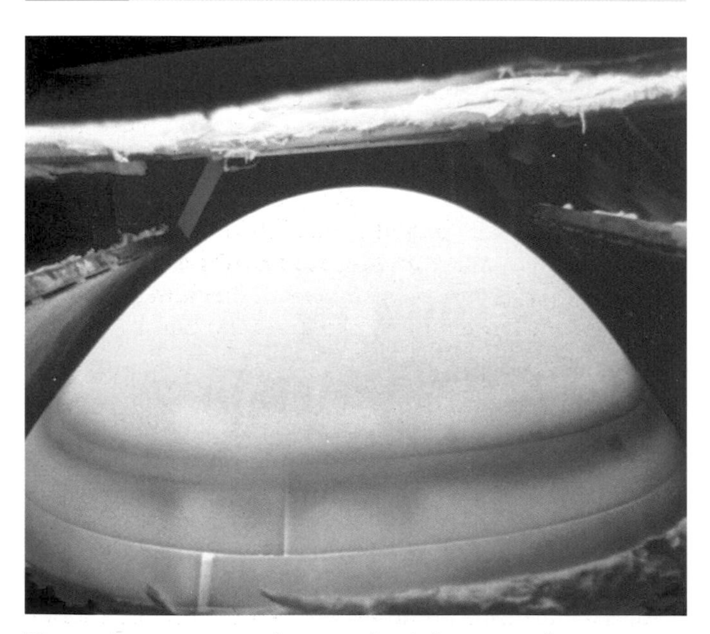

The graphite nose cone of a space shuttle being tested in a furnace. Why is graphite used for the nose cone?

One useful way of classifying elements is as metals and non-metals (section 13.4). Unfortunately, it is not easy to classify some elements in this way. Take, for example, graphite and silicon. These two elements have high melting points and high boiling points (like metals) but they have low densities (like non-metals). They conduct electricity better than non-metals but not as well as metals. Elements with some properties like metals and other properties like non-metals are called **metalloids**.

Because of this difficulty in classifying elements neatly as metals and non-metals, chemists looked for patterns in the properties and reactions of smaller groups of elements.

16.2 Mendeléev's periodic table

During the 19th century, several chemists looked for patterns in the properties of elements. The most successful of these approaches was by the Russian chemist Dmitri Mendeléev in 1869.

- Mendeléev arranged all the known elements in order of their relative atomic masses.
- He also arranged the elements in horizontal rows so that elements with similar properties were in the same vertical column.

Because of the *periodic* repetition of elements with similar properties, Mendeléev called his arrangement a **periodic table**.

Figure 16.1 shows part of Mendeléev's periodic table. Notice that elements with similar properties, such as sodium and potassium, fall in the same vertical column. Which other pairs or trios of similar elements appear in the same vertical column of Mendeléev's table?

In the periodic table,

- The vertical columns of similar elements are called **groups**.
- The horizontal rows of elements are called **periods**.

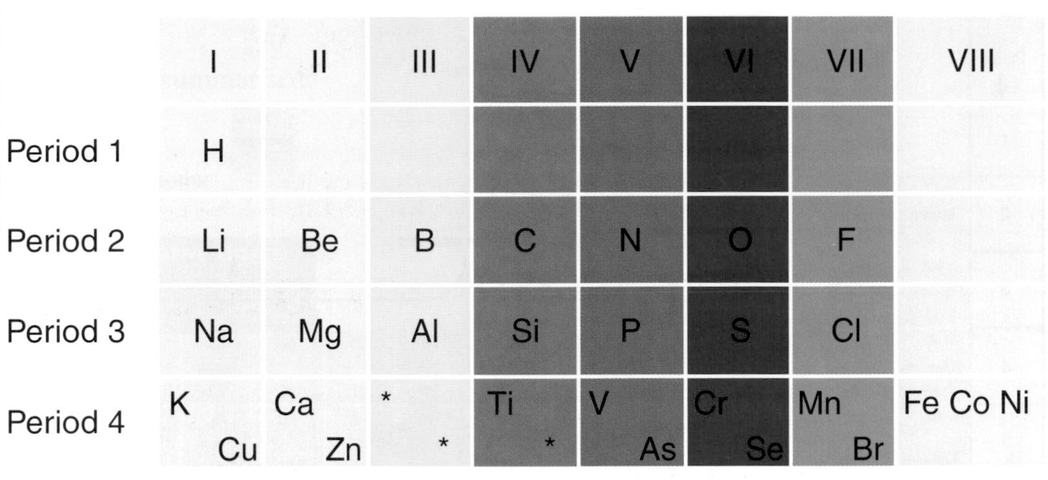

Figure 16.1 Mendeléev's periodic table

Mendeléev had some brilliant and successful ideas in connection with his periodic table.

- He left gaps in his table so that similar elements were in the same vertical group. Three of these gaps are shown as asterisks in Figure 16.1.
- He predicted the properties of the missing elements from the properties of elements above and below them in his table. Within 15 years of his predictions, the missing elements had been discovered. They were called scandium, gallium and germanium. Their properties were very similar to Mendeléev's predictions.

Dmitri Mendeléev was the first chemist to successfully arrange the elements into a pattern linking their properties and relative atomic masses

The success of Mendeléev's predictions showed that his ideas were probably correct. His periodic table was quickly accepted as an important summary of the properties of elements.

16.3 The modern periodic table

The modern periodic table, shown in Figure 16.2 over the page, is based on Mendeléev's. It shows all the known elements numbered along each period, starting with period 1, then period 2, etc. The number given to each element is called its **atomic number**. Thus, hydrogen has an atomic number of 1, helium 2, lithium 3, etc. You will learn more about atomic numbers in section 16.4.

There are several points to note about the modern periodic table.

1. The most obvious difference between modern periodic tables and Mendeléev's is the position of the **transition elements**. These have been taken out of simple groups and placed between group II and group III. Period 4 is the first to contain a series of transition elements. These include chromium, iron, nickel, copper and zinc.
2. Some groups have names as well as numbers. These are shown below the group numbers in Figure 16.2.
3. Metals are clearly separated from non-metals. The 20 or so non-metals are packed into the top right-hand corner above the thick stepped line in Figure 16.2. Some elements close to the steps are classed as metalloids.

So,

atomic number	=	number of protons
mass number	=	number of protons + number of neutrons

So, aluminium atoms, with 13 protons and 14 neutrons, have an atomic number of 13 and a mass number of 27. Sometimes the symbol Z is used for atomic number and the symbol A for mass number. So, for aluminium, Z = 13 and A = 27. Figure 16.4 shows how the mass number and atomic number are often shown with the symbol of an element.

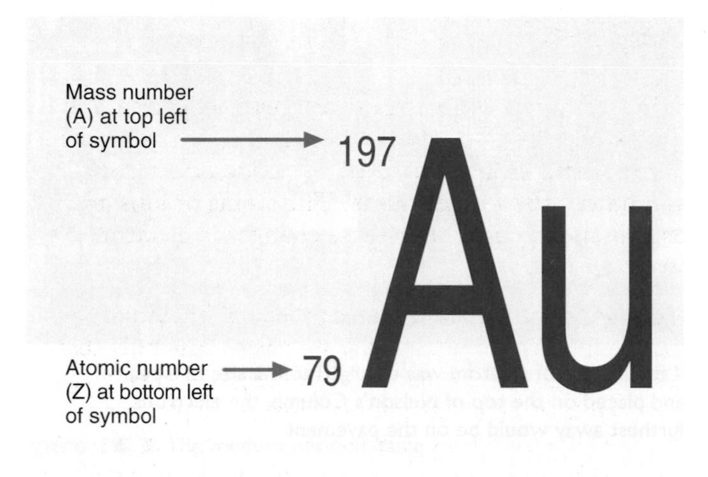

Figure 16.4 The mass number and the atomic number for gold (Au). How many protons, neutrons and electrons are there in a gold atom?

16.6 Isotopes and relative atomic mass

Several elements have relative atomic masses which are whole numbers. For example, the relative atomic mass of carbon is 12.0, that of fluorine is 19.0 and that of sodium is 23.0. This is not surprising, as the mass of an atom depends on the mass of its protons and neutrons, both of which have a relative mass of 1.0. For example, we could calculate the relative mass of fluorine as follows.

$^{19}_{9}F$ atoms have: 9 protons – relative mass = 9.0
9 electrons – relative mass = 0.0
10 neutrons – relative mass = 10.0
$\therefore$ relative atomic mass of $^{19}_{9}F$ = 19.0

Unlike fluorine, carbon and sodium, some elements have relative atomic masses that are nowhere near whole numbers. For example, the relative atomic mass of chlorine is 35.5 and that of copper is 63.5. These unexpected results were explained in 1919 when W.F. Aston built the first mass spectrometer. Using his mass spectrometer, Aston found that one element could have atoms with different masses.

These atoms of the same element with different masses are called **isotopes**. Each isotope has a relative atomic mass which is a whole number, but the average relative atomic mass for the mixture of isotopes is not always a whole number.

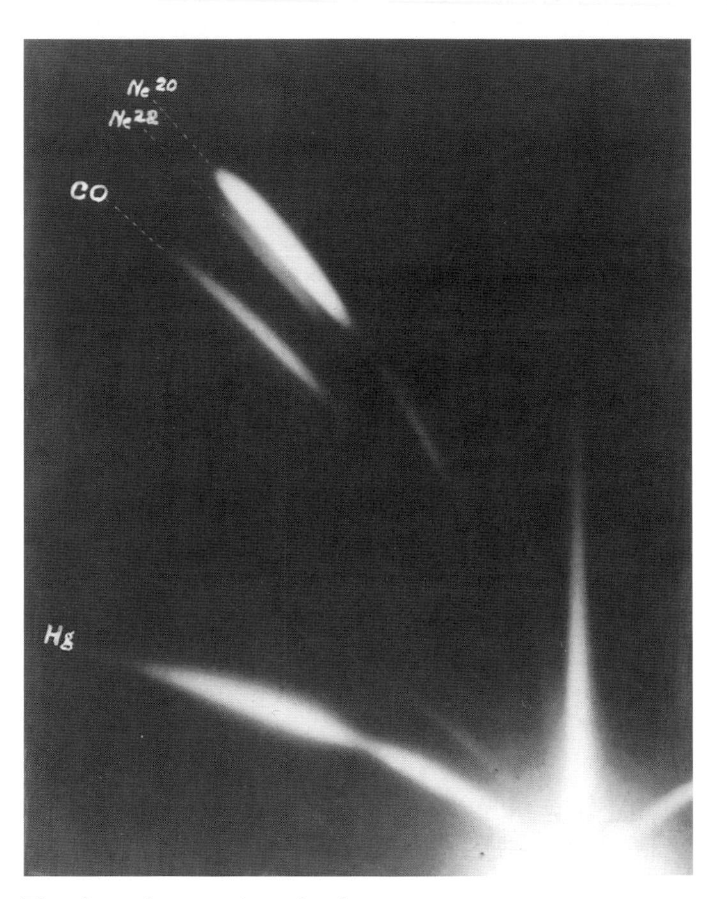

This photo shows evidence for the two isotopes in neon, neon-20 and neon-22. The trace for neon-20 is much thicker than that for neon-22. What does this tell you about the two isotopes? What does CO represent?

Chlorine is a good example of an element with isotopes. Naturally occurring chlorine contains two isotopes, $^{35}_{17}Cl$ called chlorine-35 and $^{37}_{17}Cl$ called chlorine-37. Each of these isotopes has 17 protons and 17 electrons. Therefore, both isotopes have the same atomic number and the *same chemical properties* because these are determined by the number of electrons.

However, one isotope ($^{35}_{17}Cl$) has 18 neutrons and the other ($^{37}_{17}Cl$) has 20 neutrons. Therefore, they have different mass numbers, different masses and hence *different physical properties* because these depend on the masses of atoms and molecules.

The similarities and differences between isotopes of the same element are summarised in Table 16.2.

Table 16.2 The similarities and differences between isotopes of the same element

Isotopes have the same	Isotopes have different
number of protons number of electrons atomic number chemical properties	numbers of neutrons mass numbers physical properties

Isotopes are atoms with the same atomic number, but different mass numbers.

16.7 Calculating relative atomic masses

The relative atomic mass of an element is the average mass of one atom. This can be calculated from the relative masses of its isotopes and their relative proportions.

Look closely at Figure 16.5 showing a mass spectrometer trace for chlorine. The trace shows that chlorine contains two isotopes, with mass numbers of 35 and 37.

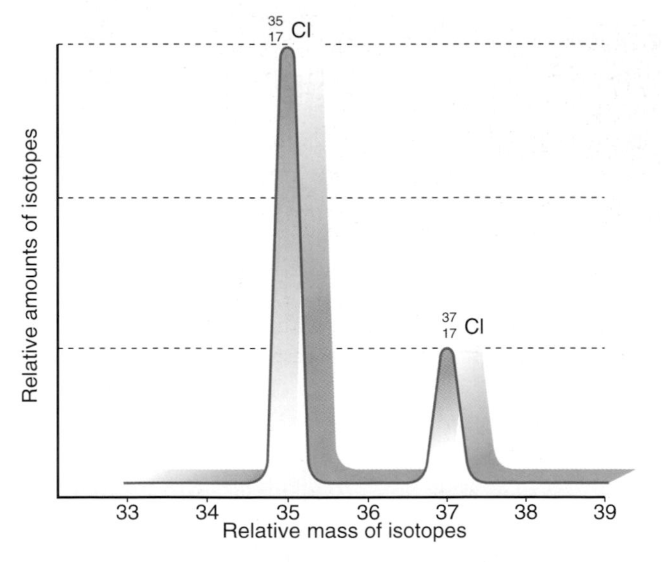

Figure 16.5 A mass spectrometer trace for chlorine. What are the relative amounts of $^{35}_{17}$Cl and $^{37}_{17}$Cl?

If chlorine contained 100% $^{35}_{17}$Cl, its relative atomic mass would be 35.
If it contained 100% $^{37}_{17}$Cl, its relative atomic mass would be 37.
If it contained 50% $^{35}_{17}$Cl and 50% $^{37}_{17}$Cl, the relative atomic mass would be:

$$\left(\frac{50}{100} \times 35\right) + \left(\frac{50}{100} \times 37\right)$$
$$= 17.5 + 18.5$$
$$= 36$$

Now, Figure 16.5 shows that chlorine contains three times as much $^{35}_{17}$Cl as $^{37}_{17}$Cl i.e. the percentages of the two isotopes are 75% to 25%. Therefore the atomic mass of chlorine

$$= \left(\frac{75}{100} \times 35\right) + \left(\frac{25}{100} \times 37\right)$$
$$= 26.25 + 9.25$$
$$= 35.5$$

16.8 Electron structures and the periodic table

When chemical reactions occur, there are changes in the number of electrons belonging to different atoms. Some atoms **gain electrons**, some **lose electrons** and others **share electrons**. The atoms or ions are usually more stable after these changes in electron structure.

However, some atoms, such as those of helium and neon, never react. This led scientists to conclude that atoms of helium and the other noble gases must have very stable electron structures. Further experiments have shown that atoms and ions have very stable electron structures if they have 2 electrons, like helium, 10 electrons like neon, 18 electrons like argon and so on. These stable electron structures are closely related to the way in which electrons are arranged in layers or **shells**. We now know that:

- the first shell is stable when it contains 2 electrons,
- the second shell is stable when it contains 8 electrons,
- the third shell is stable when it contains 8 electrons.

Thus, neon atoms are stable because they have 10 electrons, 2 in the first shell and 8 in the second shell. The electron structure of neon is written as 2,8.

Argon, the next noble gas is stable because its atoms have 18 electrons; 2 in the first shell, 8 in the second and 8 in the third. The electron structure of argon is therefore written as 2,8,8.

Figure 16.6 over the page shows the electron structures for the first twenty elements in the periodic table. When the first shell is full (at helium), electrons go into the second shell. So the electron structure of lithium is 2,1, beryllium is 2,2, boron is 2,3 and so on. When the second shell is full (at neon) electrons go into the third shell. So, the electron structure of sodium is 2,8,1, magnesium is 2,8,2 and so on.

Using these electron structures, in the next four sections we will explain the chemical properties of some of the elements.

Period 1	H							He
electron structure	1							2
Period 2	Li	Be	B	C	N	O	F	Ne
electron structure	2,1	2,2	2,3	2,4	2,5	2,6	2,7	2,8
Period 3	Na	Mg	Al	Si	P	S	Cl	Ar
electron structure	2,8,1	2,8,2	2,8,3	2,8,4	2,8,5	2,8,6	2,8,7	2,8,8
Period 4	K	Ca						
electron structure	2,8,8,1	2,8,8,2						

Figure 16.6 Electron structures of the first 20 elements in the periodic table

16.9 The alkali metals

Group I
Li lithium
Na sodium
K potassium
Rb rubidium
Cs caesium
Fr francium

Figure 16.7 The alkali metals

The elements in group I are called **alkali metals** (Figure 16.7) because they react with water to form alkaline solutions. The alkali metals are so reactive that they must be stored under oil to prevent them reacting with the oxygen and water vapour in the air. Some of the properties and reactions of the alkali metals are summarised in Table 16.3.

Property	Property/reaction
Appearance	Shiny when freshly cut, but quickly form a dull layer of oxide
Hardness	Soft metals. Li, Na and K can be cut with a knife, but Rb and Cs are not so soft.
Melting and boiling points	Low compared with other metals. Melting and boiling points fall from Li to Fr
Density	Low compared to other metals. Li, Na and K will float on water.
Reaction with air	Burn with increasing vigour from Li down the group forming white oxides, e.g. sodium + oxygen → sodium oxide $4Na + O_2 \rightarrow 2Na_2O$
Reaction with water	Li reacts steadily, sodium vigorously, potassium violently. The products are an alkaline solution of the metal hydroxide and hydrogen, e.g. sodium + water → sodium hydroxide + hydrogen $2Na + 2H_2O \rightarrow 2NaOH + H_2$

Table 16.3 Properties and reactions of the alkali metals

Notice three important points from Table 16.3.

1. All alkali metals have very similar properties and reactions.
2. The alkali metals have much lower melting points, boiling points and densities and are much softer than other metals.
3. The reactions of the alkali metals with air and water show that the elements get more reactive from lithium to potassium.

These properties illustrate an important feature of the periodic table.

> Although elements in a group have similar properties, there is a gradual change in the properties of the elements from the top to the bottom of the group.

This feature enables us to predict the properties of different elements.

Each alkali metal appears immediately after a noble gas in the periodic table. So, alkali metals have one electron in their outer shell (Figure 16.6). By losing this outer electron, their atoms form positive ions (Li^+, Na^+, K^+, etc.) with stable electron structures, like a noble gas. For example, Li^+ ions have an electron structure like helium and Na^+ ions have an electron structure like neon. Alkali metals are so reactive because they are keen and able to lose their outer electron very easily.

Compounds of the alkali metals

Compounds of the alkali metals are soluble in water. Their solid compounds are white, forming colourless solutions unless the anion present is coloured. The most important sodium compounds are sodium chloride and sodium hydroxide.

Rock salt being mined in the USA

The yellow glow from this street light is produced by a sodium vapour lamp

In hot countries, sodium chloride (salt) is left when sea water is allowed to evaporate. Sodium chloride also occurs in salt beds below the ground. Impure salt is used for de-icing roads. Pure salt is used in cooking. Sodium and chlorine are obtained by electrolysing molten salt. Liquid sodium is used as a coolant in fast nuclear reactors and sodium vapour provides the yellow glow in street lamps.

Sodium hydroxide has been added to wood pulp in this photo to remove gums and resins in the manufacture of paper.

Sodium hydroxide is used in large amounts as an industrial alkali and to make soap, paper, rayon and other cellulose fibres. Paper, rayon and cellulose fibres are all made from wood. The wood is made into pulp and soaked in sodium hydroxide solution. This removes gums and resins, leaving the natural fibres of cellulose.

Soaps and soap powders are made by boiling oils and fats with sodium hydroxide. The oils and fats are converted to complex sodium compounds which make up soap. The equation for this process is:

$$\begin{array}{l} CH_2-O-\underset{\underset{O}{\|}}{C}-\text{WW} \\ \;| \\ CH-O-\underset{\underset{O}{\|}}{C}-\text{WW} \\ \;| \\ CH_2-O-\underset{\underset{O}{\|}}{C}-\text{WW} \end{array} + 3NaOH \rightarrow 3Na^{+}\,{}^{-}O-\underset{\underset{O}{\|}}{C}-\text{WW} + \begin{array}{c} CH_2OH \\ | \\ CHOH \\ | \\ CH_2OH \end{array}$$

Oil or fat — Soap (complex sodium compound) — Glycerine (glycerol)

16.10 The halogens

The elements in group VII of the periodic table are called **halogens** (Figure 16.8). They form a group of reactive non-metals. In fact, the halogens are so reactive that they never occur as pure elements in nature, but are combined with metals in salts such as sodium chloride (NaCl), calcium fluoride (CaF_2) and magnesium bromide ($MgBr_2$). This gives rise to the name halogens which means 'salt-formers'.

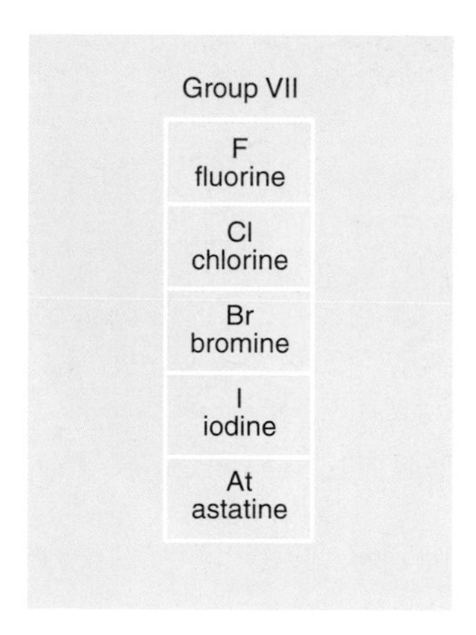

Figure 16.8 The halogens

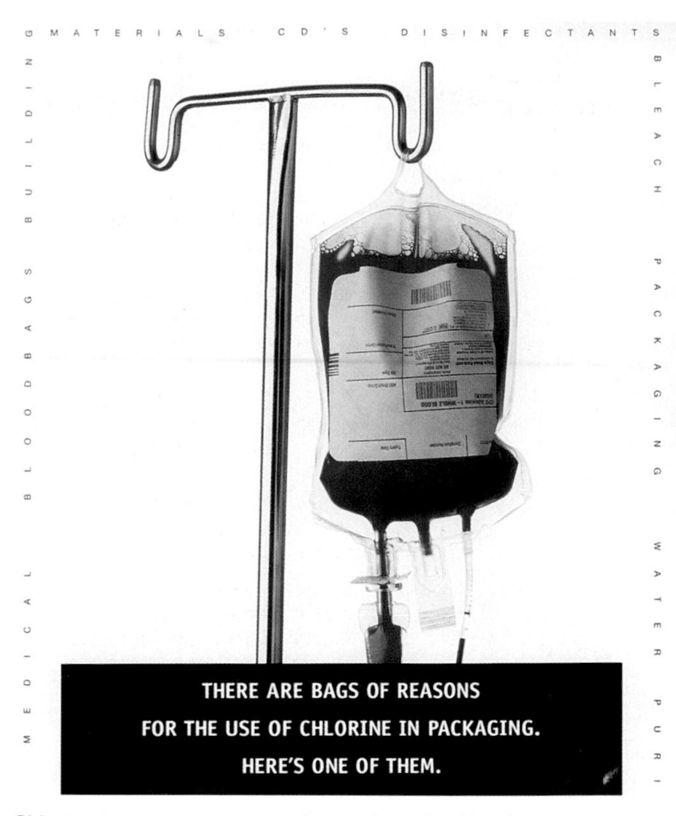

Chlorine is a very important chemical used in bleaches, degreasing agents, pesticides, disinfectants and in the manufacture of PVC. Because blood bags have to be non-biodegradable, they are made of PVC

Trends in properties

Table 16.4 lists some physical properties of the first four halogens – fluorine, chlorine, bromine and iodine. Look at Table 16.4. How do the following properties of the halogens change as their relative atomic mass increases?

(i) state at room temperature, (ii) colour,
(iii) melting point, (iv) boiling point?

All the halogens exist as **simple molecules** – F_2, Cl_2, Br_2, I_2. Strong bonds hold the two atoms together as a molecule, but the bonds between separate molecules are very weak. This means that the molecules are easily separated, so they have low melting points and low boiling points.

Moving down the group, the halogen molecules get larger and heavier. Therefore, from fluorine to iodine, they are gradually more difficult to melt and vaporize. This results in increasing melting points and boiling points.

Reactions of chlorine and the halogens

The chemical reactions of the halogens are similar, with gradual changes from one element to the next down the group. Fluorine is the most reactive of all non-metals. Chlorine is also very reactive, whilst iodine is moderately reactive.

Chlorine is the most important element in group VII. Some of the uses of chlorine are shown in Figure 16.9.

Table 16.4 Some properties of fluorine, chlorine, bromine and iodine

Element	Relative atomic mass	State at room temperature	Colour	Structure	Melting point (°C)	Boiling point (°C)
Fluorine	19.0	gas	pale yellow	F_2 molecules	−220	−188
Chlorine	35.5	gas	pale green	Cl_2 molecules	−101	−35
Bromine	79.9	liquid	red brown	Br_2 molecules	−7	58
Iodine	126.9	solid	dark grey	I_2 molecules	114	183

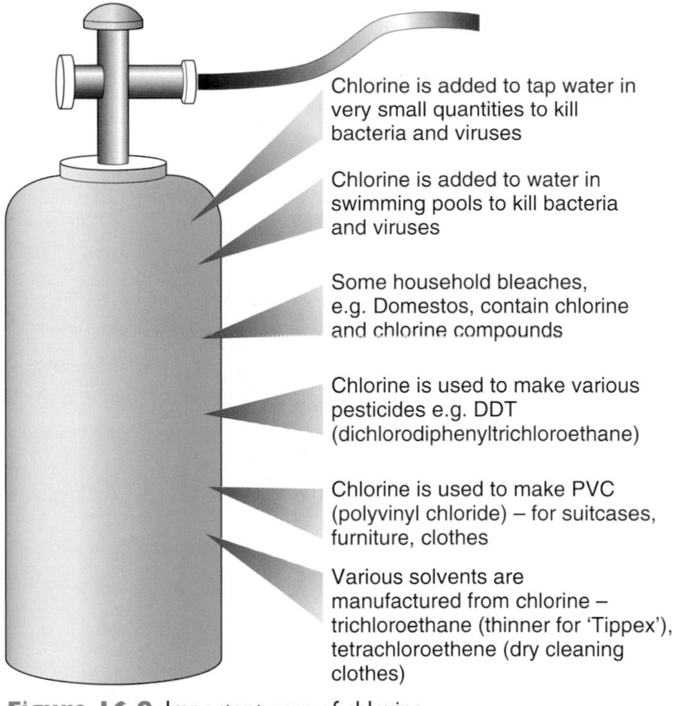

Figure 16.9 Important uses of chlorine

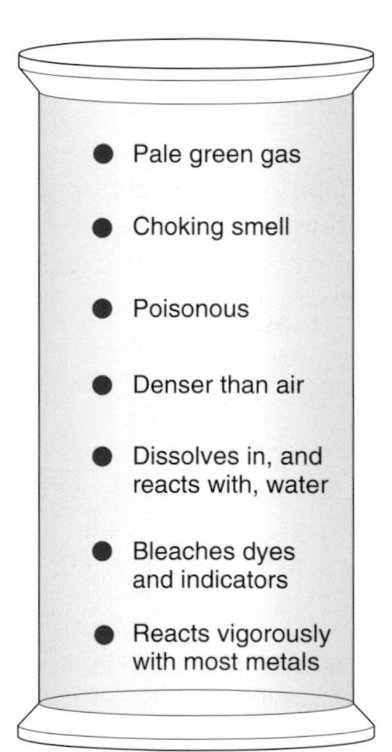

The important properties of chlorine are summarised in Figure 16.10. Remember that **chlorine is poisonous** (toxic) with a choking smell. Always treat it with care! Experiments with chlorine should always be done in a fume cupboard unless very small quantities are involved.

Figure 16.10 Important properties of chlorine

Test for chlorine

The best test for chlorine uses the fact that **chlorine bleaches moist litmus paper**. If blue litmus paper is used, the paper first turns red (because chlorine reacts with water to form an acidic solution) and then the paper goes white.

$$\text{chlorine} + \text{water} \rightarrow \text{hydrochloric acid} + \text{hypochlorous acid}$$

$$Cl_2 + H_2O \rightarrow HCl + HClO$$

Reactions with iron

The reactions between halogens and iron can be studied using the apparatus in Figure 16.11. If you try these experiments, **wear eye protection and use a fume cupboard**.

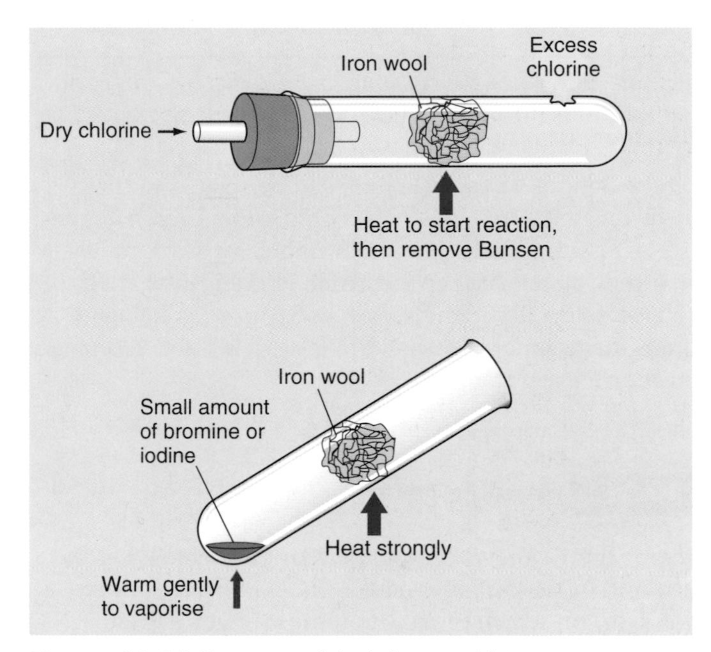

Figure 16.11 Reactions of the halogens with iron

When iron is heated in chlorine, it reacts with a bright glow forming brown iron(III) chloride.

$$\text{iron} + \text{chlorine} \rightarrow \text{iron(III) chloride}$$
$$2Fe + 3Cl_2 \rightarrow 2FeCl_3$$

When iron is heated in bromine vapour, the reaction is less vigorous forming iron(III) bromide ($FeBr_3$). With iodine, the iron reacts very slowly to form iron(II) iodide (FeI_2).

Notice that the halogens react with iron and with other metals to form salts – chlorides, bromides and iodides. Notice also that *the halogens get less reactive as their relative atomic mass increases*. This is opposite to the trend in group I where the alkali metals get more reactive with increasing relative atomic mass.

Displacement reactions

The relative reactivity of the halogens is confirmed by the fact that a more reactive halogen will displace a less reactive one from its compounds. Thus, chlorine will displace bromine from sodium bromide solution. The more reactive chlorine forms sodium chloride, displacing the less reactive bromine, i.e.

$$Cl_2 + 2NaBr \rightarrow 2NaCl + Br_2$$

However, the less reactive bromine will not displace the more reactive chlorine from sodium chloride solution, i.e.

$$Br_2 + 2NaCl \rightarrow \text{no reaction}$$

Halogen	Atom	Electron structure of atom	Halide	Ion	Electron structure of ion
Fluorine	F	2,7	Fluoride	F^-	2,8
Chlorine	Cl	2,8,7	Chloride	Cl^-	2,8,8
Bromine	Br	2,8,18,7	Bromide	Br^-	2,8,18,8

Table 16.5 Electron structures of the atoms and ions of the first three halogens

Electron structures

The reactions of the halogens can be related to their electron structures. In the periodic table, each halogen is placed immediately before a noble gas. This means that they all have seven electrons in their outer shell (Table 16.5). By gaining one electron, each halogen atom forms a negative halide ion with a stable electron structure like a noble gas.

16.11 The noble gases

Figure 16.12 shows the position of the noble gases in group 0 of the periodic table. The commonest noble gas is argon which makes up almost 1% of dry air. Neon, krypton and xenon occur in smaller quantities in the air. These four noble gases are obtained industrially during fractional distillation of liquid air. There are only minute traces of helium in the air. It is, therefore, more economical to extract helium from natural gas.

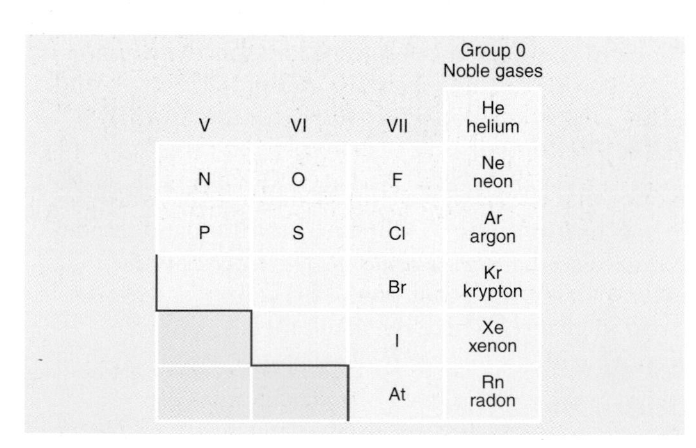

Figure 16.12 The position of the noble gases in group 0 of the periodic table

Properties of the noble gases

The noble gases are all colourless and very unreactive. As expected, their melting points, boiling points and densities show a steady change as the relative atomic mass increases. The graph in Figure 16.13 shows the trend in their boiling points and densities with increasing relative atomic mass.

The noble gases all exist as separate single atoms. Until 1962, no compounds of the noble gases were known and chemists thought they were completely unreactive. This is why they were initially called the *inert* gases. Nowadays, several compounds of them are known and the name inert has been replaced by **noble**. This compares to the naming of unreactive metals like gold, as noble metals.

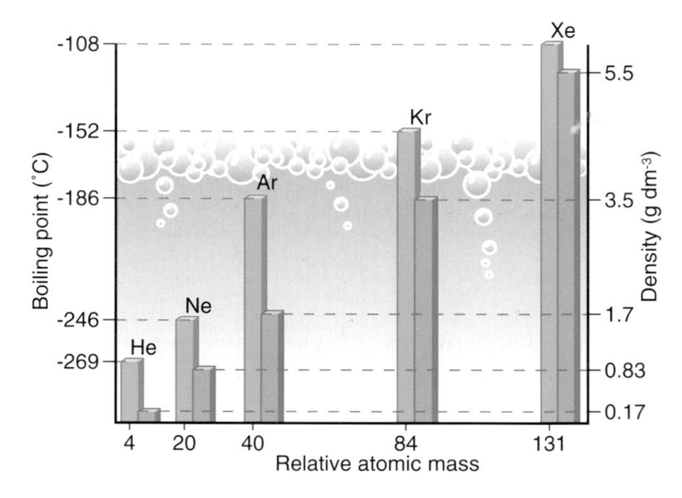

Figure 16.13 A graph showing the increase in boiling point and density of the noble gases as their relative atomic masses increase

Uses of the noble gases

A meteorological balloon being launched in Finland

Table 16.6 shows the important uses of the noble gases with reasons for these uses.

Table 16.6 Important uses of the noble gases

Noble gas	Use	Reason for the use
Helium	in weather balloons	very low density, non flammable
Neon	in neon lights	neon gives a bright glow when an electric discharge (spark) passes through it
Argon	during welding and in lasers	argon provides an inert atmosphere so that during the welding or laser process there is no reaction with oxygen in the air
Argon and krypton	in electric light bulbs	white hot tungsten filaments in electric lights will not react with argon or krypton

16.12 The transition metals

Figure 16.14 shows the position of transition metals in the periodic table. They lie between the reactive metals in groups I and II and the poor metals in groups III and IV. Symbols of the more common transition metals are also shown in Figure 16.14.

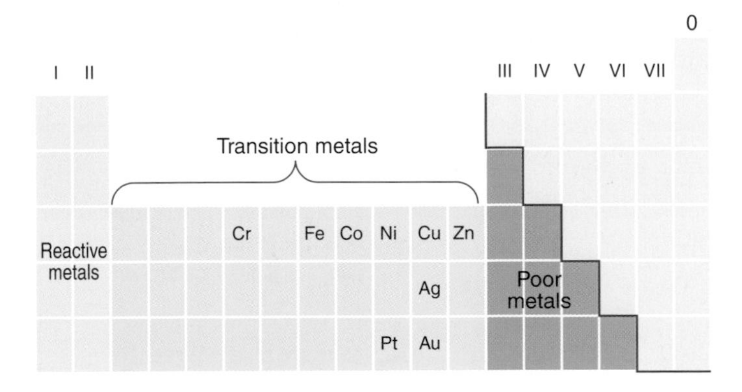

Figure 16.14 The position of transition metals in the periodic table

> The transition metals have similar properties. Unlike other parts of the periodic table, there are similarities throughout the whole block of transition metals across the periods as well as down the groups.

Uses of the transition metals

The most important transition metals are iron and copper. Iron is the most widely used metal. More than 700 million tonnes of it are manufactured each year. Almost all of this is converted to steel which is hard, strong and relatively cheap. Steel is used in girders and supports for bridges and buildings, in vehicles, in engines and in tools.

The Howrah Bridge, Calcutta, India. Large amounts of iron (steel) are used in building bridges.

After iron and aluminium, copper is the third most widely used metal. About 9 million tonnes are manufactured each year. Copper is a good conductor of heat and electricity. It is also malleable and can be made into different shapes and drawn into wires. Because of these properties, copper is used in electrical wires and cables and in hot water pipes and radiators.

The uses of copper are increased by **alloying** it with other metals. Alloys of copper and zinc produce brass which is harder and stronger than pure copper. Alloys of copper and tin produce bronze. This is stronger than copper and also easier to cast into moulds.

Why are distillery vessels made of copper?

Properties of transition metals

Some properties of iron and copper are shown in Table 16.7. Compare these with the properties of group I metals in Table 16.3.

The information in Table 16.7 illustrates the characteristic properties of transition metals and these can be summarised as follows.

- **High melting points and high boiling points** – much higher than alkali metals.
- **High densities** – much higher than alkali metals.
- **Unreactive with water**, unlike alkali metals. None of the transition metals react with cold water, but a few of them, like iron, react slowly with steam.
- **More than one ion.** Most of the transition metals form more than one stable ion, each with its own series of compounds. For example, iron forms Fe^{2+} and Fe^{3+} ions resulting in iron(II) and iron(III) compounds. Copper forms Cu^{+} and Cu^{2+} ions resulting in copper(I) and copper(II) compounds. Alkali metals form only one stable ion with a charge of 1+.
- **Coloured compounds.** Transition metals usually have coloured compounds due to the colour of their ions. In contrast, alkali metals form white compounds with colourless solutions, unless the constituent anions are coloured.
- **Catalytic properties.** Transition metals and their compounds can act as **catalysts.** For example, iron or iron(III) oxide is a catalyst in the Haber process, (see section 21.3) platinum or vanadium(V) oxide is a catalyst for the manufacture of sulphuric acid and nickel is a catalyst for the production of margarine from oils.

Table 16.7 Some properties of iron and copper

Element	Melting point (°C)	Boiling point (°C)	Density ($g\ cm^{-3}$)	Reaction with water	Formulas of oxides	Symbols of ions	Colour of salts
Iron	1540	3000	7.9	does not react with pure water but reacts slowly with steam	FeO Fe_2O_3	Fe^{2+} Fe^{3+}	Fe^{2+} salts are green Fe^{3+} salts are yellow or brown
Copper	1080	2600	8.9	no reaction with water or steam	Cu_2O CuO	Cu^{+} Cu^{2+}	Cu^{2+} salts are blue or green

1 This question is about sodium chloride (common salt) which is an important chemical. Sodium chloride can be made by burning sodium in chlorine gas.

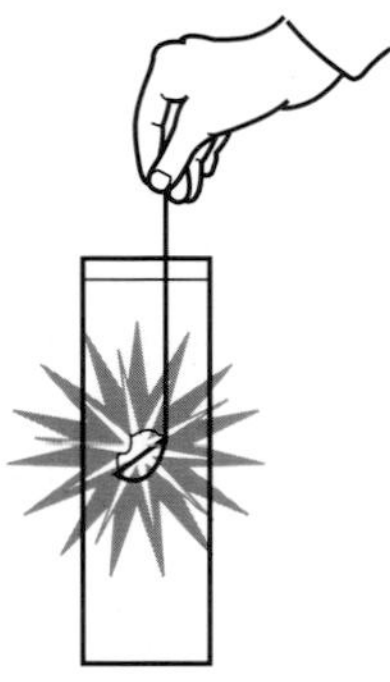

a) Copy the equation, balancing it for the reaction of sodium with chlorine.

$$\ldots\ldots\ldots\ldots Na + Cl_2 \rightarrow \ldots\ldots\ldots\ldots NaCl$$

b) i) Copy and complete the diagrams to show the electronic structure of a sodium atom and a chlorine atom.

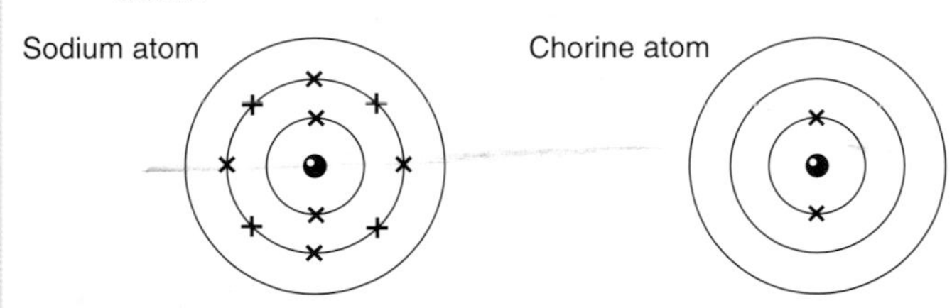

ii) How does a sodium atom change into a sodium ion?

c) The apparatus shown below can be used to electrolyse sodium chloride solution.

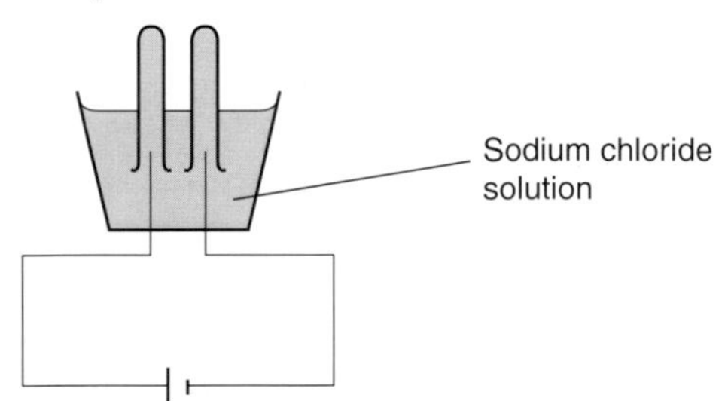

i) Name the product formed at:
1. the positive electrode,
2. the negative electrode.
ii) Give **one** large-scale use of the product formed at the negative electrode.
iii) The final solution contains Na^+ ions and OH^- ions. Name the useful chemical that could be obtained from this solution.

d) The element potassium is in the same group of the periodic table as sodium. Potassium reacts with chlorine to make potassium chloride. This is sometimes used in cooking instead of common salt.
i) Give the chemical formula of potassium chloride.
ii) By reference to the electronic structures of sodium and potassium, explain why the reaction of sodium with chlorine is similar to the reaction of potassium with chlorine.

e) Fluorine is the most reactive element in group VII of the periodic table. It reacts with all the other elements in the periodic table except some of the noble gases. It does not react with helium, argon and neon, but it does react with xenon (Xe) to form xenon fluoride.
i) Explain why fluorine is more reactive than chlorine.
ii) Explain why the noble gases are generally unreactive.
iii) Predict, with reasons, whether Radon (Rn) will react with fluorine. **NEAB**

2 a) Chlorine, bromine and iodine are three of the halogen elements that occur in group VII of the periodic table. The following table gives some data for these elements.

Element	Melting point (temperature)(°C)	Boiling point (temperature)(°C)
Chlorine	−101	−35
Bromine	−7	59
Iodine	114	184

There is another halogen element called astatine which has a higher atomic number and follows iodine in group VII and is very unstable. Would you expect astatine to be a gas, liquid or solid at room temperature (20 °C)?

b) The atoms of the elements hydrogen, chlorine and sodium, are represented by the symbols $^{1}_{1}H$; $^{35}_{17}Cl$ and $^{23}_{11}Na$, respectively.
Give the electronic structure of hydrogen, chlorine and sodium. **WJEC**

3 a) Give the symbol of:
i) A metal.
ii) A non metal.
iii) The first element in group II.
iv) An element which consists of molecules which are made up of pairs of atoms.
v) An element which forms a hydroxide which dissolves in water to give an alkaline solution.
vi) An element which forms an ion of the type X^-.

b) The salt sodium hydrogen phosphate, (Na_2HPO_4), is used as a softening agent in processed cheese. The salt can be made by reacting phosphoric acid (H_3PO_4) with an alkali.
i) Complete the name of an alkali that could react with phosphoric acid to make sodium hydrogen phosphate.

_______________ hydroxide

ii) What name is given to a reaction in which an acid reacts with an alkali to make a salt?
iii) How would the pH change when alkali is added to the phosphoric acid solution? **NEAB**

CHAPTER

The structure and bonding of materials

17.1 Studying structures

What is the general shape of salt (sodium chloride) crystals in this photo?

Look at the crystals of sodium chloride in this photograph. What do you notice about all the salt crystals? All the salt crystals are roughly the same cubic shape. Further studies show that *all the crystals of one substance have similar shapes*. This suggests that the particles in the crystals are always packed in a regular fashion to give the same overall shape. Sometimes, crystals grow unevenly and their shapes become distorted. Even so, it is usually easy to see their general shape. Solid substances which have a regular packing of particles are described as **crystalline**. The particles may be atoms, ions or molecules. Figure 17.1 shows how cubic crystals and hexagonal crystals can form. If the particles are always placed in parallel lines or at 90° to each other, the crystal will be cubic. If the particles are placed at 120° in the shape of a hexagon, the final crystal will be hexagonal.

Figure 17.1

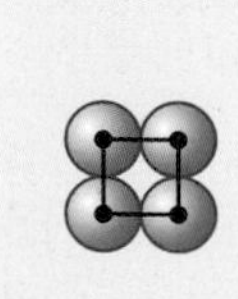

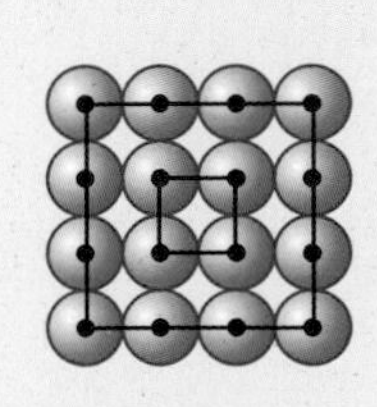

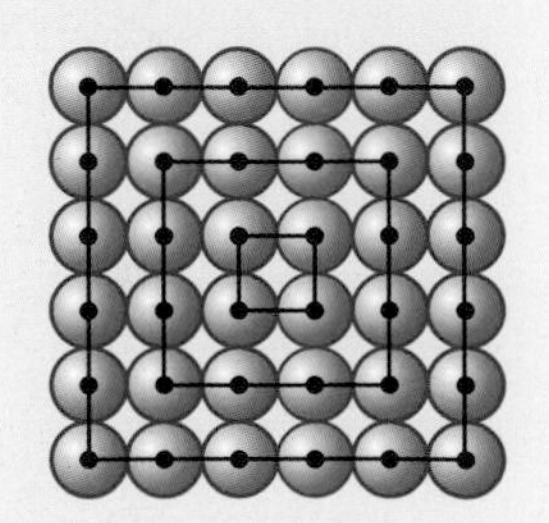

When the particles are arranged in a cubic fashion, the final crystal will be cubic

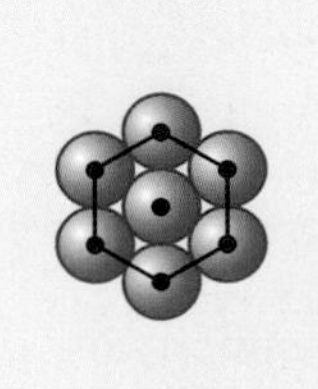

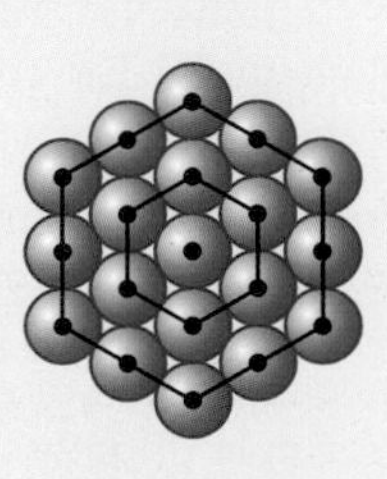

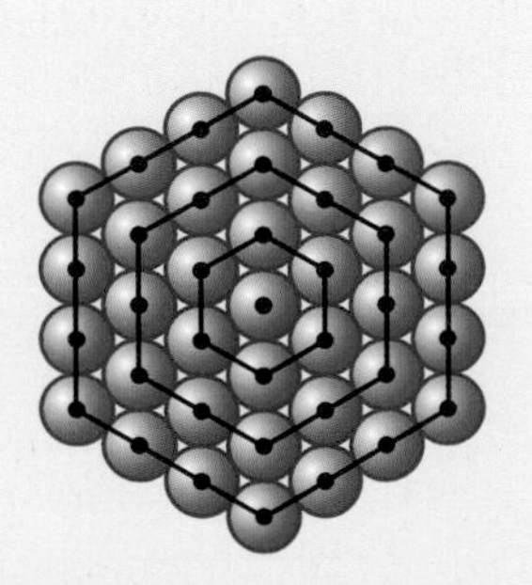

When the particles are arranged in a hexagonal fashion, the final crystal will be hexagonal

We can compare the way in which a crystal grows to the way in which a bricklayer lays bricks. If the bricklayer always places the bricks in parallel lines or at 90° to each other, then the final buildings will be like cubes or boxes. However, if the bricks are laid at 120° to make hexagons, then the final buildings will be hexagonal.

The overall shape of a crystal can only give a clue to the way in which the particles are arranged. X-rays give much better evidence.

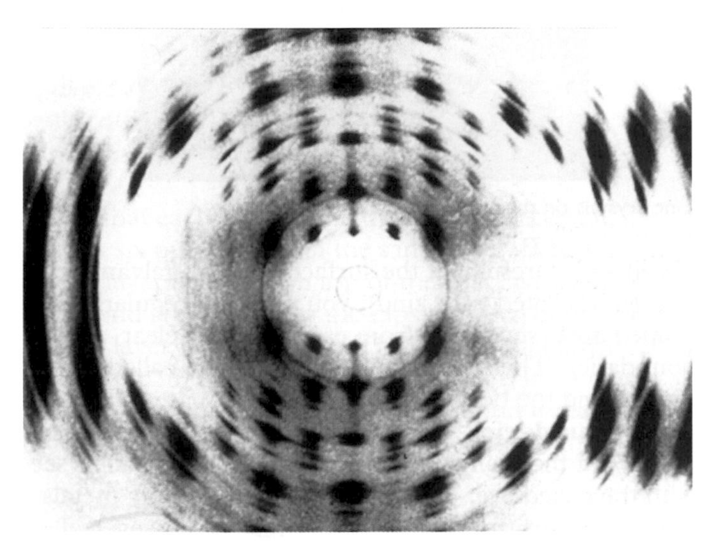

An X-ray diffraction photo of DNA, the chemical that makes up our genes. Notice the general pattern in the dots

Using X-rays to study crystals

Look through a piece of thin stretched cloth at a small bright light. The pattern you see is due to the deflection of the light as it passes through the regularly spaced threads of the fabric. This deflection of the light is called **diffraction** and the patterns produced are **diffraction patterns**. If the cloth is stretched so that the threads in the fabric get closer, then the pattern spreads further out. From the diffraction pattern which we *can* see, we can work out the pattern of the threads in the fabric which we *cannot* see. The same idea is used to work out how the particles are arranged in a crystal.

A narrow beam of X-rays is directed at a well-formed crystal (Figure 17.2). Some of the X-rays are diffracted by particles in the crystal onto X-ray sensitive film. When the film is developed, a regular pattern of spots appears. This is the diffraction pattern for the crystal. From the diffraction pattern which we *can* see, it is possible to work out the pattern of particles in the crystal which we *cannot* see. A regular arrangement of spots on the film indicates a regular arrangement of particles in the crystal. This regular arrangement of particles in the crystal is called a **lattice**.

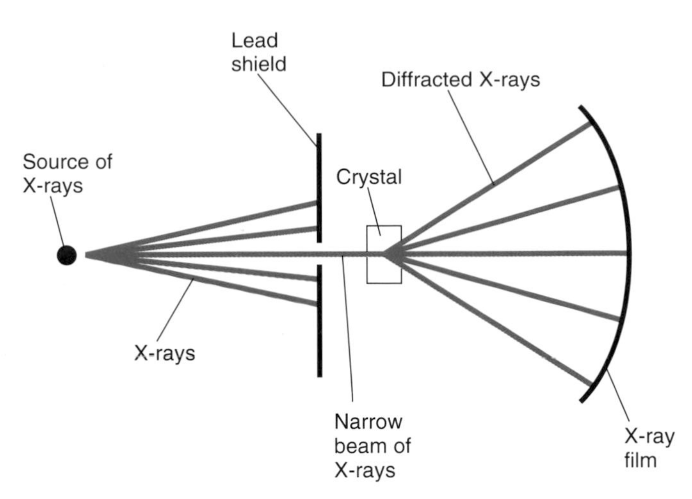

Figure 17.2 The diffraction of X-rays by a crystal

X-rays have been used in this way to study the structure of thousands of different solids. Beams of electrons can also be used, like X-rays, to study the way in which particles are arranged in crystals.

17.2 The structure of substances

What properties must the material in nappies have? What do you think the structure of nappy material is like?

Why was metal used to make suits of armour in the Middle Ages?

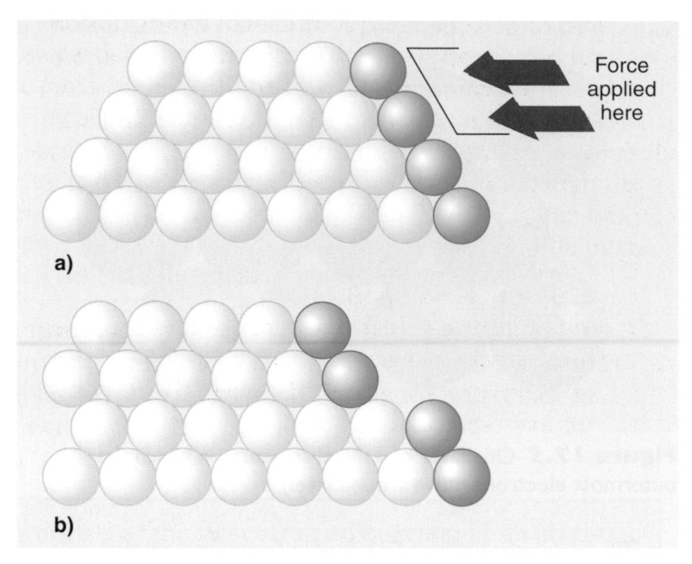

Figure 17.7 The positions of atoms in a metal crystal, (a) before and (b) after slip has occurred

17.4 Giant covalent structures

In giant covalent structures, like diamond, polythene and sand (silicon dioxide), strong covalent bonds (see section 17.9) join one atom to another in very large molecules containing thousands or even millions of atoms.

In diamond, each carbon atom is joined to four others (Figure 17.8). Each carbon atom is at the centre of a tetrahedron with four other carbon atoms at the corners of the tetrahedron.

Figure 17.8 A model of the structure of diamond

Every carbon atom shares its four outer electrons, one with each of its four neighbours forming strong covalent bonds. The covalent bonds extend through the whole diamond forming a three-dimensional giant covalent structure. Thus, a diamond is a **single giant molecule** or a **macromolecule**. Only a small number of atoms are shown in the model in Figure 17.8. In a real diamond, there are billions of atoms.

The properties of diamond

- **Diamond is very hard** because its carbon atoms are linked by very strong covalent bonds. Another reason for its hardness is that the atoms are *not* arranged in layers so they cannot slide over one another like the atoms in metals. In fact, diamond is the hardest known natural substance. Most of its industrial uses depend on this hardness.

Diamonds which are not good enough for gems are used as glass cutters and in diamond-studded saws. This photo shows a craftsman using a diamond cutter to add decorative stars to the pattern on this dish

- **Diamond has a very high melting point** because of the strong covalent bonds linking carbon atoms in a giant structure. This means that the atoms cannot vibrate fast enough to break away from their neighbours until very high temperatures are reached.
- **Diamond does not conduct electricity.** Unlike metals, diamond has no free electrons because all four electrons in the outer shell of each carbon atom are held firmly in covalent bonds. So in diamond there are no free electrons to form an electric current.

17.5 Giant ionic structures

Chalk cliffs are composed of an ionic compound, calcium carbonate. This contains calcium ions (Ca^{2+}) and carbonate ions (CO_3^{2-})

Ionic compounds form when metals react with non-metals. For example, when sodium burns in chlorine, sodium chloride is formed.

2Na	+	Cl_2	→	$2Na^+Cl^-$
two sodium atoms	+	one chlorine molecule	→	two sodium ions two chloride ions

In solid ionic compounds, the ions are held together by the attraction between positive ions and negative ions. Figure 17.9 shows how the ions are arranged in one layer of sodium chloride and Figure 17.10 is a three-dimensional model of the structure of sodium chloride. Notice that Na^+ ions are surrounded by Cl^- ions and vice versa.

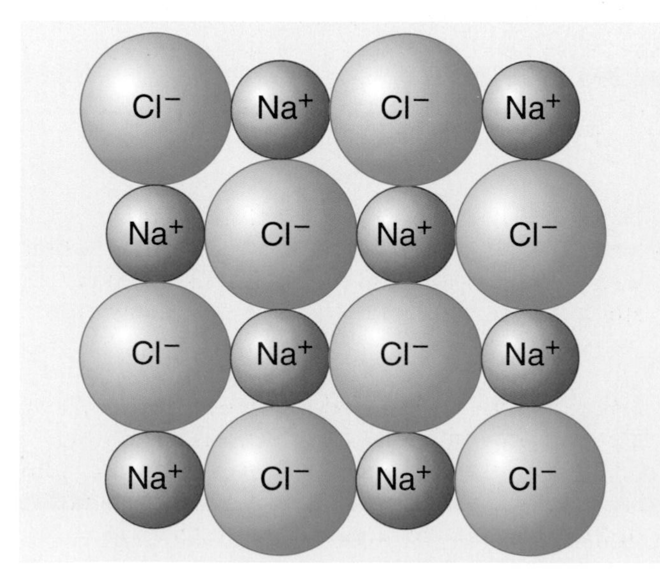

Figure 17.9 The arrangement of ions in one layer of a sodium chloride crystal

This kind of arrangement in which large numbers of ions are packed together in a regular pattern is another example of a **giant structure**. The force of attraction between oppositely charged ions is called an **ionic** or **electrovalent bond**. The strong ionic bonds hold the ions together very firmly. This explains why ionic compounds:

- are **hard** substances,
- have **high melting points**,
- **cannot conduct electricity** when solid because the ions cannot move freely, but conduct electricity when molten or aqueous as the charged ion can then move (see sections 15.8 and 15.9).

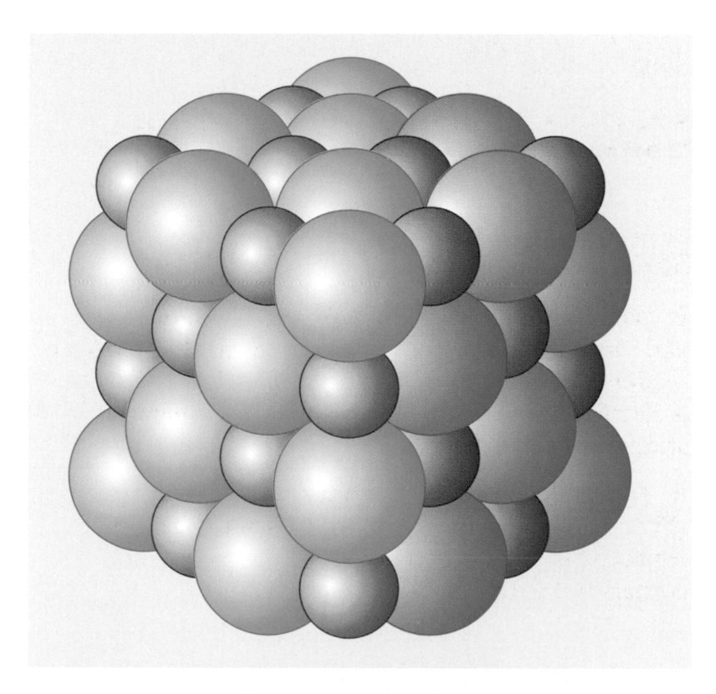

Figure 17.10 A three dimensional model of the structure of sodium chloride. The larger green balls represent Cl^- ions ($A_r = 35.5$). The smaller red balls represent Na^+ ions ($A_r = 23.0$). In the middle of the structure, six Na^+ ions surround each Cl^- ion and six Cl^- ions surround each Na^+ ion. Make sure you can appreciate this from Figures 17.9 and 17.10

17.6 Simple molecular substances

Oxygen and water are good examples of simple molecular substances. They are made of simple molecules each containing a few atoms. Their formulas and structures are shown near the top of Table 17.2 over the page. Most other non-metals and non-metal compounds are also made of simple molecules. For example, hydrogen is H_2, chlorine is Cl_2, iodine is I_2, carbon dioxide is CO_2 and tetrachloromethane is CCl_4. Sugar ($C_{12}H_{22}O_{11}$) has much larger molecules than these substances, but it still counts as a simple molecule.

Table 17.2 Formulas and structures of some simple molecular substances

Name	Molecular formula	Model of structure
Hydrogen	H_2	
Oxygen	O_2	
Water	H_2O	
Methane	CH_4	
Hydrogen chloride	HCl	
Iodine	I_2	
Carbon dioxide	CO_2	

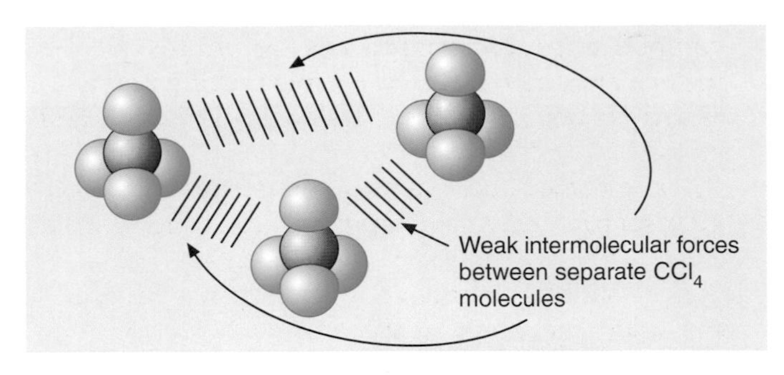

Figure 17.12 Intermolecular bonds (Van der Waals forces) in tetrachloromethane

In these simple molecular substances, the atoms are held together in each molecule by strong covalent bonds (Figure 17.11) (see section 17.9). But there are only weak forces between the separate molecules (Figure 17.12). These weak forces between the separate molecules are called **intermolecular bonds** or **Van der Waals** forces.

The properties of simple molecular substances

The properties of simple molecular substances can be explained in terms of their structure. The molecules in these substances have no electrical charge (unlike ions in ionic compounds or electrons in metals). So there are no electrical forces holding them together. But some simple molecular substances, like iodine, sugar, tetrachloromethane and water do exist as liquids and solids so there must be some intermolecular forces holding their molecules together.

- **Simple molecular substances are soft.** The separate molecules in simple molecular substances are usually further apart than atoms in metal structures and ions in ionic structures. The forces between the molecules are only weak and the molecules are easy to separate. Because of this, crystals of these substances, like iodine and sugar, are usually soft.
- **Simple molecular substances have low melting points and boiling points.** It takes less energy to separate the molecules in simple molecular substances than to separate ions in ionic

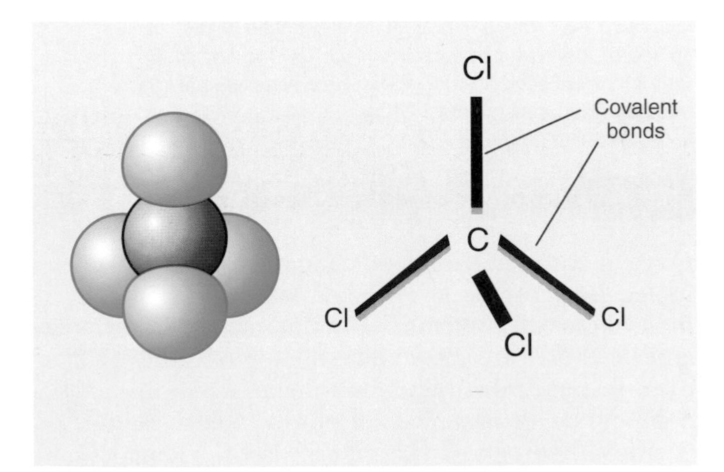

Figure 17.11 Tetrachloromethane (carbon tetrachloride) is a simple molecular substance. In tetrachloromethane, the carbon atom and four chlorine atoms are held together by strong covalent bonds

This butcher is using 'dry ice' (solid carbon dioxide) to keep meat cool and bacteria free during mincing. After mincing, the 'dry ice', which is a simple molecular substance, will evaporate rapidly without spoiling the meat

compounds, or atoms in metals. So, simple molecular compounds have lower melting points and lower boiling points than ionic compounds and metals.
- **Simple molecular substances do not conduct electricity.** They have no mobile electrons like metals. They do not have any ions either. This means that they cannot conduct electricity as solids, as liquids or in aqueous solution.

Notice the following key points from the last four sections.

- Substances with giant structures are often hard with high melting points and boiling points.
- Substances with simple molecular structures are usually soft with low melting points and boiling points.
- There are three types of strong force between particles in giant structures;
  - metallic bonds between metal atoms,
  - covalent bonds between non-metal atoms,
  - ionic bonds between positive metal ions and negative non-metal ions.
- In simple molecular substances there are relatively weak forces between the separate molecules.

17.7 Electron structure and chemical bonding

In section 16.8, we studied the electronic structures of the first twenty elements in the periodic table (Figure 16.6 on page 148).

When elements react we now know that they try to gain, lose or share electrons in order to get a more stable electron structure. In many cases, this more stable electron structure is the same as that of a noble gas.

The simple ideas expressed in this statement form the basis of the **electronic theory of chemical bonding**.

Look carefully at Table 17.3. This shows the electron structures of the atoms and ions of elements in period 3.

Notice three important points from Table 17.3.

1. The first three elements in period 3 (sodium, magnesium and aluminium) *lose* the electrons in their outer shell to form positive ions (Na^+, Mg^{2+}, Al^{3+}) with an electron structure like the previous noble gas, neon.
2. Elements in groups VI and VII (sulphur and chlorine), which are near the end of period 3, *gain* electrons to form negative ions (S^{2-}, Cl^-) with an electron structure like the next noble gas, argon.
3. Elements in the middle of the period (silicon and phosphorus) do *not* usually form ions. They get stable electron structures when they react by *sharing* electrons with other atoms instead of gaining them or losing them. This sharing of electrons results in covalent bonds between atoms which is the usual type of bonding in compounds of non-metals.

Table 17.3 Electron structure of the atoms and ions of elements in period 3

Element	Symbol	Electron structure of atoms	Number of atoms in outer shell	Symbol of common ions	Electron structure of ions
Sodium	Na	2,8,1	1	Na^+	2,8
Magnesium	Mg	2,8,2	2	Mg^{2+}	2,8
Aluminium	Al	2,8,3	3	Al^{3+}	2,8
Silicon	Si	2,8,4	4	–	–
Phosphorus	P	2,8,5	5	–	–
Sulphur	S	2,8,6	6	S^{2-}	2,8,8
Chlorine	Cl	2,8,7	7	Cl^-	2,8,8
Argon	Ar	2,8,8	8	–	–

17.8 Ionic bonding – transfer of electrons

Figure 17.13 shows what happens in terms of electrons when sodium chloride (Na^+Cl^-) is formed from sodium and chlorine atoms.

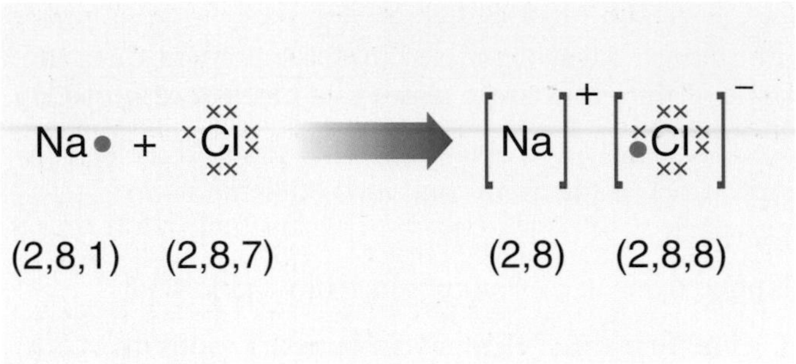

Na• + ×Cl → [Na]$^+$ [Cl]$^-$

(2,8,1) (2,8,7) (2,8) (2,8,8)

Figure 17.13 Electron transfer during the formation of sodium chloride

The number of electrons in the outer shell of each atom is shown by dots or crosses around its symbol. So diagrams like this are sometimes called 'dot-cross' diagrams. Full electron structures are also shown below the symbols. Each sodium atom loses the one electron in its outer shell to form a Na^+ ion with the same electron structure as neon. The electrons given up by sodium atoms are taken by chlorine atoms. Each chlorine atom gains one electron to form a Cl^- ion with the same electron structure as argon. So, the formation of NaCl involves the *complete transfer* of an electron from a sodium atom to a chlorine atom, forming Na^+ and Cl^- ions.

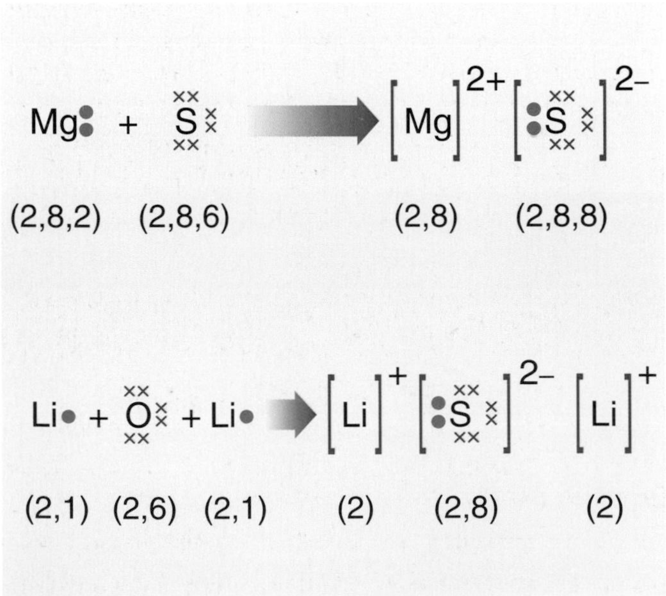

Mg + S → [Mg]$^{2+}$ [S]$^{2-}$

(2,8,2) (2,8,6) (2,8) (2,8,8)

Li• + O + Li• → [Li]$^+$ [S]$^{2-}$ [Li]$^+$

(2,1) (2,6) (2,1) (2) (2,8) (2)

Figure 17.14 Electron transfer in the formation of magnesium sulphide and lithium oxide

Ionic (electrovalent) bonds result from the transfer of electrons from metal atoms to non-metal atoms forming positive and negative ions. The electrical forces between these oppositely charged ions produce strong ionic bonds.

Figure 17.14 shows the electron transfers that take place in the formation of magnesium sulphide and lithium oxide. As a result of the electron transfers, all the ions have electron structures like a noble gas. Li^+ ions have an electron structure like He (2), Mg^{2+} and O^{2-} ions have an electron structure like Ne (2,8) and S^{2-} ions have an electron structure like Ar (2,8,8).

17.9 Covalent bonding – sharing electrons

A chlorine atom is very unstable. Its outer shell contains only seven electrons. At normal temperatures, chlorine atoms join up in pairs to form Cl_2 molecules. Why is this? If two chlorine atoms come close together, the electrons in their outer shells can overlap so that one pair of electrons is shared by each atom (Figure 17.15). Both atoms then have a full outer shell.

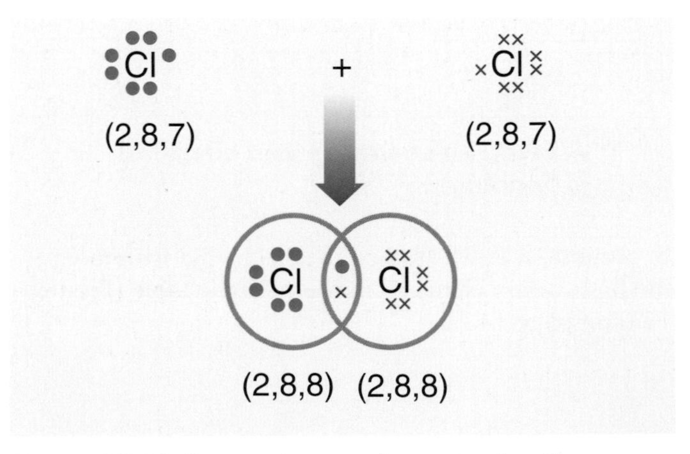

Figure 17.15 Electron sharing in the covalent bond in a chlorine molecule

The shared pair of electrons is attracted by the positive nucleus of each atom forming a **covalent bond**. Circles are used in diagrams like Figure 17.15 to enclose the electrons in the outer shell of each chlorine atom. So:

A covalent bond is formed by the sharing of a pair of electrons between two atoms. Each atom contributes one electron to the bond.

The atoms in simple molecular compounds and in giant covalent structures are joined by covalent bonds. The electron structures of some common molecular substances are shown in Figure 17.16. Notice the following points.

1 All the atoms have an electron structure like a noble gas.
2 Double covalent bonds result from the sharing of two pairs of electrons as in oxygen and carbon dioxide. Triple covalent bonds with three pairs of electrons are also known.
3 Each covalent bond can be shown as a line between the two atoms which it bonds (e.g. H—O—H).

When this is done:

- hydrogen and chlorine atoms form one covalent bond,
- oxygen atoms form two bonds,
- nitrogen atoms form three bonds and
- carbon atoms form four bonds.

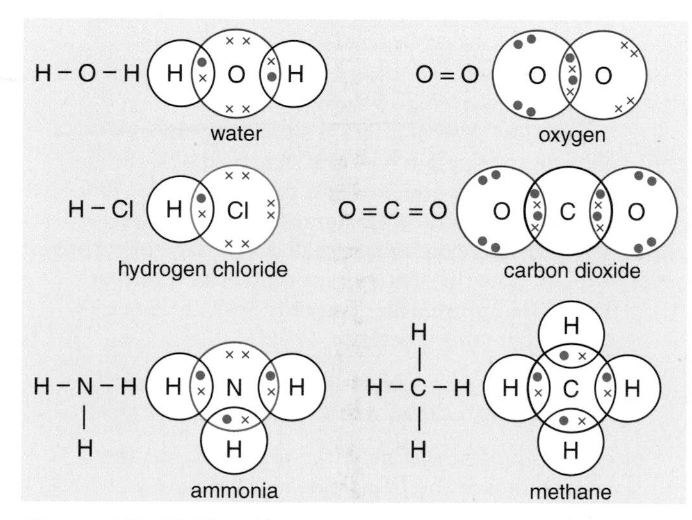

Figure 17.16 The electron structures of some simple molecular substances

17.10 Making and breaking chemical bonds

When a chemical reaction occurs, one substance changes to another. This means that bonds in the reactants must first be *broken* and then new bonds must be *made* in the products.

Now breaking bonds involves pulling atoms apart and this requires energy. On the other hand, making bonds helps to make atoms more stable and this gives out energy. So,

- bond breaking is **endothermic**,
- bond making is **exothermic**.

Figure 17.17 shows the bond breaking and bond making when hydrogen reacts with chlorine to form hydrogen chloride. The bonds between hydrogen atoms in H_2 molecules and those between chlorine atoms in Cl_2 molecules must first be broken. These bond breaking processes require energy. New bonds are then made between the H and Cl atoms as they form hydrogen chloride. This process gives out energy. We can work out the energy changes in these processes using **bond energies**. These tell us the amount of energy taken in or given out when a particular bond is broken or made.

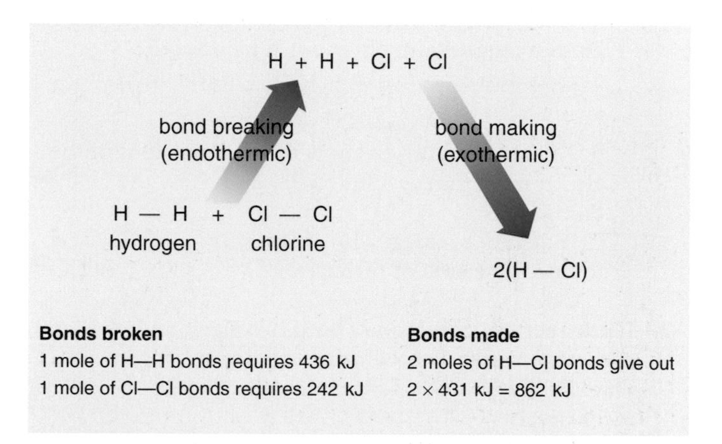

Figure 17.17 Bond breaking and bond making when hydrogen and chlorine react

From Figure 17.17 you will see that:

the total energy required for bond breaking = 678 kJ
the total energy given out on bond making = 862 kJ
∴ overall energy change = 184 kJ given out

The energy changes in chemical reactions, like those in Figure 17.17 can be summarised in **energy level diagrams**. Figure 17.18 shows the energy level diagram for the reaction of hydrogen with chlorine. In this case, the reaction is exothermic. Energy is lost during the reaction, so the products are at a lower level than the reactants.

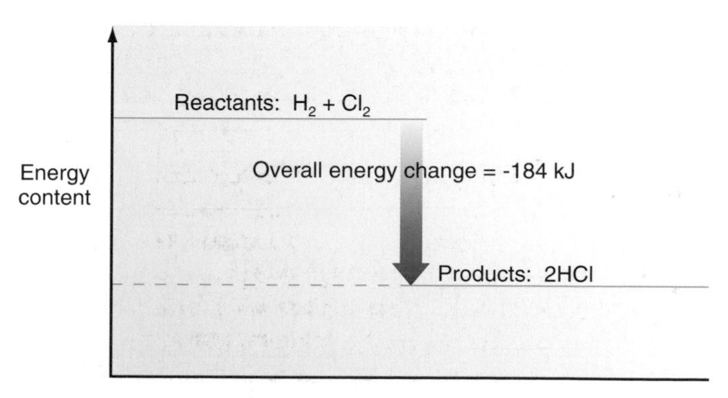

Figure 17.18 An energy level diagram for the reaction between hydrogen and chlorine

1 This question is about magnesium and its compounds.

a) The bonding in magnesium is metallic.
i) Draw a diagram to illustrate metallic bonding.
ii) Explain why magnesium is a good conductor of electricity.
iii) Use your understanding of metallic bonding to explain why metals can be pulled into wires.

b) Magnesium chloride, $MgCl_2$, is a white crystalline salt similar to sodium chloride.
i) Using your knowledge of the reactions of metals, bases and carbonates with acids, write balanced formula equations to illustrate **two different** methods of making magnesium chloride.
ii) Magnesium is manufactured by the electrolysis of molten magnesium chloride. Why does the magnesium chloride have to be molten and not solid?

c) i) Draw a diagram to show the arrangement of the electrons in magnesium oxide, and show the charges on the ions.
ii) Suggest a reason why magnesium oxide is used to line the inside of furnaces. **NICCEA**

2 Under certain conditions, both hydrogen and sodium react vigorously with chlorine to form hydrogen chloride (HCl) and sodium chloride (NaCl) respectively.

a) Give the electronic structure of hydrogen chloride.

b) Show the electronic changes that take place during the formation of sodium chloride, including the structure of the product.

c) Write down with a reason, whether the structure of sodium chloride is classified as a simple molecule or a giant structure. **WJEC**

3 The diagram shows part of a layer of particles in pure iron.

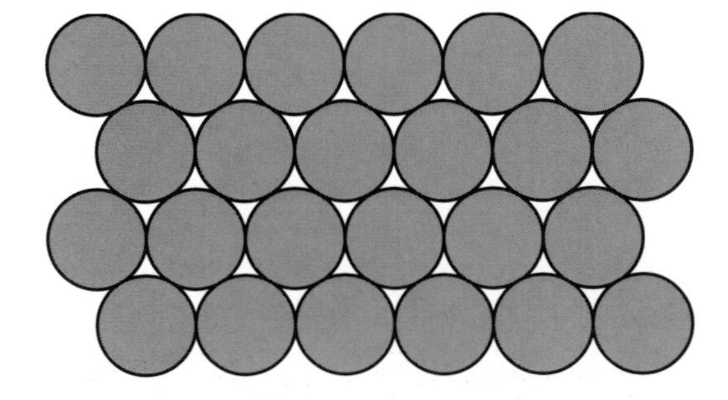

a) Explain how the structure of pure iron makes it a good conductor of electricity.
The table at the bottom of the page contains information about pure iron, cast iron, mild steel and stainless steel.

b) Cast iron is made when molten iron is poured into moulds and cooled. Cast iron contains a number of impurities including silicon, carbon, sulphur and phosphorus. The percentages of the impurities are variable.
i) Suggest why cast iron is cheaper than other forms of iron and steel.
ii) Suggest a reason why the structure of cast iron gives it a low tensile strength.

c) The following data may be useful in answering this question.

Particle	Radius of particle (nm)
Iron	0.74
Carbon	0.08
Chromium	0.63

Steels are made by adding elements such as carbon and chromium to pure iron.
i) Stainless steel contains a small amount of chromium. Explain how the structure of stainless steel gives it greater strength than pure iron. You may wish to include a diagram in your answer.
ii) Mild steel contains a small percentage of carbon. Explain how the structure of mild steel makes it harder than pure iron. You may wish to include a diagram in your answer.

d) Suggest **one** example of a use where pure iron would be better than mild steel or stainless steel.

e) Using your knowledge of the materials and the information in the first table, suggest **two** advantages of using stainless steel for car exhausts rather than the cheaper mild steel. **MEG**

4 a) Some power stations produce energy by the combustion of fossil fuels such as methane.

$$CH_4 + 2O_2 \rightarrow CO_2 + 2H_2O$$

Explain clearly **why** the combustion of methane is exothermic. **NICCEA**

Table for question 3

Metal	Density (kg per m³)	Melting point (K)	Tensile strength (MPa)	Relative electrical conductivity
Pure iron	7870	1810	300	10
Cast iron	7150	1500	100	10
Mild steel	7860	1700	300	6
Stainless steel	7930	1800	600	1

CHAPTER

Reaction rates

18.1 How fast? – different reaction rates

Every day our lives are affected by the rates of chemical reactions. We need to know how quickly or how slowly toast is made, how long it takes to cook rice, how fast the rust appears on bikes and cars and how long we can keep milk before it tastes sour.

There is a great variation in the rates at which different reactions take place. Some reactions, like explosions, are so fast that they are almost instantaneous. Other reactions, like the rusting of steel and the weathering of limestone on buildings, happen so slowly that it may be months or even years before we notice their effects.

The reactions which take place when food is cooked occur at a steady rate

Limestone is weathered very slowly. Carbonic acid in rainwater has reacted with the limestone over many years to 'eat' away at these gravestones

An explosion is an extremely fast reaction and has been used to break up rocks in this quarry

The rates of most reactions do, however, fall somewhere between those just described. The reactions which take place when fuels burn, when food is cooked and when metals react with acids, are good examples of reactions which occur at steady rates.

18.2 Why are reaction rates important?

Any cook knows the importance of reaction rates and how they are affected by **temperature**, and nowadays, most kitchens will be furnished with a gas or electric cooker and a refrigerator. Cooking involves lots of complicated reactions which have to be well synchronized throughout the preparation of a meal. The cooker can be used to speed up some reactions whilst the fridge is used to slow others down.

We are aware that food will cook faster if it is cut up into small pieces and there is a larger **surface area**. Rashers of bacon cook much faster than a bacon joint and scones bake faster than a large fruit cake.

Most of the reactions in our bodies would never happen if the reacting substances were just mixed together. Fortunately, every reaction in our bodies is helped along by its own special **catalyst** (section 18.7). Catalysts allow substances to react more easily and the catalysts in living things are called **enzymes**. Without enzymes, the reactions in your body would stop and you would die.

One of these enzymes is amylase. Amylase is present in saliva. It speeds up the first stage in the breakdown of starch in foods such as bread, rice and potatoes.

Industrial chemists cannot be satisfied with just turning one substance into another. They must look for ways to carry out reactions faster and more cheaply. In industry, speeding up reactions makes them more economical because saving time usually saves money.

The key reaction in the manufacture of ammonia is the Haber process (section 21.3). This involves converting nitrogen and hydrogen to ammonia.

$$N_2(g) + 3H_2(g) \rightarrow 2NH_3(g)$$

At room temperature, this reaction will not happen. But, chemical engineers have found that the reaction occurs rapidly at 400°C, using an iron catalyst and if the **concentrations** of nitrogen and hydrogen are increased by raising the pressure to about 200 atmospheres.

The examples mentioned in this section illustrate the four important factors which affect the rates of chemical reactions:

- temperature
- surface area
- catalysts
- concentration

These factors will be studied in more detail later in this chapter.

A workman fitting a special catalyst section to the exhaust system of a car. The catalyst removes nitrogen oxide from the car's exhaust fumes

18.3 Measuring reaction rates

During a reaction, reactants are being used up and products are forming. So, the amounts of the reactants fall and the amounts of the products rise. We can use these changes to measure the reaction rate *by calculating how much reactant is used up or how much product forms in a given time.* Therefore;

$$\text{reaction rate} = \frac{\text{change in amount of reactant or product}}{\text{time taken}}$$

For example, when 0.1 g of magnesium was added to dilute hydrochloric acid, the magnesium reacted and disappeared in 10 seconds.

∴ reaction rate

$$= \frac{\text{change in mass of magnesium}}{\text{time taken}}$$

$$= \frac{0.1}{10} \text{ g magnesium used up per second}$$

$$= 0.01 \text{ gs}^{-1}$$

Strictly speaking, this is the *average* reaction rate over the 10 seconds it takes for all the magnesium to react. Although reaction rates are usually measured as changes in mass (or concentration) with time, we can also use changes in volume, pressure, colour and conductivity.

Calculating reaction rates

The rate of reaction between small marble chips (calcium carbonate, $CaCO_3$) and dilute hydrochloric acid was studied using the apparatus in Figure 18.1.

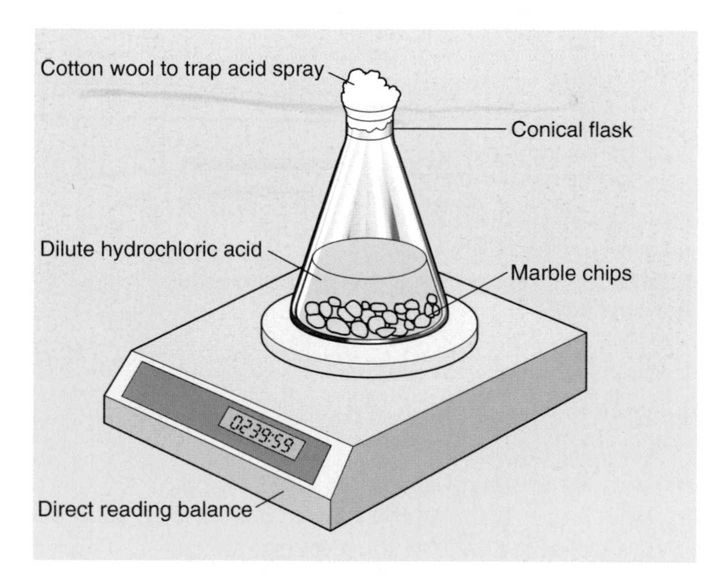

Figure 18.1 Studying the rate of reaction between marble chips and hydrochloric acid. (**Wear eye protection if you try this experiment**)

As the reaction occurs, carbon dioxide is produced. This escapes from the flask and so the mass of the flask and its contents decrease.

$$CaCO_3(s) + 2HCl(aq) \rightarrow CaCl_2(aq) + H_2O(l) + CO_2(g)$$

The decrease in mass is the mass of carbon dioxide produced. The results of one experiment are given in Table 18.1. These results have been plotted on a graph in Figure 18.2.

Table 18.1 The results of an experiment to measure the rate of reaction between marble chips and dilute hydrochloric acid

Time (minute)	Mass of flask and contents (g)	Decrease in mass (g)	Decrease in mass for each minute interval (g)
0	78.00	0	
			1.50
1	76.50	1.50	
			1.00
2	75.50	2.50	
			0.55
3	74.95	3.05	
			0.35
4	74.60	3.40	
			0.19
5	74.41	3.59	
			0.08
6	74.33	3.67	
			0.03
7	74.30	3.70	
			0
8	74.30	3.70	

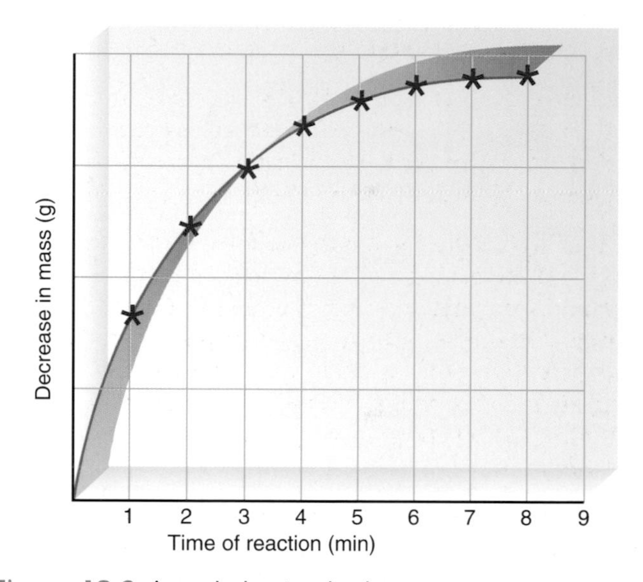

Figure 18.2 A graph showing the decrease in mass against time when marble chips react with hydrochloric acid

During the first minute, there is a decrease in mass of 1.5 g as carbon dioxide escapes.

∴ average rate of reaction in the first minute

$$= \frac{\text{change in mass}}{\text{time taken}}$$

$$= \frac{1.5 \text{ g}}{1 \text{ min}}$$

$$= 1.5 \text{ g of carbon dioxide per minute}$$

(Notice that the units for reaction rate are grams per minute this time.) During the second minute (from time = 1 minute to time = 2 minutes), 1.0 g of carbon dioxide escapes.

∴ average rate of reaction in the second minute

$$= \frac{1.0 \text{ g}}{1 \text{ min}}$$

= 1.0 g of carbon dioxide per minute

These calculations and the graph show that the reaction is fastest at the start of the reaction when the slope of the graph is steepest. During the reaction, the rate falls and the slope of the graph levels off. Eventually, the reaction stops (reaction rate = 0.0 g/min) and the graph becomes flat (gradient or slope = 0).

This mechanic is tuning the car engine to adjust the rate at which petrol burns in the cylinders. The car's performance depends on the rate of this reaction

18.4 Surface area and reaction rates

If you have ever needed to light a fire, you will know that it is easier to burn sticks than logs. The main reason for this is that the sticks have a greater surface area. There is a larger area in contact with oxygen in the air, so the sticks burn more easily.

> In general, reactions go faster when there is more surface area to react.

The reaction between marble chips (calcium carbonate) and dilute hydrochloric acid used in the last section to calculate reaction rates, can also be used to study the effect of surface area on reaction rate. The equation for the reaction is:

$$CaCO_3(s) + 2HCl(aq) \rightarrow CaCl_2(aq) + H_2O(l) + CO_2(g)$$

The results are shown in Figure 18.3. In experiment I, thirty *small* marble chips (with a total mass of 10 g) reacted with 100 cm^3 of dilute hydrochloric acid. In experiment II, six *large* marble chips (with the same total mass of 10 g) reacted with 100 cm^3 of the same hydrochloric acid. There is a large excess of marble chips in each of these experiments, so the acid will be used up first.

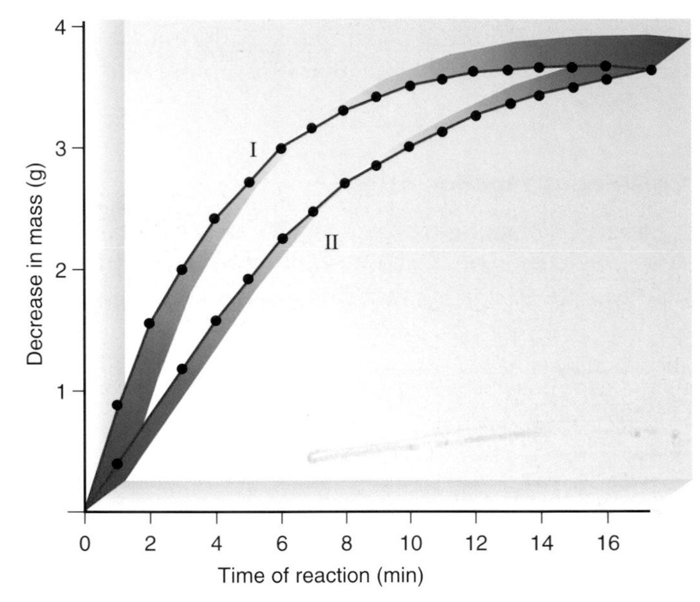

Figure 18.3 Studying the effect of surface area on reaction rate

Look closely at Figure 18.3.

- Why is the overall decrease in mass the same in both experiments?
- Why do the graphs become flat?
- Which graph shows the greater decrease in mass per minute at the start of the experiment?
- Which experiment begins at the faster rate?
- Why is the reaction rate different in the two experiments?

18.5 Concentration and reaction rates

Substances that burn in air burn much more rapidly in pure oxygen. Charcoal in a barbecue normally burns very slowly with a red glow. But, if you blow on to it so that it gets more air and therefore more oxygen, it glows brighter and may burst into flames. When red hot charcoal is put into oxygen it burns with a bright flame.

> Chemical reactions occur when particles of the reacting substances collide with each other.

Collisions between carbon atoms and oxygen molecules occur more frequently when oxygen is used instead of air. So, the reaction happens faster, gives off more heat and produces flames when the concentration of oxygen is increased by using the pure gas. Pure oxygen is also used to speed up chemical changes in the body. This can help the recovery of hospital patients such as those suffering from breathing difficulties.

In general, reactions go faster when the concentration of reactants is increased.

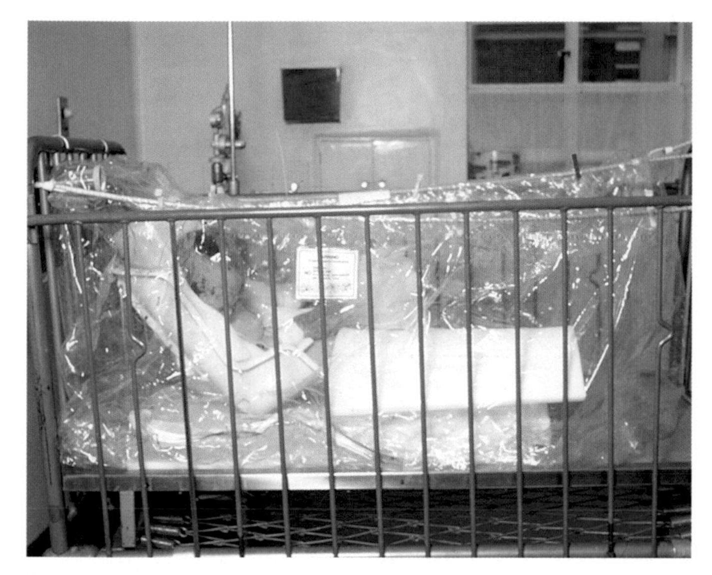

Pure oxygen is used in oxygen tents like this one to speed up the recovery of patients who have difficulty in breathing

In reactions between gases, the concentration of each gas can be increased by increasing its pressure. Some industrial processes use very high pressures. For example, in the Haber process (section 21.3), nitrogen and hydrogen are made to react at a reasonable rate by increasing the pressure to 200 times atmospheric pressure.

18.6 Temperature and reaction rates

Milk will keep for several days in a cool refrigerator, but it turns sour very quickly if it is left in the sun. Other perishable foods, like fruit and cream, also go bad more quickly at higher temperatures. This is because:

Chemical reactions go faster at higher temperatures.

The effect of temperature on reaction rates can be investigated using the reaction between sodium thiosulphate solution ($Na_2S_2O_3(aq)$) and dilute hydrochloric acid.

$$Na_2S_2O_3(aq) + 2HCl(aq) \rightarrow 2NaCl(aq) + H_2O(l) + S(s) + SO_2(g)$$

When the reactants are mixed, the solution becomes cloudy because sulphur is precipitated (Figure 18.4).

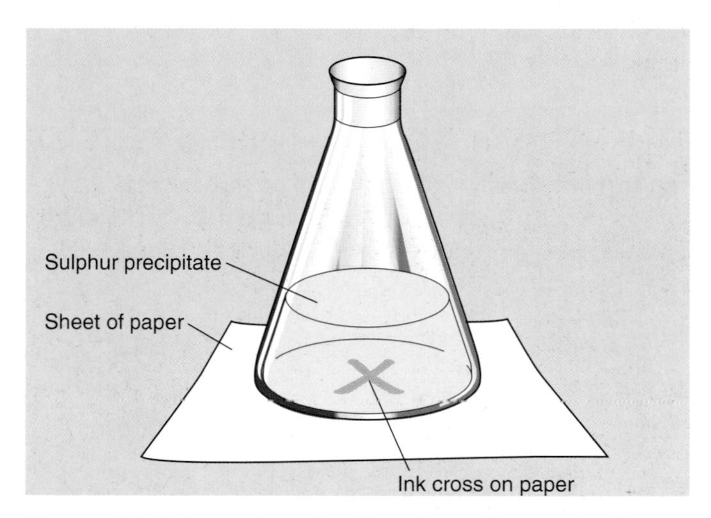

Figure 18.4 Investigating the effect of temperature on reaction rate

As the precipitate gets thicker, the ink cross on white paper below the flask slowly disappears. We can find the reaction rate by mixing 5 cm^3 of 2.0 mol dm^{-3} hydrochloric acid with 50 cm^3 of 0.05 mol dm^{-3} $Na_2S_2O_3(aq)$ and then measuring the time it takes for the cross to disappear.

Table 18.2 shows the results obtained when the reaction was carried out at different temperatures.

Table 18.2 The results of an experiment to study the reaction between sodium thiosulphate and dilute hydrochloric acid at different temperatures

Temperature (°C)	Time for cross to disappear (s)	$\frac{1}{\text{time for cross to disappear}}$ (s^{-1})
23	132	0.0076
29	90	0.0111
34	64	0.0156
44	38	0.0263
59	20	0.0500

Notice from the results in Table 18.2 that:

- the cross disappears more quickly and therefore the reaction goes faster at higher temperatures,

- if the temperature rises by 10°C, the reaction rate is about twice as fast. For example, at 23°C the cross disappears in 132 seconds whilst at 34°C it disappears in about half the time (64 seconds).

In this experiment, using the equation from section 18.3,

$$\text{reaction rate} = \frac{\text{amount of sulphur precipitated}}{\text{time for cross to disappear}}$$

As the cross disappears at the same thickness of precipitate each time, the amount of sulphur precipitated is the same at each temperature. So,

$$\text{reaction rate} \propto \frac{1}{\text{time for cross to disappear}}$$

Figure 18.5 shows a graph of this reciprocal against temperature. The graph shows clearly that the reaction rate increases as temperature increases.

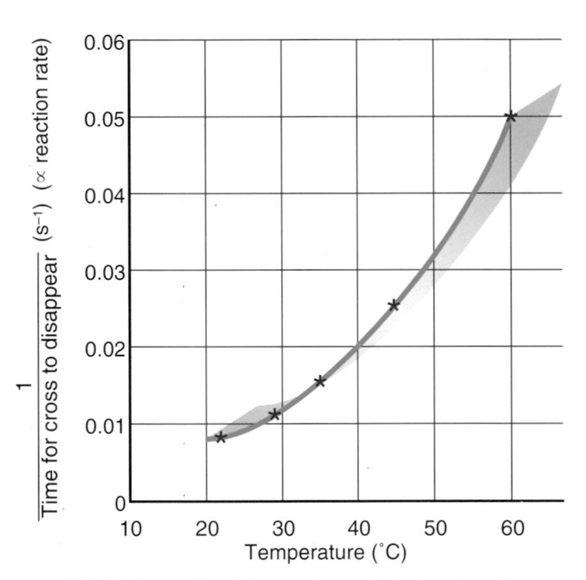

Figure 18.5 A graph of the reciprocal of the time for the cross to disappear (∝ reaction rate) against temperature

Why do chemical reactions go faster at higher temperatures?

Lots of processes can be speeded up by increasing the temperature. The chemical reactions involved in baking and cooking would never happen unless the food was heated to a high temperature.

At higher temperatures, particles are moving faster. So, they collide more often and this causes reactions to go faster. But, particles do not always react when they collide. Sometimes, they collide too gently. The particles do not collide with enough energy for bonds to stretch and break. In some reactions, only the molecules with high energy can react. So there are two reasons why reactions go faster at higher temperatures:

- The particles move faster at higher temperatures and collide more often.
- The particles collide with more energy at higher temperatures, so more collisions result in a reaction.

Perishable foods can be stored in a freezer for long periods of time because reaction rates are halted at a temperature of −18°C

18.7 Catalysts and reaction rates

Hydrogen peroxide solution ($H_2O_2(aq)$) decomposes slowly into water and oxygen at room temperature.

$$2H_2O_2(aq) \rightarrow 2H_2O(l) + O_2(g)$$

When manganese(IV) oxide, MnO_2, is added, the hydrogen peroxide decomposes rapidly. The manganese(IV) oxide helps the H_2O_2 to decompose, but it is not used up in the reaction. The manganese(IV) oxide left at the end weighs exactly the same as that at the start of the reaction. So, the MnO_2 has acted as a **catalyst**.

Catalysts are substances which change the rate of chemical reactions without being used up during the reaction.

Most catalysts are used to speed up reactions, but a few are used to slow reactions down. These substances are called negative catalysts or **inhibitors**. For example, glycerine is added to hydrogen peroxide as an inhibitor to slow down its rate of decomposition during storage. Hydrogen peroxide is used in industry to bleach textiles, paper and pulp.

Substances called antioxidants are an important group of inhibitors. **Antioxidants** prevent substances and materials being oxidized. They are added to paints to slow down rusting. Antioxidants are often added to fatty foods like cakes, margarine and crisps. They prevent the fats in the food being oxidized by oxygen in the air to form unpleasant, rancid products and as a consequence, they extend their shelf life.

This cheesecake contains a small amount of antioxidant which acts as a negative catalyst to prevent the cake reacting with oxygen in the air

Different reactions do, of course, need d
catalysts and catalysts play an importa different
chemical industry. Petrol (section 20. nt part in the
(section 20.7) and ammonia (sectio 6), margarine
produced by processes involvin on 21.3) are all
catalysts in many of these imp s catalysts. The
processes are transition meta ortant industrial
Iron is the catalyst for the m als or their compounds.
in the Haber process, nick manufacture of ammonia
the production of margar el is the catalyst in
are used in the catalytic ine, and platinum alloys
cars. converters fitted to modern

Catalysts that speed
react more readily. up reactions allow substances to
break more easily. They do this by allowing bonds to
react and the rea So, the particles need less energy to
those mentione action is faster. Solid catalysts, like
reactions by br d in the last paragraph, speed up
surface. This a inging the reactants together on the
new bonds to allows bonds to stretch and break and
form more easily.

Catalysts can
(motorway) be compared to motorways. Catalysts
reaction (jou provide a faster, easier path (route) for the
than the un rney) which needs less energy (petrol)
roads). catalysed reaction (winding and narrow

18.8 Enzymes – biological catalysts

One of the most important discoveries of this century is that all chemical reactions in living things need catalysts.

> The catalysts for biological processes are called **enzymes** and every chemical reaction in living things has its own specific enzyme.

Enzymes in our bodies catalyse the breakdown of our food. They also catalyse the reactions which synthesize important chemicals like proteins in our muscles and DNA in our genes. Almost all enzymes are themselves proteins and these are also synthesized in our bodies using other enzymes as catalysts.

So, enzymes are cruc al to the control of life. Without enzymes, the rea ctions in your body would stop and you would d e.

More and more industrial processes are being developed which use enzymes. These processes include baking, brewing (section 3.6), yogurt and cheese making and the manufacture of fruit juices, vitamins and pharmaceuticals. The enzymes for these processes are often extracted from living material such as animal tissues, plants, yeast and fungi.

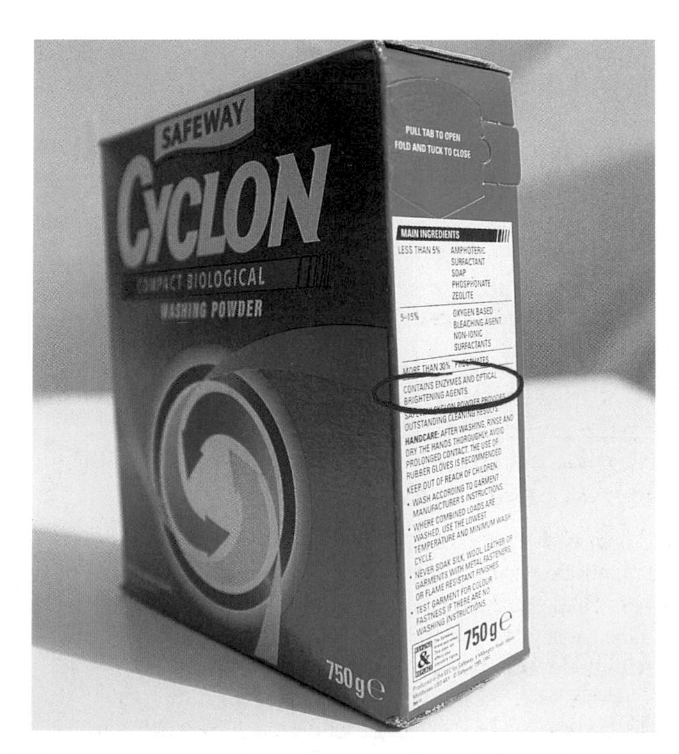

Biological washing powders contain enzymes which break down the chemicals in food, grease or dirt stains on clothing

Another very important use of enzymes is in biological detergents and washing powders. These break down fabric stains of food, blood and other biological substances. The advantage of a biological washing powder is that it works at a relatively low temperature. So, they are useful for washing delicate fabrics and electricity is saved. However, some people have a skin allergy to them and also they only work in a narrow temperature range.

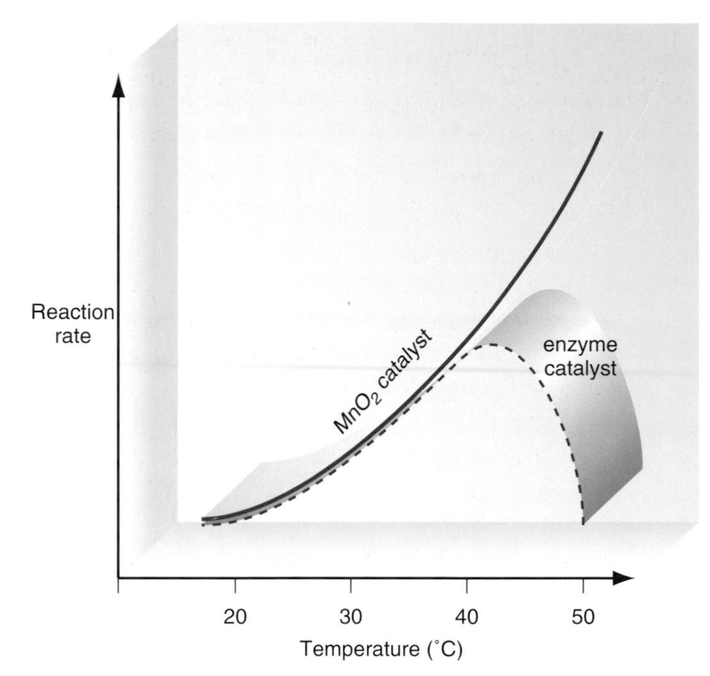

Figure 18.6 The rate of decomposition of H_2O_2 at different temperatures with (i) MnO_2 as catalyst, (ii) the enzyme, catalase, as catalyst

The effect of temperature on enzyme-catalysed reactions

The sketch graph in Figure 18.6 shows how the rate of decomposition of hydrogen peroxide changes with temperature (i) using a manganese dioxide catalyst and (ii) using the enzyme, catalase, as the catalyst.

Most catalysed reactions go faster as the temperature rises like the graph for MnO_2 in Figure 18.6. This means that the reaction rate steadily increases with temperature as expected.

The reaction rate of enzyme-catalysed reactions also increases at fairly low temperatures, but above about 40°C their reaction rate decreases rapidly as the temperature rises.

This is because the enzymes are proteins and their structure is damaged as the temperature rises above 40°C. This damage to the protein structure is called **denaturation**. As the protein is denatured, it becomes less and less effective as a catalyst. So, the enzyme-catalysed reaction goes slower and eventually stops.

This explains why enzyme (biological) washing powders, which clean by catalysing the breakdown of grease and other stains, cannot be used with water at temperatures above 40°C.

QUESTIONS

1 Calcium carbonate reacts with hydrochloric acid. The word equation for this reaction is:

calcium carbonate + hydrochloric acid → calcium chloride + water + carbon dioxide

Clare investigated the rate of this reaction at 20 °C using the apparatus shown.

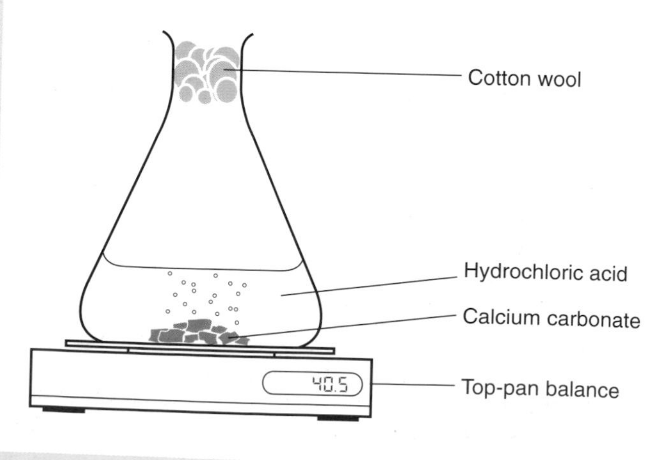

This is a graph of her results.

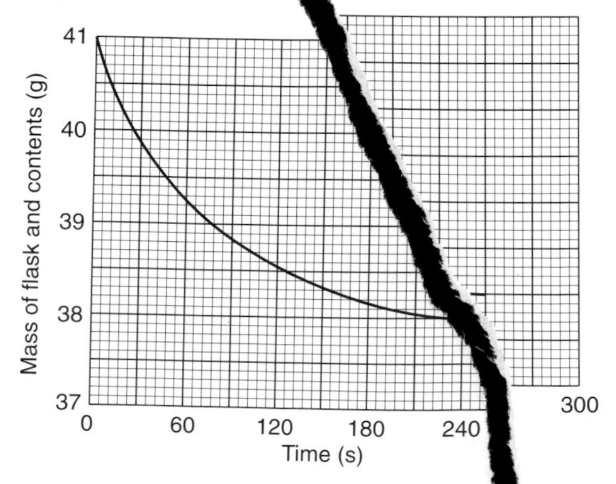

a) i) Why did the mass decrease?
   ii) Why did the graph level off?

b) Clare repeated the experiment at 40 °C. Copy the graph and draw on the line you think she will get at this temperature.
Edexcel

2 Chemical reactions in our bodies produce hydrogen peroxide, which is poisonous and must be removed. The enzyme catalase speeds up the decomposition of hydrogen peroxide to form water and oxygen.

a) Copy and complete the balanced symbolic equation for the decomposition of hydrogen peroxide.

$$H_2O_2(aq) \rightarrow \_\_\_\_\_\_\_\_$$

Manganese(IV) oxide is a non-biological substance which also speeds up the decomposition of hydrogen peroxide. The following experiments were carried out to compare catalase and manganese(IV) oxide as catalysts.

Experiment 1 0.1 g of catalase was added to 2 cm^3 of hydrogen peroxide and the volume of gas produced was measured every 5 seconds.

Experiment 2 0.1 g of manganese(IV) oxide was added to 2 cm^3 of hydrogen peroxide and the volume of gas produced was measured every 5 seconds.

The results collected from the experiments were plotted on the grid and two graphs drawn.

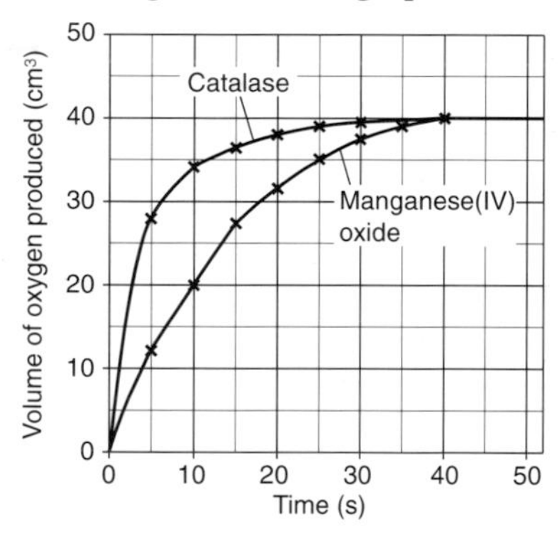

b) What can be concluded from the graphs about the effectiveness of catalase and manganese(IV) oxide as catalysts? Explain your conclusion.

c) A solution of hydrogen peroxide of concentration 15% produces a total volume of 10 cm^3 of oxygen on decomposition from each cubic centimetre of solution. What was the concentration of the hydrogen peroxide used in experiments 1 and 2?

d) At the end of the experiment, the solid manganese(IV) oxide was separated from the solution.
i) How would you separate manganese(IV) oxide from the solution?
ii) What mass of manganese(IV) oxide would you expect to recover?

e) Catalase contains iron which can be removed chemically. When 0.1 g of iron-free catalase was added to 2 cm^3 of hydrogen peroxide the results in the following table were obtained.

Time (s)	5	10	15	20	30	40	50
Volume of gas produced (cm^3)	2	4	6	8	12	16	20

Copy the grid from part a) and plot these results on it. Draw the best line through the points and label this line 'iron-free catalase'.
ii) What can be concluded from these results?

f) The substances in the list are all catalysts in chemical reactions.

iron manganese(IV) oxide nickel platinum vanadium(V) oxide

Use the periodic table to identify what these catalysts all have in common. **MEG**

3 The rate at which magnesium metal reacts with hydrochloric acid can be measured using the apparatus shown below.

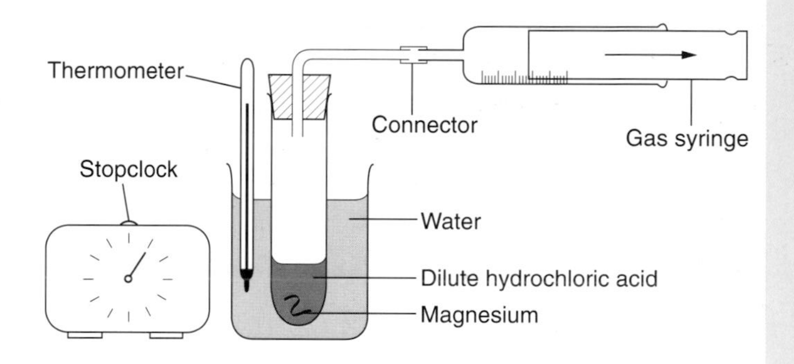

a) Name the **two** substances formed when magnesium reacts with hydrochloric acid.

b) The volume of gas made by the reaction of magnesium with the acid is plotted against time. This is shown in the graph below.

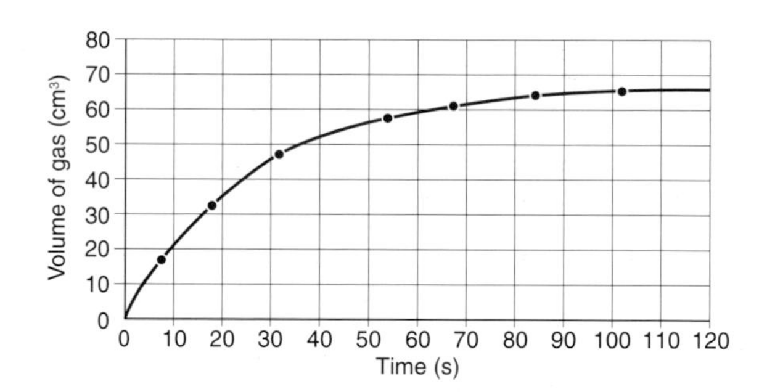

i) What was the **total** volume of gas produced by the reaction?
ii) Another experiment was carried out, using the same amounts of magnesium and hydrochloric acid. However, in this experiment the acid was heated before use. Copy the graph above and sketch a second curve to show the graph that would be obtained.
iii) Explain why the rate of reaction is changed when the acid is heated.

c) You have been told that copper powder will act as a catalyst in the reaction between magnesium and hydrochloric acid. How could you show whether or not this suggestion is true? **SEG**

CHAPTER

Earth science

19.1 The Earth – our source of raw materials

Look around you and notice the variety of useful materials:

- wood to make chairs, tables and desks,
- fibres to make curtains, carpets and clothes,
- glass for windows, jars and ornaments,
- metals to make cutlery, vehicles and girders.

All of these useful materials have been manufactured from raw materials found in or on the Earth. Most of the manufacturing processes employ chemical reactions to convert the raw materials into useful products.

Figure 19.1 shows the major sources of our raw materials:

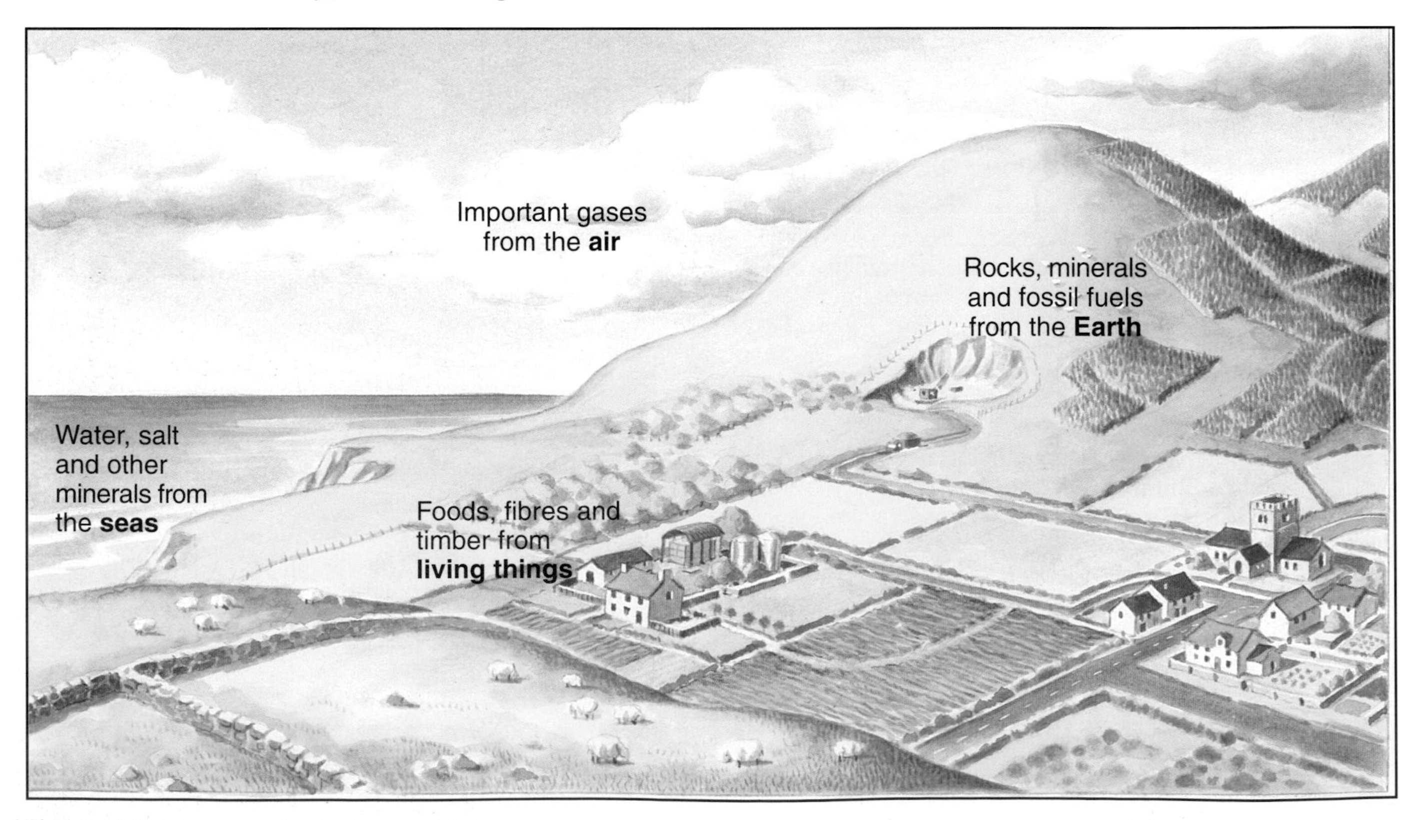

Figure 19.1 Sources of our raw materials

- the Earth's crust provides minerals, rocks and fossil fuels,
- the seas provide water, salt and other minerals,
- the atmosphere provides important gases,
- living things provide timber, food and fibres.

Most of our really important raw materials come from the Earth as fossil fuels, minerals and rocks.

A mineral is a single substance which has a chemical name and a formula.

Limestone and rock salt are good examples of minerals. Pure limestone is calcium carbonate, $CaCO_3$ and pure rock salt is sodium chloride, NaCl.

Minerals are mostly impure when they are first dug from the Earth. Mixtures of crystals or grains of different minerals form rocks. So impure limestone and impure rock salt are normally classed as rocks.

A rock is a mixture of different minerals.

This cottage is built from two important rocks. It has limestone walls and a slate roof

19.2 Exploiting the Earth's resources

This area was once a huge gravel pit. Nowadays, Thorpe Park attracts crowds of people who come to enjoy its many leisure activities

Pitlochry Dam and salmon fish ladder (below) is a tourist attraction. Hydro-Electric recognises the need to safeguard the salmon as well as generating electricity

Stockley Park Golf Course (below), near Uxbridge, was built on a landfill site

Exploitation of the Earth's resources raises important issues for society. Forestry, farming, quarrying, mining and industry provide vital materials for us all. Everyday, we depend on farms for food, forests for wood and paper, quarries for rocks and minerals and industrial plants for fertilizers, metals, alloys and plastics.

In choosing the site for forestry, mining and industry, there are three kinds of issue to consider – social, environmental and economic.

1 *Social issues*

- Will it benefit or disrupt the present community?
- Will it provide employment?
- Is there a readily available and skilled workforce?
- Are the existing social, medical and community services adequate?

2 *Environmental issues*

- Will it disfigure the landscape with unsightly roads, buildings, quarries, etc.?
- Will it destroy habitats and endanger wildlife?
- Will it produce large amounts of waste or pollution?
- Will it cause high levels of noise from traffic, machinery and blasting?
- How will waste be disposed of?
- Are there any dangers to health and safety?

3 *Economic issues*

- Is there cheap land available for any necessary developments?
- What are the costs of the raw material and fuel?
- Will the depletion of the raw material be serious?
- Will the activity/processing provide sufficient profits?
- Will the profits benefit the local community?

19.3 Evolution of the Earth, atmosphere and oceans

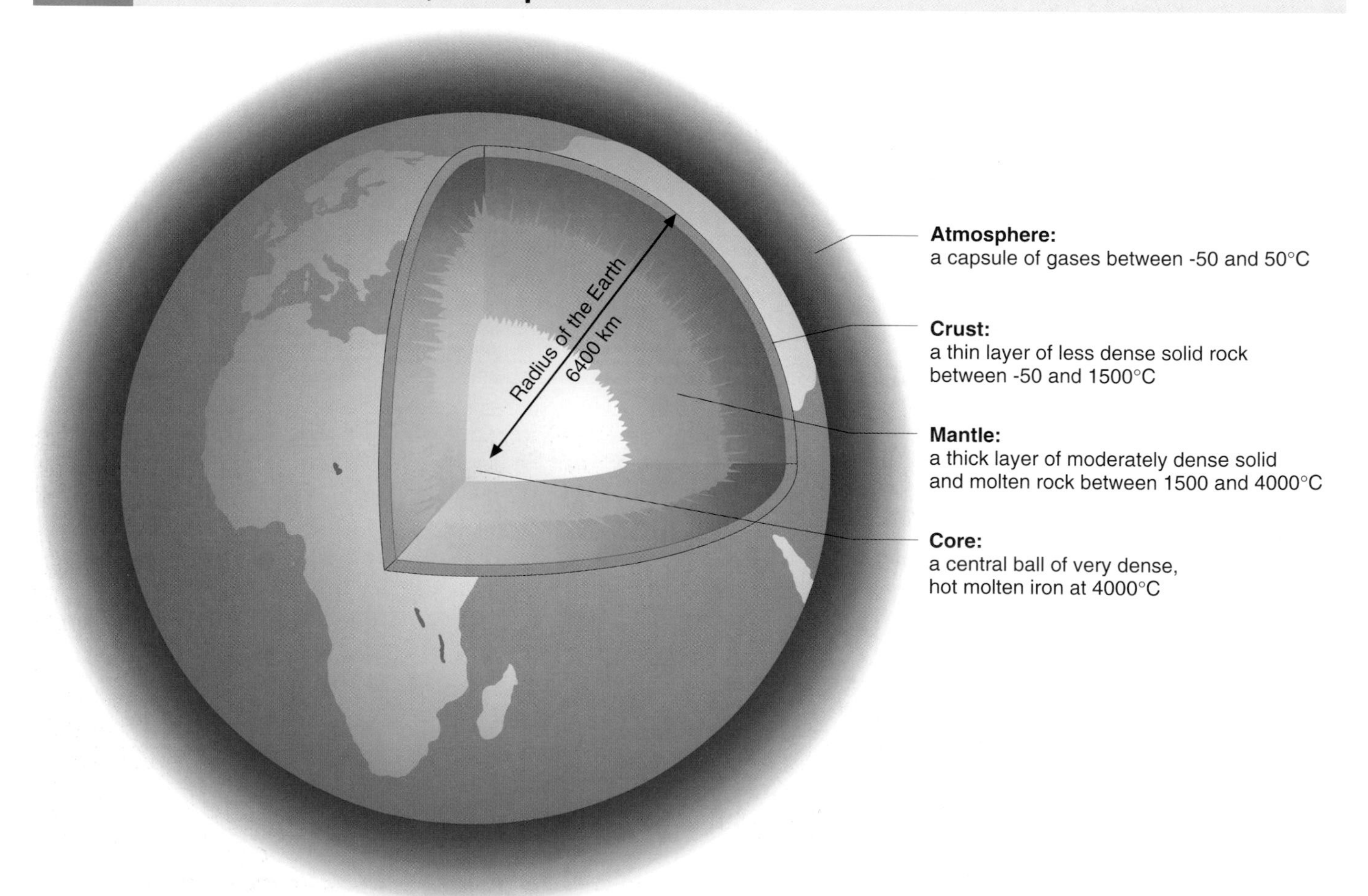

Figure 19.2 Layers of the Earth. Notice how the temperature and density increase from the atmosphere to the core

4500 million years ago, the Earth was a ball of molten rock which has slowly cooled. During this period, heavier metals sunk to the centre of the Earth forming a **core** of dense, partly solid, partly molten iron, nickel and sulphur at about 4000°C (Figure 19.2). The core is surrounded by a thick layer of moderately dense solid and molten rock in the **mantle** at temperatures between 1500 and 4000°C.

Less dense material collected on the surface of the Earth forming a thin, solid **crust** about 50 km thick. Where the crust is thickest, its surface is above sea level. The rocks in the crust are mainly carbonates and silicates.

Outside and above the Earth is the **atmosphere** – a layer of gases about 100 km deep. At about 6 km above sea level, the air becomes too thin for us to survive. The oceans are about 6 km deep, so almost all life on Earth exists in a relatively thin band about 12 km thick.

While the Earth was still forming, the atmosphere was mainly hydrogen and helium. These gases had such small molecules that they escaped from the Earth's gravitational attraction into outer space. Once volcanic activity started, other gases were added as rocks decomposed and elements reacted.

Aerial photos such as this one of Glen Canyon in Arizona, USA, help us to understand the effect of water on the Earth's surface and the formation of features like rivers, lakes and deltas

These gases included water vapour, carbon dioxide, hydrogen sulphide, methane and nitrogen. As the temperature dropped still further, water vapour condensed to form rivers, lakes and oceans.

When plants appeared, 3500 million years ago, they formed oxygen from water and carbon dioxide by photosynthesis. At the same time, plants used up the oxygen during respiration reforming water and carbon dioxide. Flammable gases such as hydrogen, methane and hydrogen sulphide burnt in this oxygen forming water, carbon dioxide and sulphur dioxide. In time, animals evolved and used the oxygen for respiration. This further helped to keep a balance between the production and removal of oxygen and carbon dioxide in the atmosphere. The composition of the atmosphere has remained more or less constant for the last 500 million years. The main constituents in dry air are nitrogen (about $\frac{4}{5}$ths), oxygen (about $\frac{1}{5}$th) and argon (about 1%) with traces of other noble gases and carbon dioxide.

The carbon cycle (section 9.6) also helps to maintain the composition of the atmosphere. Carbon dioxide and water vapour are removed from the atmosphere by photosynthesis, but returned to the atmosphere when animals and plants respire and during the combustion of carbon compounds.

Coral reefs such as this occur as a result of the formation of shells by marine organisms. The shells, composed mainly of calcium carbonate, are formed from Ca^{2+} and CO_3^{2-} ions in sea water

During the last century, an increase in the burning of fossil fuels has led to a small but steady increase in the amount of carbon dioxide in the atmosphere. This increase in the amount and concentration of carbon dioxide has been recognised as a possible cause of global warming, although the evidence is far from conclusive.

The composition of the oceans, like that of the atmosphere, is also more or less constant. In this case there is a balance between the input of dissolved salts in river water from the weathering of rocks and the removal of dissolved salts by processes such as:

- the formation of shells by marine organisms,
- the crystallization of salt from sea water in hot countries,
- the deposition of sea-floor sediments.

19.4 The weathering of rocks

When rocks are exposed to the weather, they slowly break up and wear away.

> The breaking up of rocks by wind, rain, ice and water is called **weathering**.

There are two main types of weathering – physical and chemical.

I Physical weathering

This involves the cracking, loosening and breaking up of rocks by physical means. There are two ways in which this happens.

i) By expansion and contraction of the rock

If you pour really hot water into a jam jar, it cracks. This is because the glass on the inside has expanded much faster than that on the outside. This sets up stresses and strains in the glass causing it to crack. In the same way, the exposed surface of a rock will expand and contract much more rapidly than the inside. As the rock surface heats up and expands during the day and then cools and contracts at night, stresses will be set up causing the rock to crack.

Large amounts of limestone rock scree have collected below Gordale Scar near Malham, Yorkshire

ii) By the freezing of water

Once a rock is cracked, water will seep into the joint. Some rocks will absorb water. If the water filling a crack or absorbed in a rock freezes, it expands pushing the pieces or grains of rock apart. When the rock warms up, the ice melts and seeps further into the rocks. Figure 19.3 shows what happens in this freeze/thaw weathering.

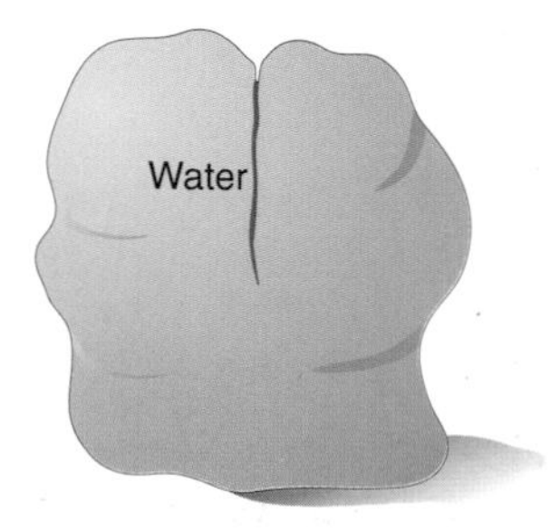

a) Water seeps into a crack in the rock

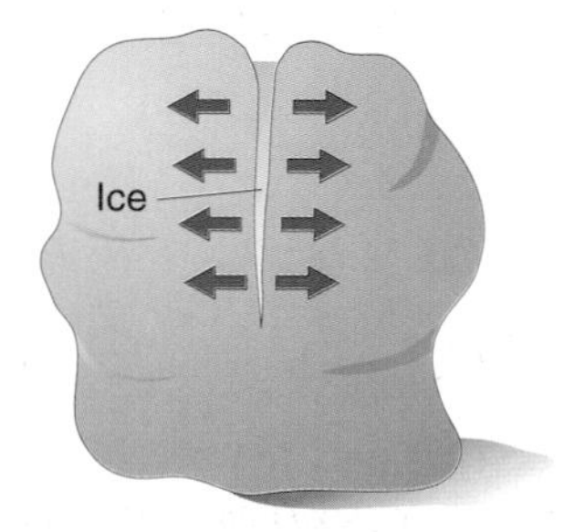

b) The water freezes, expands and forces the pieces of rock apart

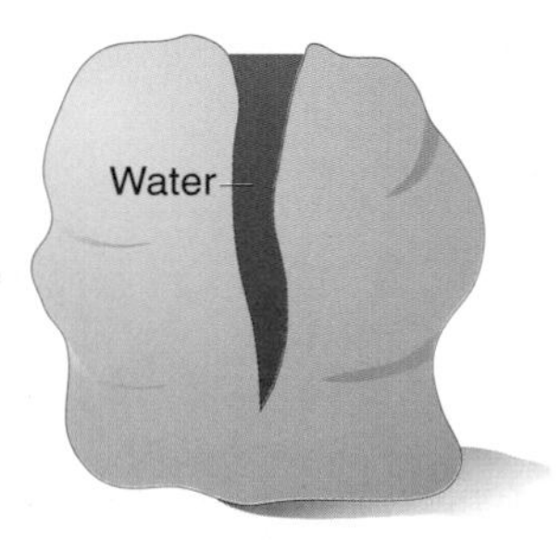

c) The ice thaws and water seeps further into the crack

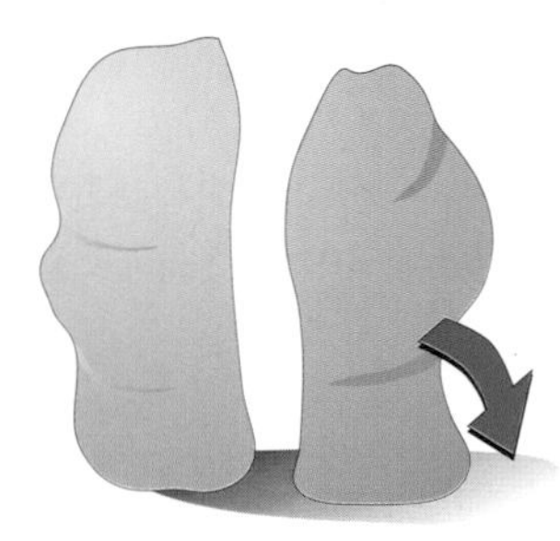

d) Further freezing and thawing occurs and eventually a piece of rock breaks off

Figure 19.3 The physical weathering of a rock by repeated freezing and thawing of water

2 *Chemical weathering*

This involves chemical reactions between the rocks and water. This can happen in two ways.

i) By reaction with the water itself
Sometimes, the minerals which make up a rock slowly change by reaction with water. This is what happens to one of the minerals called feldspar in granite. Feldspar slowly reacts with water to form clay, leaving behind quartz and mica.

In the hot wet climate of the past, feldspar in the granite of this china clay pit has reacted with warm water to form clay. The soft wet clay is easily removed and is used to manufacture high quality crockery

ii) By reaction with substances dissolved in the water
Some rocks react with chemicals in rainwater and river water. The most important example of this is limestone (Figure 19.4).

As rain falls, it reacts with carbon dioxide in the atmosphere to form a dilute solution of carbonic acid.

$$H_2O(l) + CO_2(g) \rightarrow \underset{\text{carbonic acid}}{H_2CO_3(aq)}$$

This carbonic acid in rainwater reacts very slowly with insoluble calcium carbonate (limestone) to form soluble calcium hydrogencarbonate. This dissolves in the water and is washed away.

$$H_2CO_3(aq) + \underset{\text{insoluble limestone}}{CaCO_3(s)} \rightarrow \underset{\text{soluble calcium hydrogencarbonate}}{Ca(HCO_3)_2(aq)}$$

Figure 19.4 These limestone statues on Rheims Cathedral, France, have been weathered over the centuries by acid in rain water

19.5 Weathering, transport and erosion

In mountainous areas, rocks and soil fall down steep slopes into gulleys. Fast moving streams pick up this material and carry it away. *This carrying away of weathered material is called* **transport**. *The process of wearing away rocks and then carrying the weathered material away is called* **erosion**.

Erosion = weathering + transport

However, fast moving streams are not the only methods of transport for weathered material. The most important agents of erosion are:

- moving water
- wind
- glaciers

These agents pick up particles of weathered material and transport them from one place to another.

Eventually the eroded material is deposited as mud, sand or pebbles. This happens when the water, wind or ice which is carrying the weathered material slows down. Gravity takes over and the weathered material falls to the ground. You can actually see this happening in a dust storm or a sand storm.

19.6 Rocks in the Earth

This is a smoothly polished sample of granite – an igneous rock. 'Igneous' means formed by fire. Igneous rocks are formed from the cooling of very hot, molten rock. Granite forms when molten rocks cool slowly below the Earth's surface so the crystals in it are large

This is a sample of red sandstone – a sedimentary rock. It has formed from the erosion of granite

Figure 19.5 The formation of igneous, sedimentary and metamorphic rocks

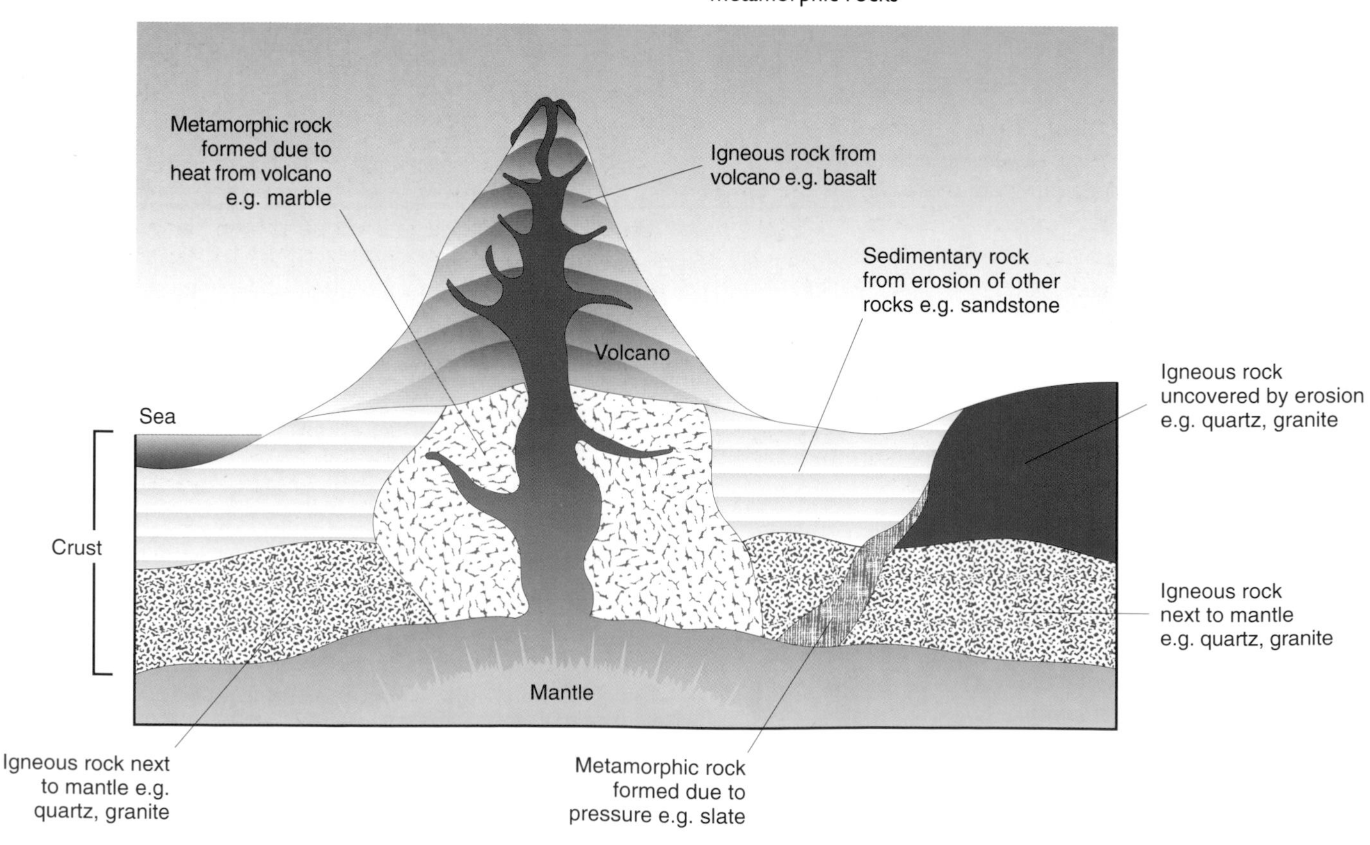

Rocks in the Earth are usually mixtures of different substances. When the Earth first cooled, its molten crust solidified to form **igneous rocks**. Initially, there were no other types of rock. Over millions of years, two other types of rock were created – **sedimentary rocks** and **metamorphic rocks**. Figure 19.5 shows how these three types of rock are being formed today.

Igneous rocks

Igneous rocks are formed by the cooling and solidifying of hot molten material called **magma** in the Earth's mantle. As the molten magma cools, it solidifies as a mixture of different minerals with interlocking crystals.

The size of crystals in the igneous rock depends on the rate at which the magma has cooled and crystallized. Some igneous rocks are produced when volcanoes erupt and the lava cools quickly in a matter of days or weeks. This produces rocks, such as basalt, with small crystals which are usually dark in colour. These igneous rocks which form relatively quickly at or near the Earth's surface are described as **extrusive**.

Other igneous rocks are formed deep in the Earth's crust next to the mantle. Here the magma cools very slowly, possibly over centuries. This results in rocks with much larger crystals such as granite and quartz which are usually light in colour. These igneous rocks which form slowly and deeply below the Earth's surface are described as **intrusive**.

Sedimentary rocks

When rocks are eroded, they form sediments and fragments such as sand and gravel. These sediments may be carried by rivers or ocean currents and deposited elsewhere. As the layers of sediment build up over millions of years, the material below is compressed and consolidated forming new, soft rocks such as mudstone, sandstone and shale.

Other sedimentary rocks, such as coal, chalk and rock salt have formed by processes involving the decay of living things or crystallization from sea water.

If the layers are buried deeper, the soft sediments such as chalk and mudstone get converted to harder sedimentary rocks like limestone. All of these rocks are known as sedimentary rocks because they are formed by the build up of sediments.

The texture of most sedimentary rocks consists of small fragments bound together by cementing material. The size, shape and composition of the fragments and the cementing material provide evidence about the environment when deposition of the fragments occurred. In some cases, these fragments include fossils which provide valuable records of the past.

Metamorphic rocks

This is a sample of impure marble – a metamorphic rock. It has formed from sedimentary rocks at high temperature and pressure

Table 19.1 Identifying igneous, sedimentary and metamorphic rocks

Property	Igneous	Sedimentary	Metamorphic
• Is the rock hard or soft?	hard	soft (grains can be rubbed off)	usually hard
• What is the structure?	interlocking crystals	separate grains	grains or crystalline
• Might the rock have layers?	no	yes	yes
• Might the rock have fossils?	no	yes	no
• Might the rock fizz (give off CO_2) with dilute HCl?	no	yes	yes

Sometimes, igneous and sedimentary rocks are changed into harder rocks by enormous pressure or very high temperatures. The new rock has a different structure from the original rock. It is therefore called 'metamorphic rock' from a Greek word meaning 'change of shape'. Slate and marble are good examples of metamorphic rocks. Slate is formed when clay and mud are subjected to very high temperatures. Marble is formed when limestone comes into contact with very hot igneous rock.

Table 19.1 on page 183 summarises the properties of igneous, sedimentary and metamorphic rocks. It will help you to decide whether a rock is igneous, sedimentary or metamorphic.

19.7 The rock cycle

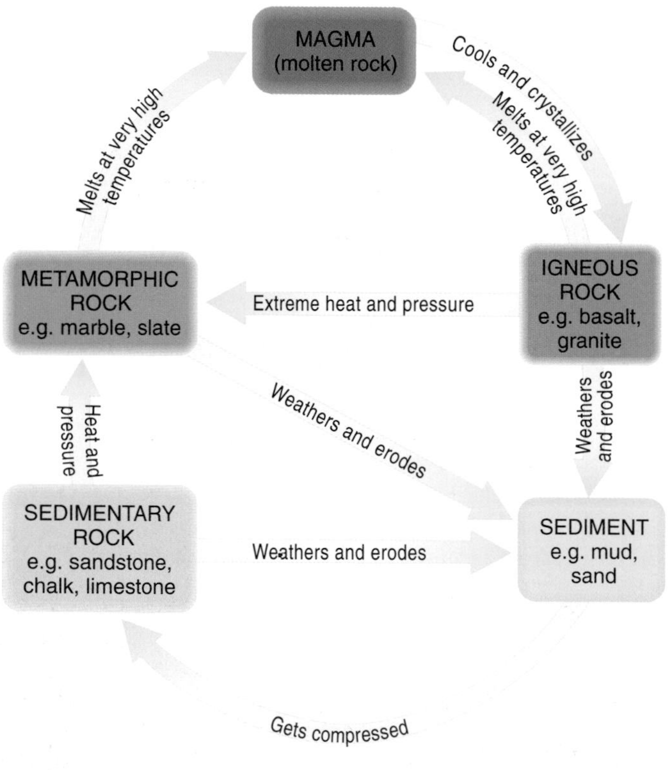

Figure 19.6 The rock cycle

Deep inside the Earth, the temperature reaches 4000°C and the temperature in the mantle ranges from 1500 to 4000°C. Therefore, as rocks are buried below the Earth's surface, they eventually melt to form magma. In due course, this magma may solidify as igneous rock beginning the sequence of events in the **rock cycle** again.

Figure 19.6 shows the main stages of the rock cycle. The complete cycle lasts hundreds of millions of years. Notice that there are both short cuts and extensions to the full cycle. So, extreme heat and pressure can change igneous rock into metamorphic rock and sedimentary and metamorphic rocks can be weathered and eroded like igneous rock.

The rock cycle involves igneous, sedimentary and metamorphic processes which take place over very different **timescales**. Landslides, volcanoes erupting and flash floods occur on *short* timescales of hours, days or possibly weeks. Other processes, such as the cooling of lava and the transport of fragments by rivers and glaciers have *moderate* timescales measured in months, years or even decades.

Some processes such as the burial and consolidation of fragments to form sedimentary rock, the cooling and solidification of magma deep inside the Earth and the effects of heat and pressure to form metamorphic rocks are measured in *long* timescales of thousands if not millions of years. These geological processes operating to very different timescales are all part of the rock cycle which has been happening since the Earth was formed 4500 million years ago.

19.8 Layers of the Earth

The Earth's **shape** is like an orange, spherical but slightly flattened at the poles. The Earth's **structure** is like a badly cracked egg. The 'cracked shell' is like the Earth's thin crust, the egg 'white' is the mantle and the 'yolk' is the core. (See Figure 19.7.)

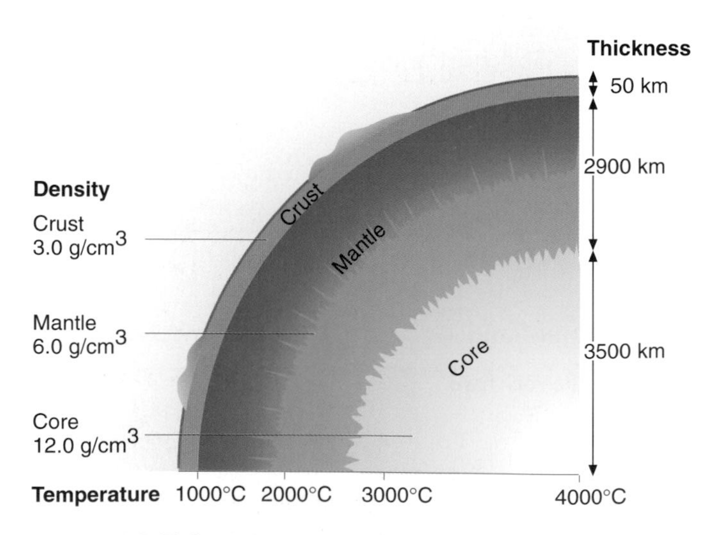

Figure 19.7 A cross section of the Earth showing the thicknesses, densities and temperatures in its internal structure

Notice in Figure 19.7 that the three concentric layers of crust, mantle and core increase in thickness, in density and in temperature towards the centre.

The deepest mines below the Earth go to a depth of about 3 km and the temperature here is about 50°C.

Deep below the Earth's surface, these miners work in the heat

Below the crust, temperatures in the outer mantle are about 1000°C. This is hot enough to soften the rocks so that they behave like stiff plasticine. Within the mantle, temperatures are between 2000 and 3000°C and the material is liquid.

In the Earth's core, the temperatures are even higher reaching 4000°C. The main constituents in the core are iron and nickel. Under normal conditions, both these metals would be gases at 4000°C, but the pressure from rocks above and the force of gravity are so great that the core is forced into the smallest possible space as a solid. The outer parts of the core and the mantle experience lower pressures and the material here is liquid.

There are two reasons why the core of the Earth is so hot.

1 When the Earth was first formed, the temperatures inside were very high. Since then the outer layers of the Earth have insulated the core and trapped the original thermal energy.

2 Rocks in the Earth and particularly igneous rocks, like granite, contain significant quantities of radioactive elements such as uranium, thorium and potassium. As the nuclei of these elements decay (break up), energy is released as heat and electromagnetic radiation. This energy helps to maintain temperatures inside the Earth and on its surface.

19.9 Earth movements

The Earth's crust is cracked and broken into huge sections called **plates**. These vast plates float on the denser mantle below. The continents and oceans sit on top of these plates which fit together like a gigantic spherical jigsaw. Over long periods of geological time, sometimes lasting millions of years, the plates move very slowly due to the convection currents in the liquid mantle.

When the plates slide past each other, move apart or push towards each other, various things can happen.

Two plates slide past each other

Stresses and strains build up in the Earth's crust. This may cause the plates to bend. In some cases, the stresses and strains are released suddenly. The Earth moves, the ground shakes violently in an **earthquake** and breaks appear in the ground. These breaks in the ground, when plates slide past each other horizontally, are called **tear faults** (Figure 19.8). The San Andreas Fault in California and the Great Glen Fault in Scotland are examples of tear faults.

Figure 19.8 A tear fault results when plates move past each other horizontally

There is severe disruption and often loss of life when an earthquake occurs. Roads are torn up and buildings damaged

Plates move apart

The crust is being stretched under tension and cracks appear in the Earth's surface. As the plates move further apart, surface rocks sink forming vertical faults. These faults produced by stretching forces are referred to as **normal faults**. When two vertical faults occur alongside each other, rift valleys are formed (Figure 19.9).

In some cases, hot molten rock (**lava**) escapes through cracks in the Earth's surface and erupts as a **volcano**.

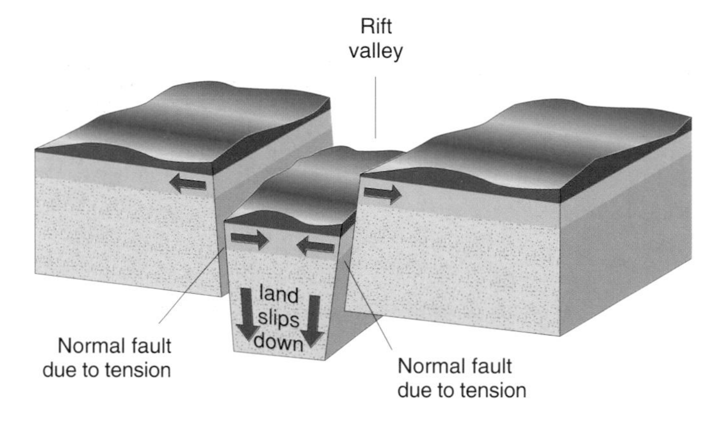

Figure 19.9 As plates move apart, the land on one side may sink into the crack. If there are two normal faults near each other, a rift valley may form

After nearly four centuries of dormancy, the volcano on the Caribbean island of Monserrat erupted in the summer of 1997. The capital, Plymouth, had to be abandoned because of the destructive flows of hot rock, ash and gases

Plates push towards each other and collide

Rocks are squeezed together. When this happens, forces of compression push layers of the Earth's crust over each other into a **fold**.

Hell's Gate, a volcanic rift valley in Kenya

Folds in the outer layer of the Earth's crust at Stair Hole near Lulworth Cove in Dorset

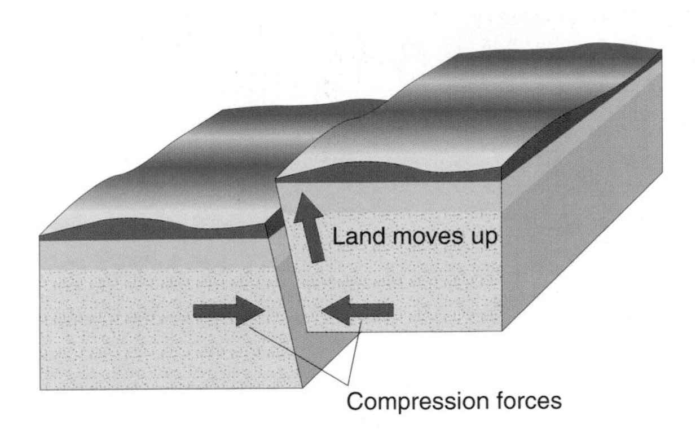

Figure 19.10 As plates move towards each other and collide, the Earth may crack with one layer moving over another as a reverse fault

Sometimes, the compression forces cracks to appear and one layer is pushed upwards and above another. This is called a **reverse fault** (Figure 19.10).

19.10 Plate tectonics

The study of the movement and interaction of the giant plates on the Earth's surface is called **plate tectonics**. Plate tectonics explains many of the large-scale geological features on the Earth.

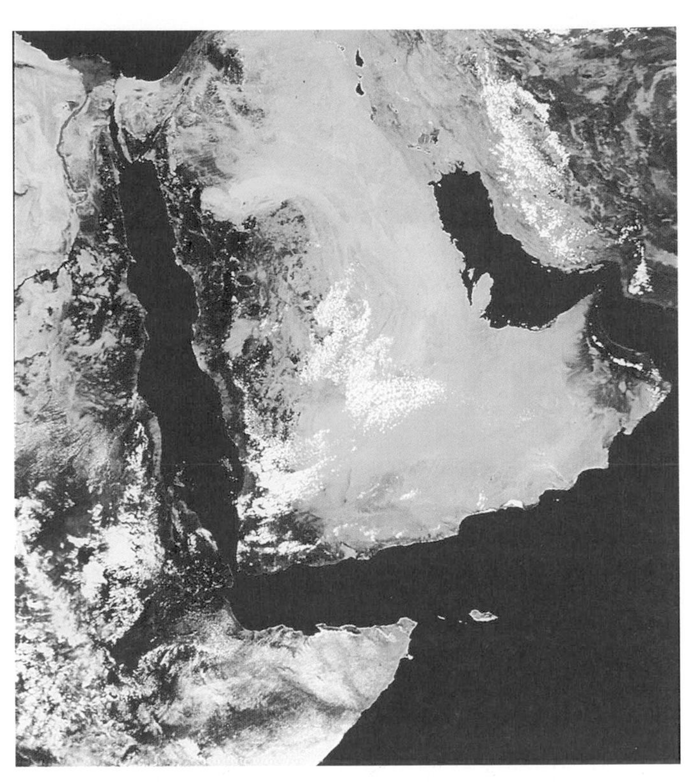

Many years ago, Africa and Arabia used to be joined together. They are now slowly drifting apart as the Red Sea gradually widens. This sea may eventually become as large as the Atlantic Ocean

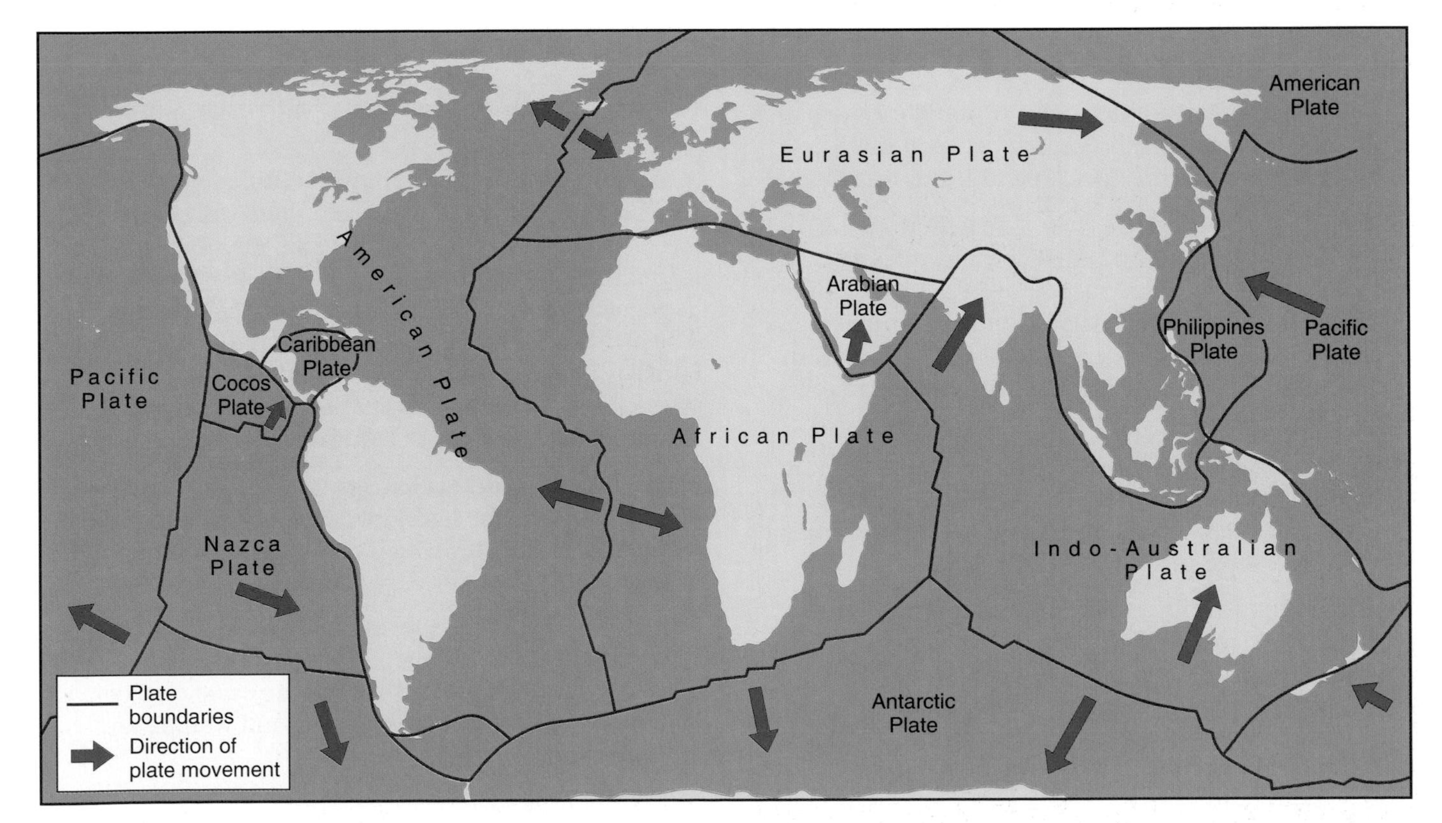

Figure 19.11 A map of the world showing the main plates and their direction of movement

The Earth's core is as hot as the surface of the Sun. This causes slow convection currents in the liquid mantle which result in slow movements of the plates in the Earth's crust. Figure 19.11 on page 187 shows how these plates are moving.

The convection currents circulating in the liquid mantle may take millions of years to rise to the surface. If currents of hot, molten rock rise in places where the Earth's crust is thin, they form **'hot spots'** of intense volcanic activity. The Hawaiian Islands lie above one of these hot spots. On these islands, magma rises to the crust surface forming **volcanoes**. In areas where the crust is thicker, rising convection currents push the crust up into a dome, causing tension and cracking (Figure 19.12a). If this causes sections of the plates on the crust to move apart, large slabs of rock sink and rift valleys form. Volcanoes appear where the magma escapes from cracks in the rift valley (Figure 19.12b). This is typical of the rift valleys in East Africa.

The Red Sea and Gulf of Aden illustrate the next stage of the **tectonic cycle**. If the floor of the rift valley widens and deepens, a **linear sea** forms (Figure 19.12c). If the African and Arabian plates continue to move apart, this region will gradually form a new ocean.

The Atlantic is an example of a **young ocean**, slowly expanding as the continents drift apart. It is widening as fast as your fingernails grow. A **mid-ocean ridge** punctuated with volcanoes has formed down the centre of the Atlantic from the Arctic to the Antarctic (Figure 19.12d). This submerged ridge has resulted from the vast amounts of lava pouring out from volcanoes along its length. It follows the boundary between the American Plate and the African and Eurasian Plates (Figure 19.11).

In some parts of the world, this volcanic activity is so great that an ocean ridge rises above the surface of the ocean as volcanic islands. Islands of this nature extend all the way down the western side of the Pacific from the Aleutian Islands in the north, through Japan to Tonga in the south.

Over millions and millions of years, thick deposits of sediment collect and spread across the ocean floor. The layers of sediment are particularly thick near continents where rivers have carried silt into the ocean.

As the Atlantic grows, the American continents are moving westwards. Along the western edge of South America, the Nazca Plate is moving towards the American Plate (Figure 19.11). Here the ocean crust is being forced under the advancing landmass. This is pushing up the continental crust and creating the Andes mountains.

These mountains were formed from horizontal beds of sediment at the bottom of an ancient ocean. As the continents moved together and the ocean closed up, this sediment was pushed up into a fold mountain range

A tectonic cycle ends when two continental landmasses converge and the ocean between them disappears. The layers of oceanic crust are squeezed into tight folds forming high mountains (Figure 19.12e). This is what happened when India moved north to collide with Asia. The ancient Tethys Ocean disappeared and the Himalayan mountains were formed. The crust here is so thick that volcanic activity has effectively stopped, although earthquakes are common.

Plenty of geological evidence supports the theory of plate tectonics. The latest evidence comes from lasers mounted on satellites which can measure continental drift of only 2–3 centimetres per year.

If the Earth was filmed from outer space using time lapse photography over a period of 1000 million years, you would be able to see the tectonic cycle clearly: continents moving, oceans waxing and waning and mountains forming and eroding.

a)

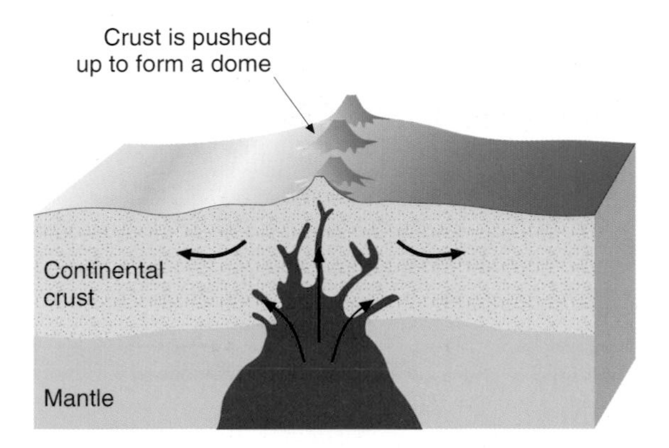

b)

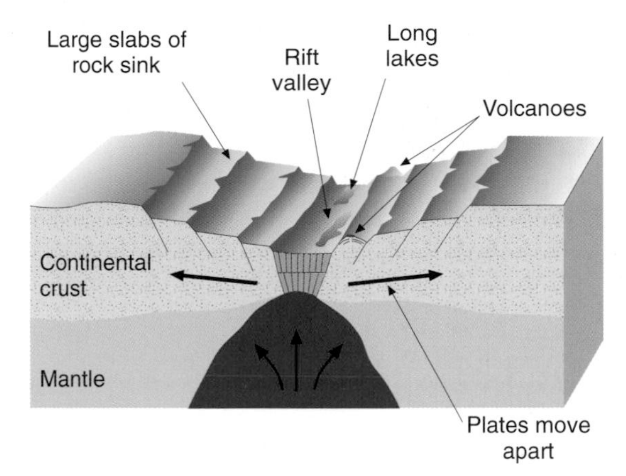

c)

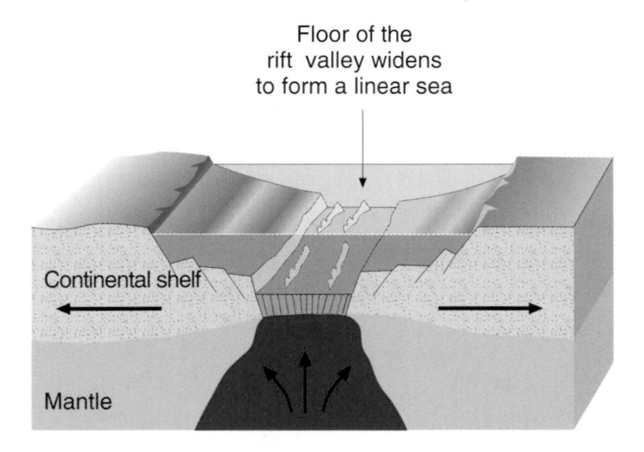

d)

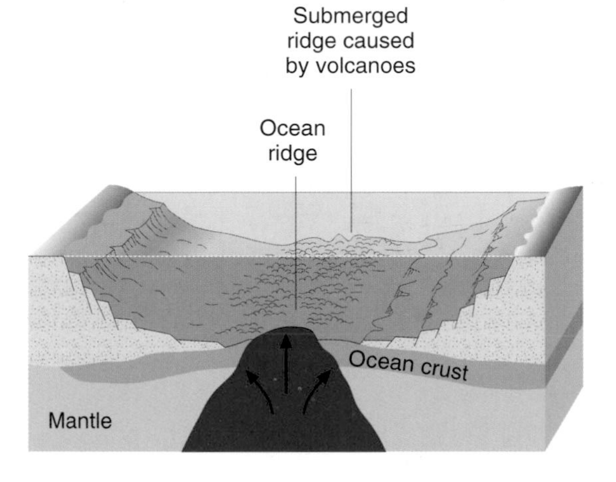

e)

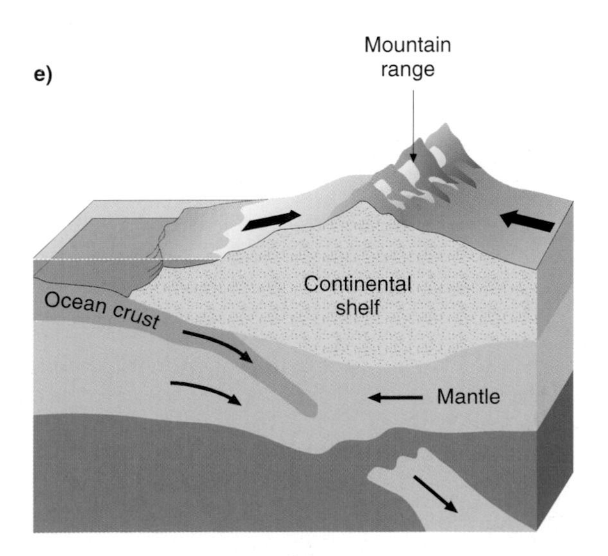

Figure 19.12 a) The continental crust is pushed up into a dome by convection currents, b) when plates move apart, rift valleys form with volcanoes, c) a linear sea develops as the rift valley widens, d) an ocean ridge forms as the ocean widens, e) as plates move towards each other and continents collide, the ocean closes and a larger land mass with high mountains is formed

1 On a hill walk a student found a rock in a stream. The student thought that the rock might be limestone.

a) Suggest a test for limestone.

b) Further tests back at school showed that the rock contained a compound with the formula $CaCO_3$.
i) Name the **three** elements that are in $CaCO_3$.
ii) Calculate the relative formula mass (M_r) of $CaCO_3$ (Relative atomic masses: Ca = 40, C = 12, O = 16)

c) The rock was limestone. Limestone is a sedimentary rock which contains the shelly remains of living organisms. Describe how these shelly remains were changed into limestone.

d) Marble is a metamorphic rock formed from limestone. Describe how marble is formed from limestone **NEAB**

2 An underwater ridge called the Mid-Atlantic Ridge runs along the floor of the Atlantic Ocean. The sea floor on either side of the Ridge is slowly spreading apart. This spreading can be explained by the theory of plate tectonics. According to this theory, materials near the Ridge moved as shown in the diagram below.

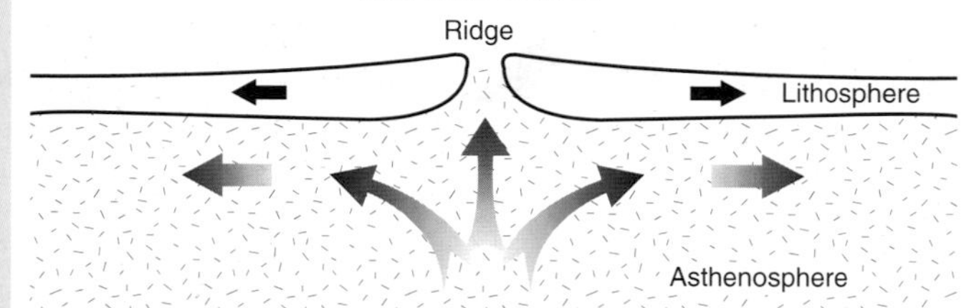

a) How can the theory of plate tectonics explain the formation of volcanoes at ocean ridges?

Plate boundaries like the Mid-Atlantic Ridge are called constructive boundaries. At some other boundaries, where plates move together, rocks are driven downwards at **subduction zones**.

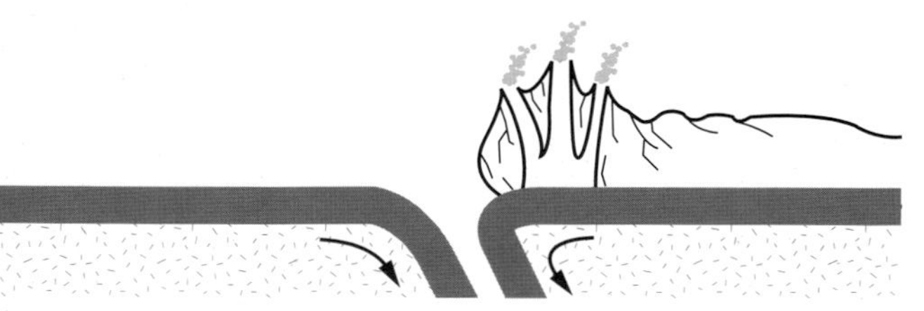

b) Use the information in the two diagrams to explain the major stages in the rock cycle.

c) How can modern ideas about plate tectonics and the rock cycle explain the movement of land masses which is called 'continental drift'? **MEG**

3 a) Limestone, basalt, slate, marble, sandstone and granite are all examples of different rocks.
From the above list of rocks choose and give the name of:
i) a sedimentary rock, ii) a metamorphic rock, iii) an igneous rock.

b) Which type of rock, sedimentary, metamorphic or igneous, is formed when molten rock (magma) is forced up from inside the Earth?

c) Describe how sedimentary rock is formed over long periods of time.

d) Rocks are widely used in the construction industry. Copy and complete the table below by choosing from the uses listed:

building cement concrete road making

Each use can only be selected once. One use has been completed for you.

Rock	Use
slate	building
granite	
limestone	
sand	

WJEC

4 a) Alistair used a key to help him identify his rock collection.
He tried a series of tests and used the results to make this key.

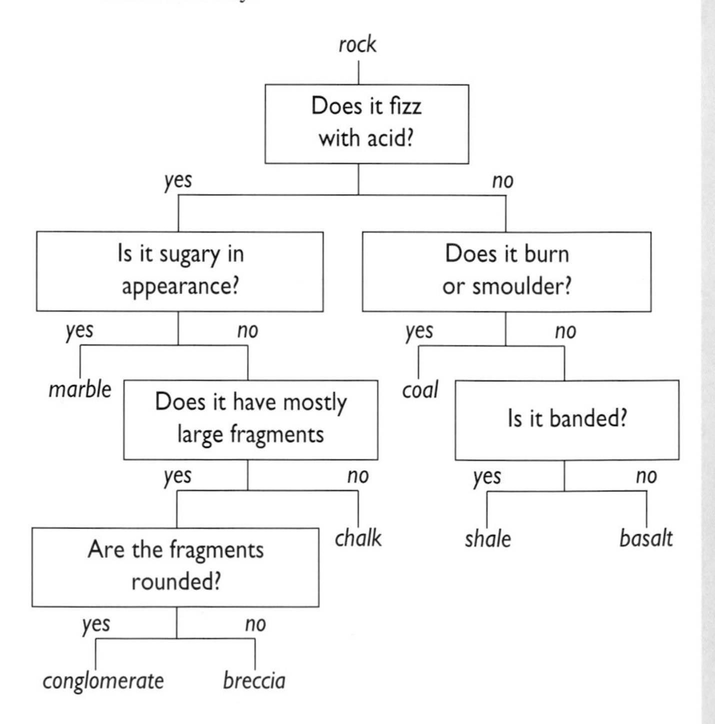

i) Use the key to identify the rock which Alistair described as having 'large, jagged fragments and it does react with acid.'
ii) Use the key to describe the piece of shale in Alistair's collection in as much detail as possible.

b) Describe how deposits of mud on the sea floor could, after a long period of time, become a metamorphic rock such as slate. **MEG**

CHAPTER

Chemicals from crude oil

20.1 The formation of coal, oil and natural gas

The most commonly used fuels are coal, oil and natural gas. These are called **fossil fuels** because they have formed from the remains of dead animals and plants.

Three hundred million years ago, the Earth was covered in forests and the seas were teeming with tiny organisms. As these plants and animals died, huge amounts of decaying organic material piled up.

When the decaying material was in contact with air and exposed to breakdown by bacteria, it rotted away completely. During this process, compounds containing carbon, hydrogen, oxygen and nitrogen in the rotting material reacted with oxygen in the air forming carbon dioxide, water and nitrogen.

In some areas, however, the decaying material was covered by the sea, by sediment from rivers, by additional deposits of dead material or by rocks from earth movements. In these places, the material continued to decay, but in the absence of oxygen. It was still attacked by bacteria. It was subjected to high temperatures as it rotted exothermically and was compressed by the water and rocks above.

Over millions of years, this led to the slow formation of coal from plants and to the formation of oil and natural gas from sea creatures.

This fossilized fern in a piece of slate mixed with coal is one of the millions of plants from which seams of coal were formed

Figure 20.1, over the page, shows how oil and natural gas are obtained by **drilling**. The gas collects above the oil which is trapped between non-permeable rocks or soaked into permeable rocks.

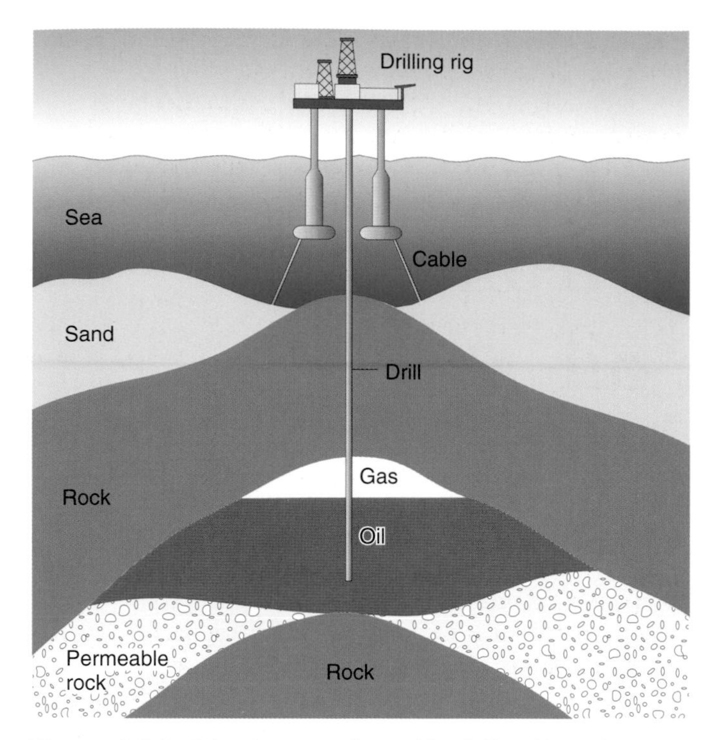

Figure 20.1 Oil and gas are obtained by drilling. First of all, a rig is built. This may be on land, on the sea bed or anchored to the sea bed. Long drills are then used to bore through rock layers to the gas and oil. The pressure of the gas forces oil through pipes to the surface

20.2 Crude oil

Crude oil (petroleum) is the main source of fuel and organic chemicals in the UK. The crude oil comes to our refineries from the North Sea and the Gulf area in the Middle East. It is a sticky, smelly, dark brown liquid. Crude oil is a mixture containing hundreds of different compounds, from simple substances like methane (CH_4) to complicated substances with long chains and rings of carbon atoms. Most of the substances in crude oil are **hydrocarbons** – substances containing only hydrogen and carbon.

A century ago oil was almost unknown. Now we could hardly survive without it. It is almost as important to our lives as air and water. In the UK, 70% of all organic chemicals come from oil. Antifreeze, brake fluid, lipstick, nylon, explosives and paint are all made from it. You may be dressed entirely in oil-based textiles, like terylene or nylon. Without oil, most transport would come to a standstill and any machine larger than a toy car would seize up from lack of lubricant.

An aerial view of the oil refinery at Pernis, Holland. Notice the fractionating column, the storage tanks and the jetties at which tankers can berth

20.3 Separating crude oil by fractional distillation

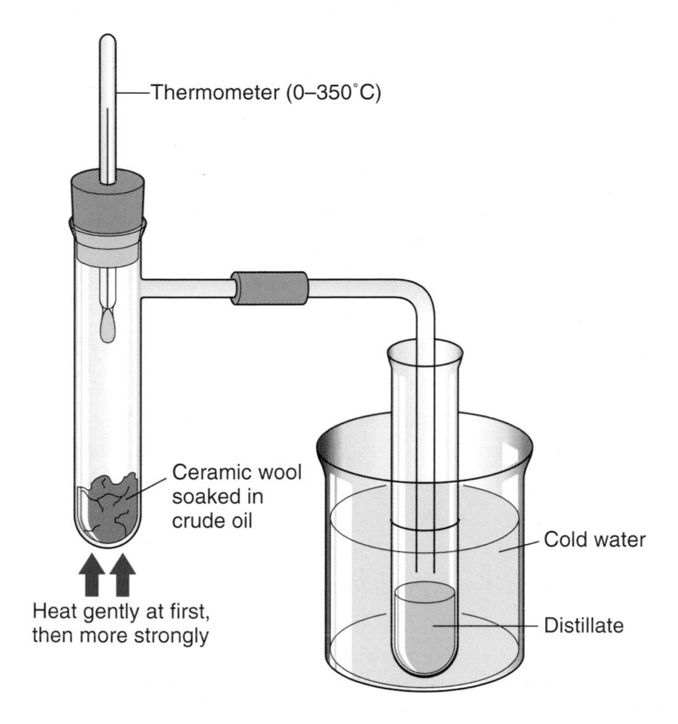

Figure 20.2 The small scale fractional distillation of crude oil

The hydrocarbons in crude oil have different boiling points so the mixture can be separated using fractional distillation. Most of the fractions from crude oil can be used as fuels for different purposes. Figure 20.2 shows the small scale fractional distillation of crude oil. The ceramic wool, soaked in crude oil, is heated very gently at first and then more strongly so that the distillate slowly drips into the collecting tube. Four fractions are collected, with the boiling ranges and properties shown in Table 20.1. Table 20.1 also shows the industrial names of the fractions to which our fractions correspond. Notice how the properties of the fractions gradually change in colour, viscosity and burning.

Figure 20.3, over the page, shows the temperatures and products at different heights in an industrial fractionating column. Inside the column there are horizontal trays with raised holes. The crude oil is heated in a furnace from which the vapours pass into the lower part of the fractionating column. As the vapours rise up the column through the holes in the trays, the temperature falls. Different vapours condense at different heights in the tower and are tapped off and used for different purposes. Liquids like petrol, which boil at low temperatures, condense high up in the tower. Liquids like lubricating oils, which boil at higher temperatures, condense lower down in the tower.

The fractions from crude oil contain similar substances with roughly the same number of carbon atoms. Figure 20.3 also shows what each fraction contains and what it is used for. The uses of the fractions depend on their properties. Petrol vaporizes easily and is very flammable, so it is ideal to use in car engines. Lubricating oil, which is very viscous, is used in lubricants and in central heating. Bitumen, which is solid but easy to melt is used for waterproofing and asphalting on roads.

Bitumen is mixed with stone chippings and used to surface roads

Boiling range (°C)	Name of fraction	Colour	Viscosity	How does it burn?
20–70	petrol (gasoline)	pale yellow	runny	easily with a clean yellow flame
70–120	naphtha	yellow	fairly runny	quite easily, yellow flame, some smoke
120–170	paraffin (kerosene)	dark yellow	fairly viscous	harder to burn, quite smoky flame
170–270	diesel oils	brown	viscous	hard to burn, smoky flame

Table 20.1 The properties of fractions obtained by the small scale fractional distillation of crude oil

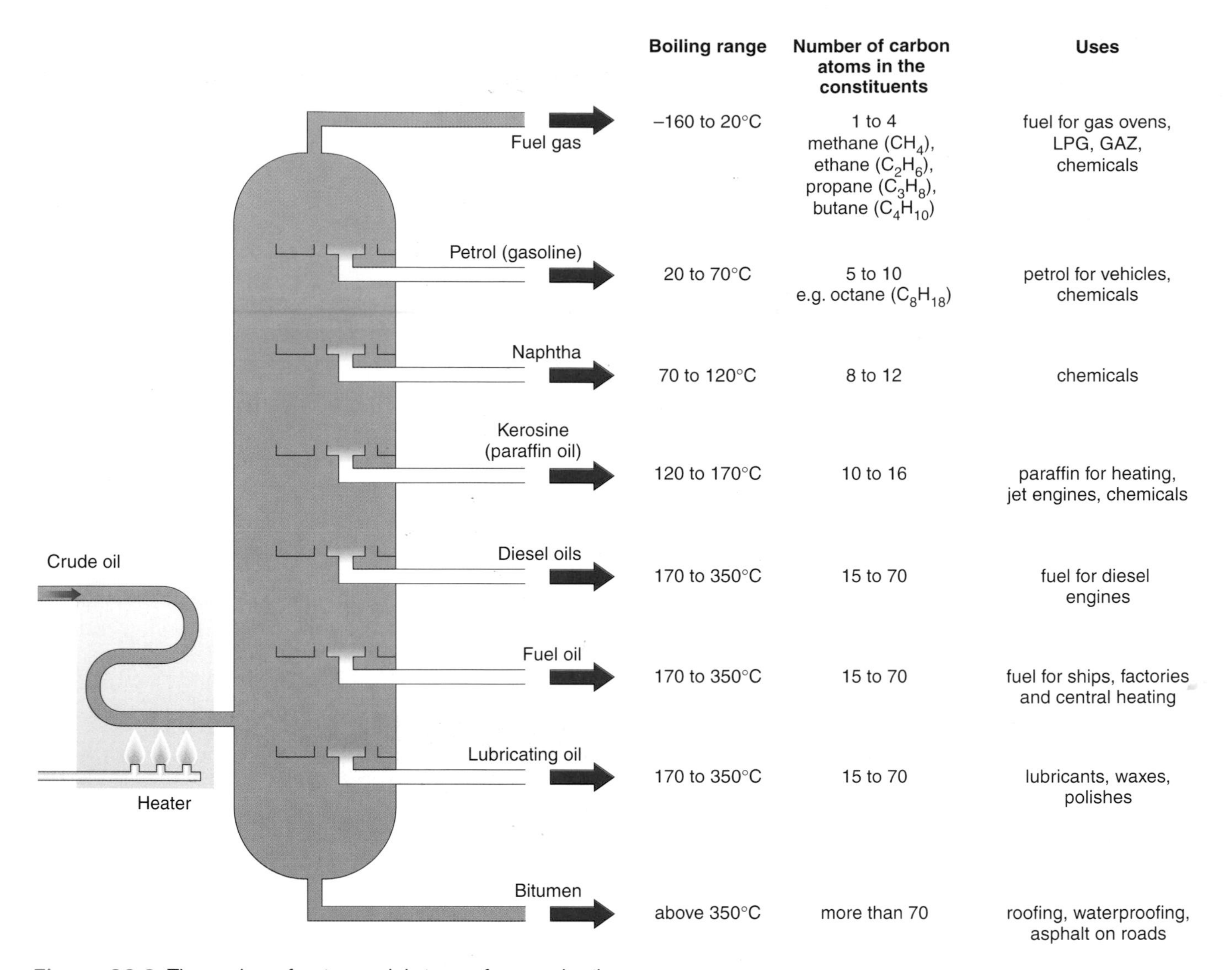

Figure 20.3 The products, fractions and their uses from crude oil

20.4 Hydrocarbons

Carbon forms millions of different compounds because carbon atoms can form strong covalent bonds with each other. Unlike carbon, the atoms of most other elements cannot do this. Because of these strong C—C bonds, carbon forms molecules containing chains of carbon atoms. In fact there are thousands of hydrocarbons alone.

The four simplest hydrocarbons are methane, ethane, propane and butane. Table 20.2 shows the molecular formulas, structural formulas and molecular models for these four hydrocarbons. The structural formulas show which atoms are attached to each other, but cannot show the correct three-dimensional structure of the molecules. The three-dimensional structures are shown in the molecular models. There are four covalent bonds to each carbon atom. Each of these bonds consists of a pair of electrons shared by two atoms (section 17.10). The four pairs of electrons around a carbon atom repel each other as far away as possible. So, the bonds around each carbon atom spread out tetrahedrally (Figure 20.4).

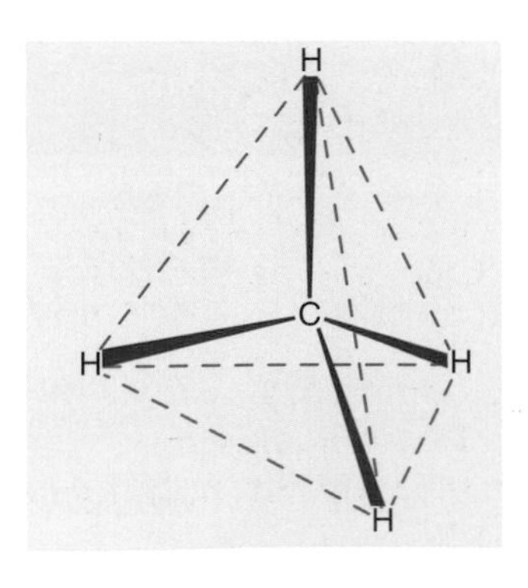

Figure 20.4 The tetrahedral arrangement of bonds in methane

Table 20.2

<table>
<tr><th>Name</th><th>Methane</th><th>Ethane</th><th>Propane</th><th>Butane</th></tr>
<tr><td>Molecular formula</td><td>CH_4</td><td>C_2H_6</td><td>C_3H_8</td><td>C_4H_{10}</td></tr>
<tr><td>Structural formula</td><td>H
|
H–C–H
|
H</td><td>H H
| |
H–C–C–H
| |
H H</td><td>H H H
| | |
H–C–C–C–H
| | |
H H H</td><td>H H H H
| | | |
H–C–C–C–C–H
| | | |
H H H H</td></tr>
<tr><td>Molecular models</td><td></td><td></td><td></td><td></td></tr>
</table>

20.5 Alkanes

Methane, ethane, propane and butane are members of a series of compounds called **alkanes**. All other alkanes are named from the number of carbon atoms in one molecule. So C_5H_{12} is *pen*tane, C_6H_{14} is *hex*ane, C_7H_{16} is *hep*tane and so on. The names of all alkanes end in -ane.

Look at the formulas of methane, CH_4, ethane, C_2H_6, propane, C_3H_8 and butane C_4H_{10}. Notice that the difference in the number of carbon and hydrogen atoms between methane and ethane is CH_2. The difference between ethane and propane is CH_2 and the difference between propane and butane is also CH_2. This is an example of a **homologous series** – a series of compounds with similar properties in which (in this case) the formulas differ by CH_2.

A small blue butane cylinder used in camping

Volatility

Alkanes are typical molecular (non-metal) compounds with low melting points and low boiling points. Alkanes with up to four carbon atoms in each molecule are gases at room temperature. Methane (CH_4) and ethane (C_2H_6) are the main constituents of natural gas. Propane (C_3H_8) and butane (C_4H_{10}) are the main constituents of 'liquefied petroleum gas' (LPG). The best known uses of LPG are 'Calor gas' and GAZ for camping, caravans and boats.

Alkanes with 5 to 17 carbon atoms are liquids at room temperature. Mixtures of these liquids are used in petrol, in paraffin and in lubricating and engine oils.

Alkanes with 18 or more carbon atoms per molecule are solids at room temperature. Tar and bitumen contain alkanes with a relative molecular mass of more than 500. Even so, they begin to melt on very hot days.

Notice that the alkanes become less volatile and change from gases to liquids and then to solids as their molecular size increases.

Insoluble in water

Alkanes are insoluble in water, but they dissolve in organic solvents such as tetrachloromethane (carbon tetrachloride) and petrol.

Cracking is important because it helps to produce more petrol. Larger alkanes are cracked to produce alkanes with about eight carbon atoms like octane, the main constituent in petrol. The petrol obtained in this way is better quality than that obtained by the distillation of crude oil. Cracked petrol is therefore blended with other petrols to improve their quality.

The BP oil refinery in Singapore

In nature, ethene acts as a trigger for the ripening of fruit, particularly bananas

20.7 Ethene

Ethene is an important industrial chemical. It is manufactured by cracking the heavier fractions from crude oil. It can be made on a small scale by cracking paraffin oil using the apparatus in Figure 20.6.

Wear eye protection if you are preparing ethene in this way and remember that both paraffin and ethene are flammable. Beware of suck back and take care that the delivery tube does not get blocked. Heat the middle of the tube below the porous pot or aluminium oxide. Heat will be conducted along the tube to vaporize the paraffin. The main gaseous product is ethene. Figure 20.7 shows some important properties of ethene.

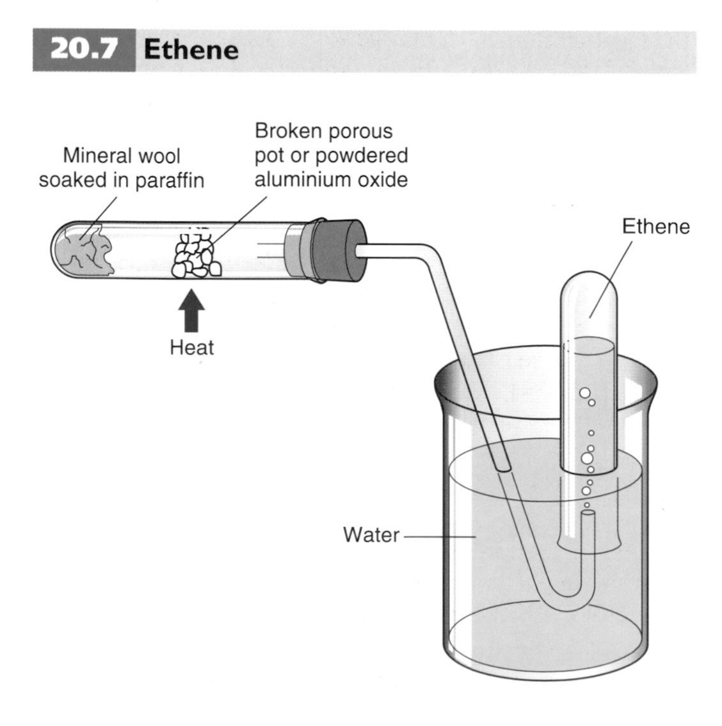

Figure 20.6 Preparing ethene by cracking paraffin oil

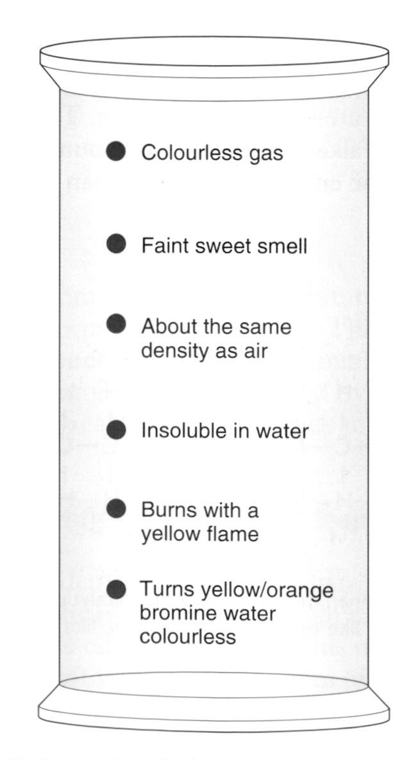

Figure 20.7 Properties of ethene

Reactions of alkenes

Alkenes, such as ethene, are much more reactive than alkanes. The most stable arrangement for the four bonds to a carbon atom is a tetrahedral one with four *single* bonds. This means that a C=C bond is unstable. Other atoms can add across the double bond to make two single bonds. So, alkenes readily undergo **addition reactions**.

This explains why yellow/orange bromine water decolorizes on shaking with an alkene, like ethene. The bromine molecules add across the double bond in ethene forming, 1,2-dibromoethane.

$$H_2C{=}CH_2 + Br_2 \rightarrow H{-}\underset{Br}{\overset{H}{C}}{-}\underset{Br}{\overset{H}{C}}{-}H$$

ethene + bromine → 1, 2-dibromoethane

Ethene also has an addition reaction with hydrogen at 150°C using a nickel catalyst. The product of this reaction is ethane.

$$H_2C{=}CH_2 + H_2 \xrightarrow[150^\circ C]{\text{Ni catalyst}} H{-}\underset{H}{\overset{H}{C}}{-}\underset{H}{\overset{H}{C}}{-}H$$

ethene + hydrogen → ethane

This process is known as **catalytic hydrogenation**. It is important in making lard and margarine from vegetable oils in palm seeds and sunflower seeds. The vegetable oils are liquids containing alkenes. During hydrogenation, these alkenes are converted to alkanes. This change in structure can turn an oily liquid into a harder, fatty solid that can be used to make margarine. By controlling the amount of hydrogen added to vegetable oil, margarine can be made as hard or as soft as required.

Compounds, like ethene, which contain double bonds are described as unsaturated. Compounds, like ethane, with only single bonds are saturated hydrocarbons. Which of the products in this photo is the most unsaturated?

Comparing alkanes with alkenes

Ethane and other alkanes are fairly unreactive. They react with oxygen (burning) and other reactive non-metals, like chlorine and bromine, but they do not react with bromine water.

In comparison, ethene and alkenes are very reactive because of the addition reactions that take place across their C=C bonds. They readily decolorize bromine water and this reaction is used to test for alkenes.

20.8 Polymers from alkenes

Ethene is one of the most valuable feedstocks for the chemical industry. Ethene and other alkenes are used to make a number of important polymers because they are so reactive. They are used in the manufacture of polythene, PVC, polypropene, polystyrene and perspex.

Polythene

In the last section, we saw that elements such as bromine and hydrogen can undergo addition reactions with ethene. If the conditions are right, molecules of ethene will also add to each other. Polythene is made by heating ethene at high pressure with special catalysts.

1 Methane CH_4 contains the elements carbon and hydrogen only. A student wanted to find out which new substances are produced when methane is burned.

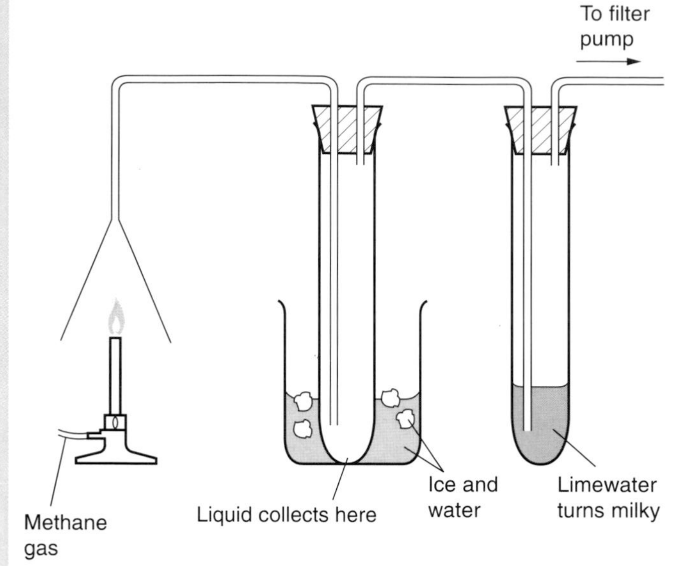

a) Which gas in the air reacts with methane when it burns?

b) Name the liquid collected.

c) Name the gas which turns limewater milky.

d) When methane burns an exothermic reaction takes place. What is meant by an exothermic reaction?

NEAB

2 a) Crude oil (petroleum) and natural gas both contain *hydrocarbons* and are both used to make important *fuels*.
Explain the meaning of the following terms used above:
i) hydrocarbon;
ii) fuel.

b) Describe how crude oil (petroleum) was formed in the geological past.

c) i) Copy and complete the word equation to represent the complete combustion of the main chemical present in natural gas, which is methane.

Methane + oxygen → ____________ + ____________

ii) Why is it important that, in natural gas heaters and cookers, **complete** combustion takes place?

d) Most plastics are made from crude oil (petroleum). Plastic products are now in widespread use and often replace traditional materials e.g. plastic gutters rather than iron, plastic window frames rather than wood, plastic containers and wrapping rather than paper and cardboard. Compare the advantages and disadvantages of the widespread use of plastics.

WJEC

3 Naphtha is produced by the fractional distillation of crude oil. It boils over a temperature range of 65 °C to 170 °C and is very flammable. When it burns in a plentiful supply of air or oxygen it produces carbon dioxide and water.

a) i) Name the **two** main elements present in naphtha.
ii) How can you tell that naphtha is a mixture of compounds rather than a single compound?

b) Naphtha is converted into ethene by a process of cracking. Give **two** conditions necessary for cracking.

c) Ethene is used to make poly(ethene) by a process of polymerisation.
i) What is **polymerisation**?
ii) Copy and complete the equation showing the changes in bonding which occur when polymerisation of ethene takes place.

$$n\begin{pmatrix} H & & & & H \\ & \rangle C & = & C\langle & \\ H & & & & H \end{pmatrix} \rightarrow$$

ethene poly(ethene)

d) There are two types of poly(ethene). They are shown as **X** and **Y** in the following table.

	X	Y
Date of discovery	1933	1953
Method of manufacture	high pressure/ catalyst	low pressure/ catalyst
Hardness	soft	hard
Density (g per cm^3)	0.92 – 0.94	0.94 – 0.96
Relative cost	low	high
Softening temperature	below 100 °C	above 100 °C

i) Give **two** reasons why **Y** is better than **X** for making plastic cutlery which will be washed and reused.
ii) In which form of poly(ethene) are the molecules more closely packed? Explain your answer.

MEG

4 An important process used in the petrochemical industry is called **cracking**. The products are used as fuels or chemical feedstocks (petrochemicals).
An example of this process is given in the following equation, for the cracking of decane to give octane and ethene, which are described as being saturated and unsaturated hydrocarbons respectively.

$$C_{10}H_{22}(l) \rightarrow C_8H_{18}(l) + C_2H_4(g)$$

a) What do the symbols (l) and (g) mean in the above equation?
b) What is the difference between a saturated and an unsaturated hydrocarbon?

WJEC

CHAPTER

Ammonia and fertilizers

21.1 Reversible reactions

Baking a cake is an irreversible reaction

Baking a cake, boiling an egg and burning natural gas are all one-way reactions. When a cake is baked or an egg boiled or natural gas is burnt, chemical reactions take place in the cake, the egg and the natural gas.

It is impossible to take the cake and turn it back into flour, fat, eggs and sugar. This also applies to the boiled egg and to the carbon dioxide and water produced when natural gas burns.

$$\underset{\text{methane in natural gas}}{CH_4(g)} + 2O_2(g) \rightarrow CO_2(g) + 2H_2O(g)$$

No matter what you do, carbon dioxide and water cannot be turned back into methane and oxygen.

Reactions like this which cannot be reversed are called **irreversible reactions**. Most of the reactions that we have studied so far are also irreversible, but there are some processes which can be reversed. For example, ice turns into water on heating.

$$H_2O(s) \xrightarrow{\text{heat}} H_2O(l),$$

But the ice reforms if water is cooled.

$$H_2O(l) \xrightarrow{\text{cool}} H_2O(s)$$

These two parts of this reversible process can be combined in one equation as:

$$H_2O(s) \underset{\text{cool}}{\overset{\text{heat}}{\rightleftharpoons}} H_2O(l)$$

When blue hydrated copper sulphate is heated, it decomposes to white anhydrous copper sulphate and water vapour.

$$\underset{\text{blue}}{CuSO_4.5H_2O(s)} \rightarrow \underset{\text{white}}{CuSO_4(s)} + 5H_2O(g)$$

If water is now added, the change can be reversed and blue hydrated copper sulphate reforms.

$$\underset{\text{white}}{CuSO_4(s)} + 5H_2O(l) \rightarrow \underset{\text{blue}}{CuSO_4.5H_2O(s)}$$

These two processes can be combined in one equation as:

$$CuSO_4.5H_2O(s) \underset{\text{mix reactants}}{\overset{\text{heat}}{\rightleftharpoons}} CuSO_4(s) + 5H_2O(l)$$

Fertilizers can be used as single compounds such as ammonium nitrate ('Nitram'), or as mixtures of compounds. For example, 'NPK' fertilizers contain nitrogen, phosphorus and potassium. The proportions of nitrogen, phosphorus and potassium in NPK fertilizers are usually shown as % nitrogen (N), % phosphorus(V) oxide (P_2O_5) and % potassium oxide (K_2O).

A solution of fertilizer being injected into the soil. What are the advantages and disadvantages of using a solution of fertilizer?

Nitrogen fertilizers
These are usually nitrates or ammonium salts. Ammonium nitrate ('Nitram'), NH_4NO_3, is the most widely used fertilizer because it is soluble, it can be stored and transported as a solid and it has a high percentage of nitrogen (Table 21.3). The higher the percentage of nitrogen the better, because less useless material needs to be stored and transported. Other nitrogen fertilizers are ammonium sulphate, urea and nitrochalk. Nitrochalk is a mixture of ammonium nitrate and chalk (calcium carbonate). This provides calcium for the soil as well as nitrogen and it also corrects soil acidity.

Phosphorus fertilizers
These are manufactured mainly from phosphate rocks containing calcium phosphate. This is insoluble in water, so it must be converted to soluble phosphorus compounds which plants can absorb through their roots. The phosphate rock is reacted with concentrated sulphuric acid. This converts it to 'super-phosphate' – a mixture of soluble calcium dihydrogenphosphate and insoluble calcium sulphate.

Although fertilizers are important in producing high yields of crops, problems are caused by their over use.

- They may change the soil pH and the pH of water which runs off the land.
- They may harm plants and animals in the soil.
- They allow those elements in the fertilizers that are not required by plants to accumulate in the soil.
- They get washed out of the soil and lead to the pollution of rivers.

Trials on the success of the use of fertilizers demonstrate that the application of nitrogen to the soil on the left of this field has promoted greener, taller and bushier wheat plants

Table 21.3 The percentage of nitrogen in different fertilizers

Fertilizer	Formula	Mass of one mole (g)	Mass of nitrogen in one mole (g)	% of nitrogen
Ammonium nitrate	NH_4NO_3	80	28	$\frac{28}{80} \times 100 = 35$
Ammonia	NH_3	17	14	$\frac{14}{17} \times 100 = 82$
Ammonium sulphate	$(NH_4)_2SO_4$	132	28	$\frac{28}{132} \times 100 = 21$
Urea	N_2H_4CO	60	28	$\frac{28}{60} \times 100 = 47$

1 a) Nitrogen is used to make ammonia.
The equation that represents the main reaction involved in the manufacture of ammonia is given below:

$$\text{nitrogen} + \text{hydrogen} \rightleftharpoons \text{ammonia}.$$

What does the symbol $\rightleftharpoons$ mean?

b) Ammonia is an alkaline gas, soluble in water. How would you neutralise the solution of ammonia in water to form ammonium sulphate?

c) i) Ammonium sulphate is used as a fertilizer. Give **one** advantage and **one** disadvantage of the use of fertilizers.
ii) Do the advantages of fertilizer use outweigh the disadvantages? Give a reason for your evaluation.
WJEC

2 The Haber process is used in industry to produce ammonia. The important reaction between nitrogen and hydrogen is a reversible reaction.

$$N_2(g) + 3H_2(g) \rightleftharpoons 2NH_3(g)$$

a) The graph shows the percentage of ammonia produced at different temperatures and pressures.

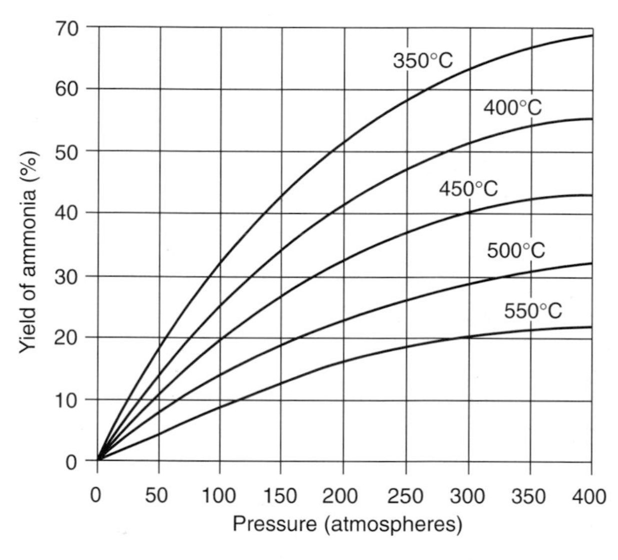

What conditions of temperature and pressure produce the highest yield of ammonia?

b) An iron catalyst is used in the Haber process.
i) What effect does an iron catalyst have on the rates of the forwards and reverse reactions?
ii) What effect does an iron catalyst have on yield of ammonia at equilibrium?
iii) State **three** conditions necessary for a reversible reaction to establish 'dynamic equilibrium'.

c) After passing over the iron catalyst, the mixture of gases contains ammonia, hydrogen and nitrogen. The following table shows the boiling point of each gas.

Gas	Boiling point (°C)
ammonia	−33
hydrogen	−253
nitrogen	−196

i) Use this information to explain how the ammonia is extracted from the mixture.
ii) Describe what happens to the hydrogen and nitrogen after the ammonia has been extracted from the mixture.

d) Much of the ammonia produced in the Haber process is used to make fertilizers such as ammonium sulphate.

$$2NH_3(g) + H_2SO_4(aq) \rightarrow (NH_4)_2SO_4(aq)$$

i) Calculate the mass of 1 mole of ammonia and 1 mole of ammonium sulphate. (Relative atomic masses: H = 1, N = 14, O = 16, S = 32)
ii) Calculate the mass of ammonium sulphate produced from 17 tonnes of ammonia.
MEG

3 a) In the blast furnace, iron(III) oxide is reduced to iron by carbon monoxide. The equation is

$$Fe_2O_3 + 3CO \rightleftharpoons 2Fe + 3CO_2$$

(Relative atomic masses: Fe = 56, O = 16, C = 12)
i) Calculate the mass of iron which could be obtained by the reduction of 800 tonnes of iron(III) oxide. Show your working.
ii) If the reaction was carried out on a small scale, what volume of carbon dioxide would be obtained by the reduction of 320 g of iron(III) oxide? Show your working.
(1 mole of gas at atmospheric pressure and temperature occupies a volume of 24 dm^3.)

b) In the Haber process, ammonia is manufactured from nitrogen and hydrogen using the reaction

$$N_2(g) + 3H_2(g) \rightleftharpoons 2NH_3(g)$$

The reaction is *reversible* and the formation of ammonia is exothermic.
i) What is meant by a *reversible* reaction?
ii) In this process what **two** conditions are normally changed to increase the **rate** of reaction between nitrogen and hydrogen?
iii) How should the manufacturing conditions be set to keep the **yield** of ammonia as high as possible?

c) Ammonia can be used to manufacture ammonium salts which are used as nitrogenous fertilizers. When fertilizers are used on farmland, some are dissolved in the ground water and eventually drain into rivers causing *eutrophication*. Explain what is meant by *eutrophication* and why it is a problem.
SEG

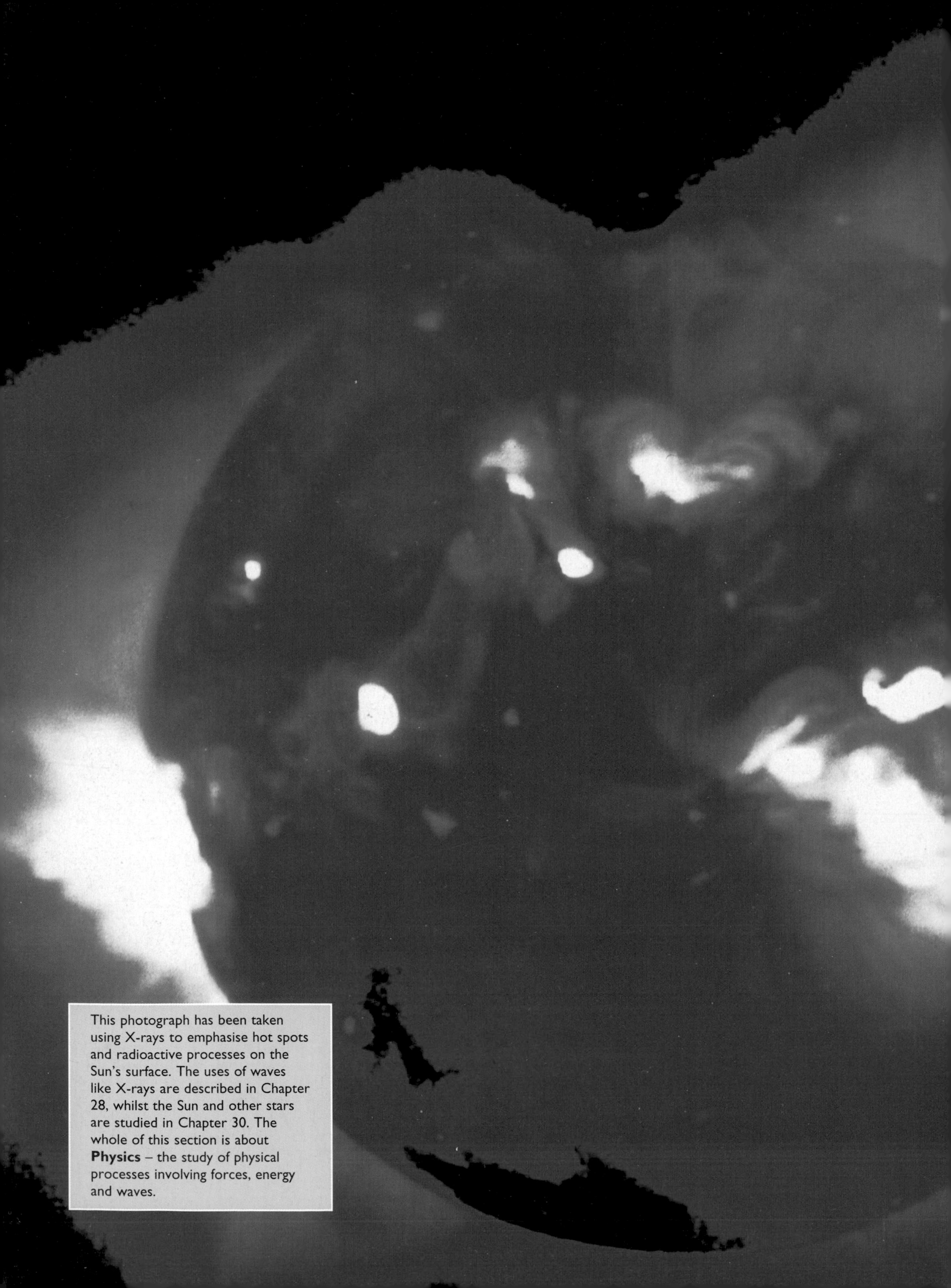

This photograph has been taken using X-rays to emphasise hot spots and radioactive processes on the Sun's surface. The uses of waves like X-rays are described in Chapter 28, whilst the Sun and other stars are studied in Chapter 30. The whole of this section is about **Physics** – the study of physical processes involving forces, energy and waves.

SECTION 3

Physical Processes

CHAPTER

22

Forces and motion

22.1 Extending materials by forces

When masses are hung on a spring, the spring extends. The force of gravity pulling on the mass causes the spring to extend. The spring can support the masses due to forces of attraction between particles in the spring. The pull in the spring which supports the weight of the masses is called a **tension**.

Figure 22.1 shows a simple experiment to investigate the extension of the spring as the stretching force increases.

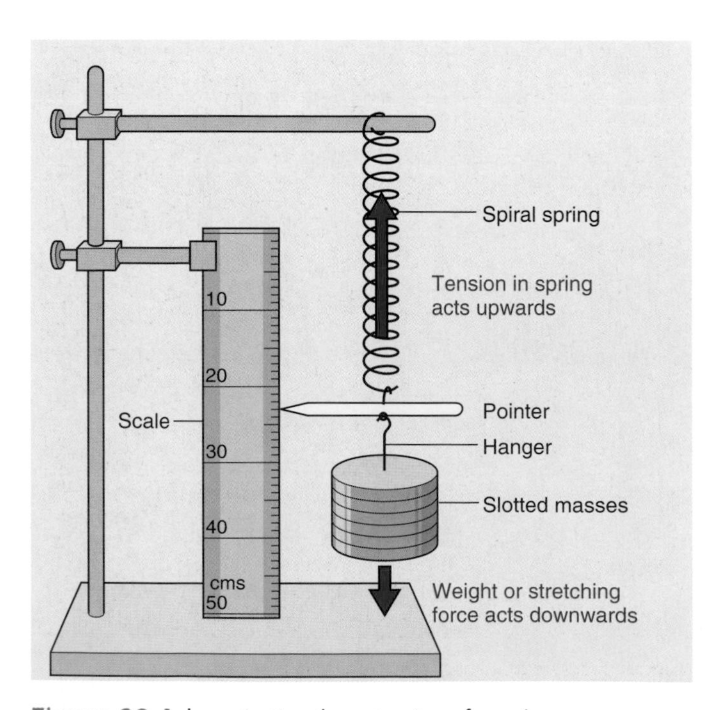

Figure 22.1 Investigating the extension of a spring

The results of one experiment are plotted on a graph in Figure 22.2.

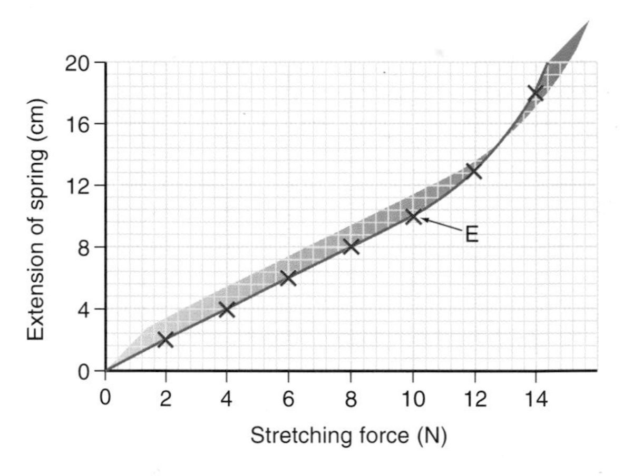

Figure 22.2 A graph of the extension of a spring against the stretching force (load)

Notice that the graph is a straight line up to point E for smaller stretching forces. After point E, the spring extends more than expected for a given stretching force. Point E is said to represent the **elastic limit** for the spring.

The results in Figure 22.2 show that:

> The extension is proportional to the stretching force provided the elastic limit is not exceeded.

This result is known as **Hooke's law**.

Provided the elastic limit is not exceeded, materials return to their original shape and size when the force is removed. Materials with this property are called **elastic**. This means that when the stretching force is removed, they return to their original shape. If their elastic limit is exceeded, they remain permanently deformed.

The elastic properties of springs are put to use in kitchen scales, in mattresses, in chairs and in car suspension systems. Although the elasticities of wood and concrete are much less than that of metals, they allow buildings to move slightly following earth movements.

A building in Mexico City after the 1985 earthquake. Wood and concrete are elastic enough to allow some movement of buildings except during an earthquake

22.2 Force and pressure

The word **pressure** is used to describe the concentration of a force. If a force acts on a small area, pressure is high. If a force acts on a large area, pressure is low.

Pressure is defined as the force per unit area.

$$\text{pressure} = \frac{\text{force (N)}}{\text{area (m}^2\text{)}}$$

So, the units of pressure are N/m^2 or pascals (Pa) where $1\ N/m^2 = 1\ Pa$

Using the idea of pressure as force per unit area, it is easy to see why:

- tractors working on boggy land need wide tyres,
- shoes with pointed stiletto heels cause damage to wooden floors,
- camels with larger hooves can move over a sandy desert easier than horses,
- woodpeckers need strong, pointed beaks.

Vehicles working on soft ground have wide tyres to reduce overall pressure

Woodpeckers need strong, pointed beaks

Example

A bridge which is being constructed will weigh 1.6×10^6 N. What minimum area must the foundations have if the bridge will sink into the soil if it creates a pressure greater than 100 000 N/m^2?

Solution

$$\text{pressure} = \frac{\text{force}}{\text{area}} \Rightarrow \text{area} = \frac{\text{force}}{\text{pressure}}$$

$\therefore$ minimum area of foundations

$$= \frac{1.6 \times 10^6 \text{ N}}{100\,000 \text{ N/m}^2} = 16 \text{ m}^2$$

22.3 Pressure in liquids – hydraulic machines

Hydraulic machines use liquid under pressure to transmit forces. Liquids have two key properties which make them ideal for use in hydraulic machines.

1 **They cannot be compressed easily**, unlike gases.

2 **They will allow pressure to be transmitted through the whole liquid and in all directions**, up, down and sideways, unlike solids.

Hydraulic machines are important in hydraulic jacks, car brakes, chair lifts and fork-lift trucks.

By using pistons with different cross-sectional areas, hydraulic machines can magnify forces.

Example

A dentist applies a force of 50 N to piston A of his hydraulic chair. What force does that provide on piston B below the patient's chair?

Solution

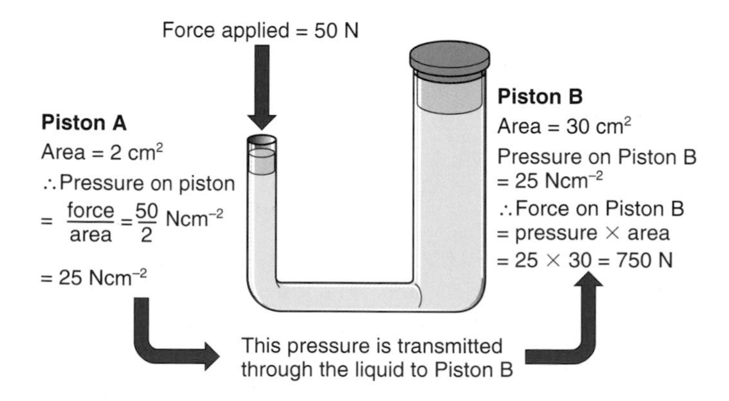

Figure 22.3 A hydraulic machine used in a dentist's chair

The force of 750 N on piston B would be sufficient to raise children and most adults in the dentist's chair. By applying different forces to piston A, the dentist can raise patients of varying weights.

22.4 Pressure in gases – Boyle's law

Gas pressure enables us to blow up balloons and tyres (Figure 22.4). When atmospheric pressure is high, clouds are driven away, skies are clear and the weather is fine. When atmospheric pressure is low, the weather is usually cloudy and wet.

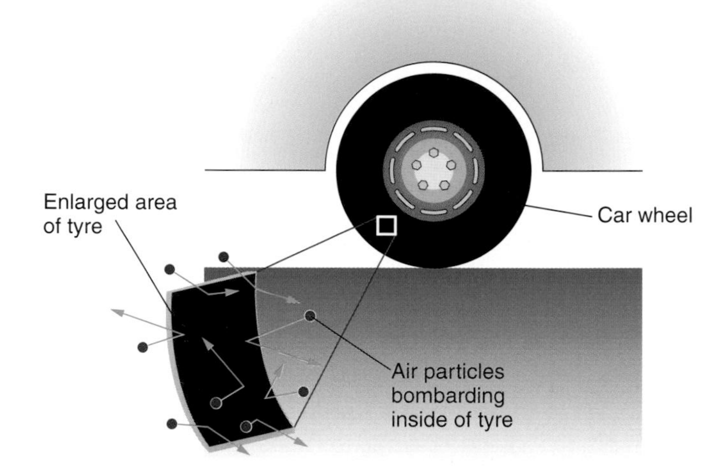

Figure 22.4 Millions of air molecules bombard the inside of the tyre every second. This produces pressure on the inside of the tyre. The tyre is also bombarded on the outside by air molecules, but the pressure on the inside of the tyre is greater than that outside, so the tyre stays inflated

If you hold your finger over the end of a bicycle pump and push in the handle, the pressure on your finger increases (Figure 22.5). The air inside the pump is pushed into a smaller volume, so the air molecules inside the pump bombard the sides of the pump and your finger more often. This causes the increased pressure on your finger.

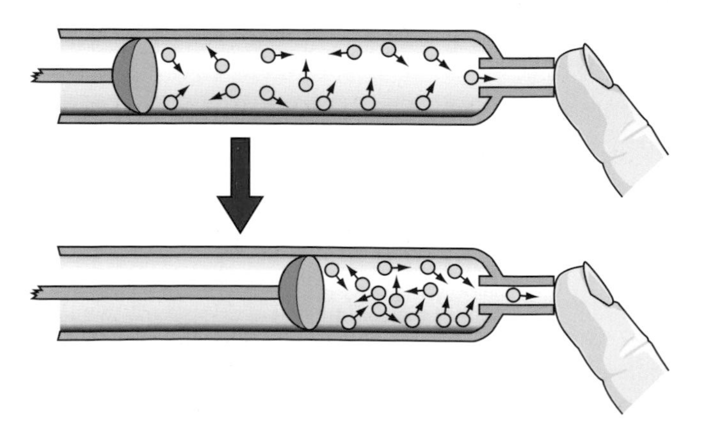

Figure 22.5

Experiments show that when the volume decreases to half, the pressure doubles and if the volume decreases to one third the pressure triples, and so on. This is an inverse relationship and we can say that:

For a fixed mass of gas at constant temperature, the volume is inversely proportional to the pressure.

i.e.

$$p \propto \frac{1}{V} \quad \text{or} \quad p = \frac{\text{constant}}{V}$$

so

$$pV = \text{constant}$$

or

initial pressure × initial volume
= final pressure × final volume.

This relationship between p and V was first discovered by Robert Boyle in the 17th century. It is called **Boyle's law**.

Example

20 m^3 of gas in the British Gas pipelines at a pressure of 750 kPa escapes into the atmosphere where the pressure is 100 kPa. What volume will the gas occupy after it escapes?

Solution

initial pressure × initial volume
= final pressure × final volume

750 (kPa) × 20 (m^3)
= 100 (kPa) × final volume after escape

$$\therefore \text{volume after escape} = \frac{750 \times 20\ \text{m}^3}{100} = 150\ \text{m}^3$$

22.5 Force, work and power

Figure 22.6 shows a crane at work on a building site. The crane is lifting a pile of bricks weighing 10 000 N through a height of 8 m.

The amount of work which the crane does depends on:

- the weight of the bricks,
- the height the bricks are lifted.

In fact, work is defined by the formula,

work	=	force	×	distance
(joules)		(newtons)		(metres)

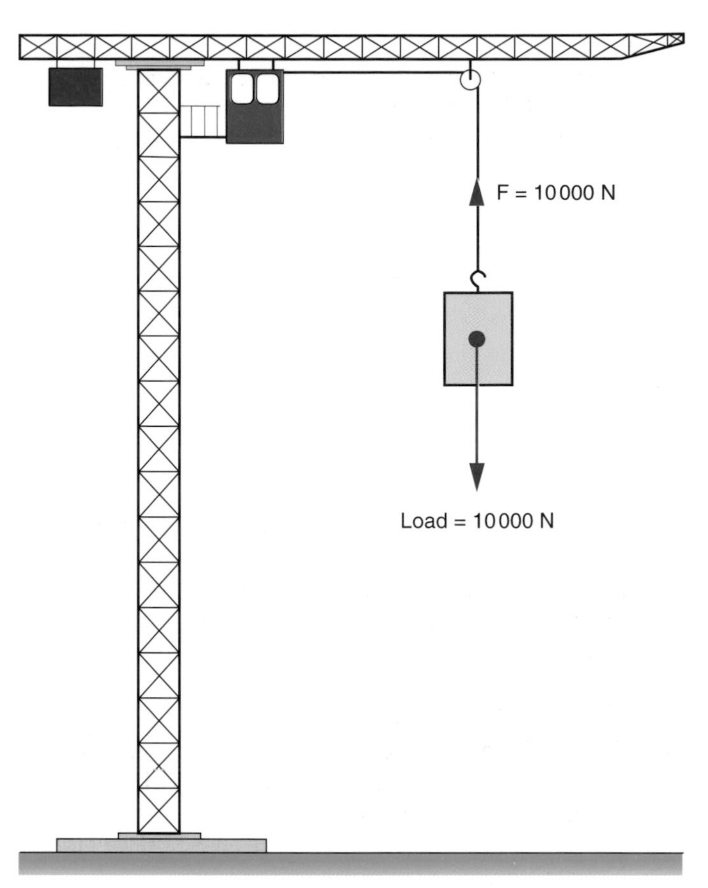

Figure 22.6 A crane working on a building site

Work is done whenever a force causes something to move. We do work when we throw a ball, turn a tap or lift a chair. Vehicles do work when they climb hills overcoming the force of gravity or when they overcome frictional forces.

Work is measured in **joules** (J). One joule of work is done when a force of 1 newton (1 N) moves something through a distance of 1 metre (1 m). (1 J ≡ 1 Nm)

We can now calculate the work done by the crane in lifting the bricks.

Work done by crane = force × distance
= weight lifted × height lifted
= 10 000 × 8 = 80 000 J
(N) (m)

It is useful to know how much work the crane has done, but this work would be useless if it took several hours to do it. In order to appreciate the usefulness of the crane, we need to know how quickly it can do the work. Now, suppose the crane lifts the bricks in 4 seconds.

Work done by crane per second

$$= \frac{80\,000\ \text{J}}{4\ \text{s}} = 20\,000\ \text{J/s}$$

The work done by the crane per second is called its **power.**

Power is the rate of working. So:

$$\text{power} = \frac{\text{work done}}{\text{time taken}}$$

Power can be given in joules per second as above, but we usually use **watts** (W) in place of joules per second. The name 'watt' is used to recognise the work of James Watt who began the scientific investigations of power in the 18th century. (1 W ≡ 1 J/s)

How will this athlete's activity change as he develops more power in his legs?

Example
Sally wants to measure the power in her arms by lifting two 25 N weights (one in each hand). She lifts them through 50 cm in 10 seconds. What is her power?

Solution

$$\begin{aligned}\text{Work done} &= \underset{(\text{N})}{\text{force}} \times \underset{(\text{m})}{\text{distance}}\\ &= \text{weight lifted} \times \text{height lifted in m}\\ &= (2 \times 25) \times 0.5\\ &= 25\ \text{J}\end{aligned}$$

So,

$$\text{power} = \frac{\text{work done}}{\text{time taken}} = \frac{25}{10} = 2.5\ \text{W}$$

22.6 Speed and velocity

John and Duane walk from Whitford to Shawley, which is exactly 20 km north-west of Whitford, in 8 hours. The actual distance which they walk is 24 km. We say that John and Duane have walked a **distance** of 24 km, but their **displacement** is only 20 km north-west.

- Distance is the total path taken.
- Displacement is the distance moved in a particular direction.

Besides knowing how far John and Duane have walked, we might want to know how fast they have moved, i.e. their speed.

Now,

$$\text{speed} = \frac{\text{distance moved}}{\text{time taken}}$$

$$\begin{aligned}&= \frac{24\ \text{km}}{8\ \text{hr}} = 3.0\ \text{km/hr}\\ &= \frac{3 \times 1000}{60 \times 60}\ \text{m/s}\\ &= 0.83\ \text{m/s}\end{aligned}$$

The S.I. units for distance and time are the metre and the second respectively, so speed is often measured in metres/second or m/s. Various other units, including km/hr and miles/hour are also used for speeds.

Although John and Duane walk 24 km in 8 hours, they are moving in different directions during that time. Their speed is 3.0 km/hr, but it has no particular direction. If we calculate their **speed in a particular direction**, this is called their **velocity**.

$$\text{velocity} = \frac{\text{displacement}}{\text{time taken}}$$

$$\text{So, John and Duane's velocity} = \frac{20\ \text{km north-west}}{8\ \text{hr}}$$
$$= 2.5\ \text{km/hr north-west}$$

Remember that whenever a velocity is given, you must show its direction as well as its size.

22.7 Distance–time graphs

Table 22.1 shows the distance that Kapil has sprinted after every 20 seconds in a 400 metres race.

Figure 22.7 shows a distance–time graph for Kapil's 400 metre sprint.

Table 22.1

Distance sprinted (m)	0	50	100	150	200	233	267	300	333	367	400
Time taken (s)	0	5	10	15	20	25	30	35	40	45	50

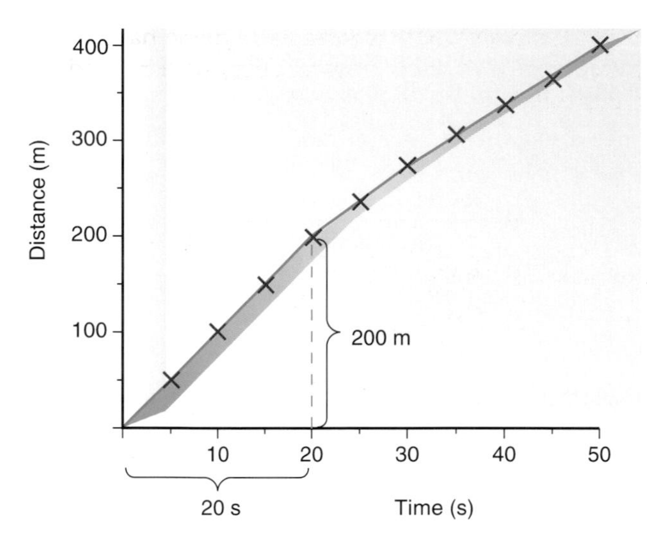

Figure 22.7 A distance–time graph for Kapil's 400 m sprint

Notice from Table 22.1 that Kapil sprints more quickly over the first 200 metres than he does over the second 200 metres.
For the first 200 metres,

$$\text{Kapil's speed} = \frac{\text{distance travelled}}{\text{time taken}}$$

$$= \frac{200\ \text{m}}{20\ \text{s}} = 10\ \text{m/s}$$

$$= \text{gradient (slope) of graph}$$

$\therefore$ speed = the gradient of the distance–time graph

What is Kapil's speed during the second 200 m of his sprint?

Although Kapil travels at two different speeds during the race, he is probably more interested in his average speed:

$$\text{average speed} = \frac{\text{total distance travelled}}{\text{total time taken}}$$

$$\therefore \text{Kapil's average speed} = \frac{400\ \text{m}}{50\ \text{s}} = 8\ \text{m/s}$$

22.8 Speed and acceleration

A sports car moves away from the traffic lights and reaches a speed of 15 m/s in 5 seconds.

We say that it has accelerated and define **acceleration** as **the rate of change of speed** or

$$\text{acceleration} = \frac{\text{change in speed}}{\text{time taken}}$$

$$\text{So, acceleration of the sports car} = \frac{15\ \text{m/s}}{5\ \text{s}} = \frac{3\ \text{m/s}}{\text{s}}$$

The car accelerates (increases its speed) by 3 m/s every second. This is usually written as 3 m/s^2.

Example
In some cases, vehicles must slow down rather than speed up. Suppose a car is travelling at 24 m/s. The driver suddenly applies the brakes and the car stops in 1.5 seconds. What is the acceleration?

Solution

$$\text{Acceleration} = \frac{\text{change in speed}}{\text{time taken}}$$

$$= \frac{-24\ \text{m/s}}{1.5\ \text{s}}$$

$$= -16\ \text{m/s}^2$$

Notice in this case that as the car is slowing down, the speed decreases and the change in speed is therefore *negative* leading to a *negative acceleration.*

Factors affecting the stopping distances of vehicles

The stopping distance of a vehicle is the sum of two distances:

1. the distance the vehicle travels during the driver's reaction time, called the '*thinking distance*',
2. the distance the vehicle travels once the brakes have been applied, called the '*braking distance*'.

Important factors affecting the thinking distance are:

- the speed of the vehicle (obviously),
- the driver's reaction (affected adversely by tiredness, alcohol or drugs),
- the general visibility due to weather conditions and road lighting.

Important factors affecting the braking distance are:

- the speed of the vehicle,
- the force applied to the brakes,
- the road conditions (dry, wet or icy, smooth or rougher surface),
- the maintenance of the vehicle, particularly the condition of the brakes and tyres.

Crash testing a van to evaluate the effectiveness of the front crumple zone and the seat belts. The van was driven into a wall at 30 mph with a dummy in the driving seat

22.9 Speed–time graphs

Look closely at Figure 22.8. This shows the speed of an athlete during the first second of a 100 m race.

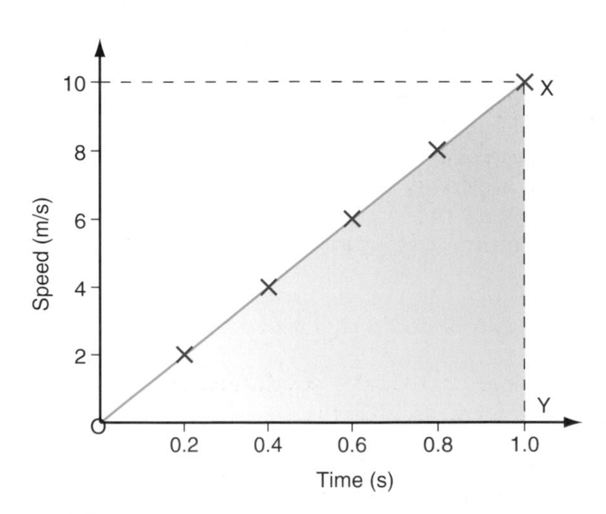

Figure 22.8

$$\text{The acceleration of the athlete} = \frac{\text{increase in speed}}{\text{time taken}}$$

$$= \frac{10 \text{ m/s}}{1 \text{ s}} = 10 \text{ m/s}^2$$

The straight line of the graph shows that the athlete's speed increases at a constant rate (i.e. acceleration is constant) during the first second.

Notice also from the graph that acceleration

$$= \frac{\text{increase in speed}}{\text{time taken}} = \frac{XY}{OY}$$

$$= \text{gradient}$$

i.e.

acceleration is equal to the gradient of a speed–time graph.

How far does the athlete run in the first second shown in Figure 22.8?

From the equation,

$$\text{speed} = \frac{\text{distance}}{\text{time}}$$

we can write,

$$\text{distance} = \text{speed} \times \text{time},$$

and if the speed is changing steadily, we must write

$$\text{distance} = \text{average speed} \times \text{time}.$$

Now the average speed of the athlete in the first second $= \frac{1}{2} \times 10$ m/s

So, the distance run in the first second

$$= \tfrac{1}{2} \times 10 \times 1 = 5 \text{ m}$$
$$= \tfrac{1}{2} \times \text{final speed} \times \text{time}$$
$$= \tfrac{1}{2} \times XY \times OY$$
$$= \text{shaded area below the graph line}$$

i.e.

distance travelled = area below the graph line in a speed–time graph.

22.10 Forces and acceleration

When an object rests on the floor, the weight of the object exerts a downward force on the floor. In response, the floor exerts an upward force on the object (Figure 22.9).

The two forces which the floor and the object exert on each other are equal in size, but opposite in direction. They are usually described as **balanced forces**. When balanced forces act on an object, the overall force is, of

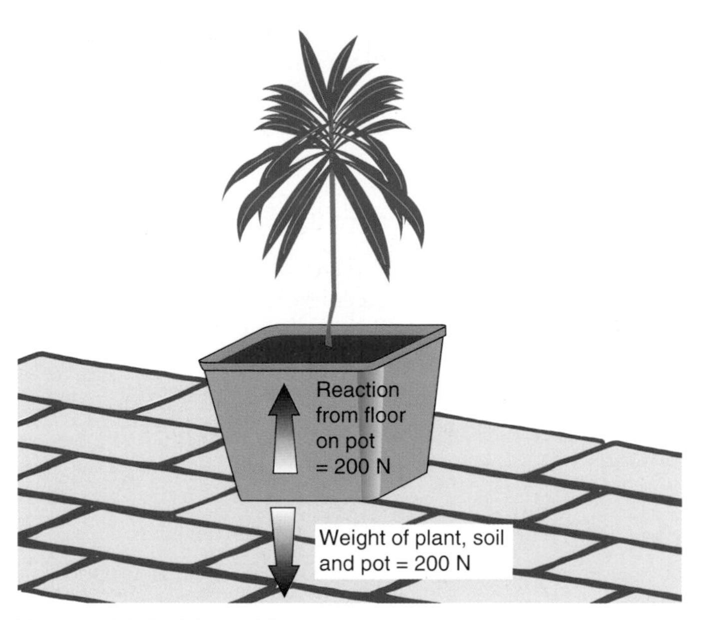

Figure 22.9 Balanced forces

course, zero and this has no effect on the movement of an object. The first scientist to appreciate this point was Isaac Newton in the 17th century. His ideas are summarised in **Newton's first law of motion**:

> If an object is stationary it will remain so, or if it is moving it will continue to move at the same speed and in the same direction unless there are unbalanced forces acting on it.

However, if you roll a ball along the floor, it does not continue to move indefinitely because unbalanced forces, due to friction, act on it. These frictional forces slow the ball down and eventually bring it to rest.

When unbalanced forces, like friction, act in the opposite direction to a moving object, the object slows down or decelerates (i.e. negative acceleration).

When an unbalanced force acts on an object in a particular direction, the object will accelerate in that direction.

If you have ever tried to push a car along a flat road, you will know that it accelerates away faster if two people push instead of one and even better with three people.

Careful experiments show that **acceleration is proportional to the force applied**.

i.e.

$$a \propto F \qquad (1)$$

You will also probably know that it is easier to push and accelerate a small car than a large lorry. Accurate experiments show that if you push larger and larger masses with the same force, the acceleration gets smaller and smaller. This is summarised by saying that **acceleration is inversely proportional to the mass**.

i.e.

$$a \propto \frac{1}{m} \qquad (2)$$

Equations (1) and (2) can be summarised to obtain

$$a = \frac{F}{m} \qquad (3)$$

$$\text{or } F = m \times a$$

$$\underset{(\text{N})}{\text{force}} = \underset{(\text{Kg})}{\text{mass}} \times \underset{(\text{m/s}^2)}{\text{acceleration}}$$

Equation 3 is the mathematical form of **Newton's second law of motion**:

> The acceleration of an object is directly proportional to the force acting on it and inversely proportional to its mass.

The acceleration of this car will be proportional to the overall force acting on it

Example
A rocket (mass = 1 tonne) accelerates from rest to 40 m/s in 5 seconds. What is:

(i) its average acceleration;
(ii) the force causing this acceleration;
(iii) the distance travelled in the 5 seconds?

Solution

(i) Average acceleration $= \dfrac{\text{change in speed}}{\text{time taken}}$

$= \dfrac{40 \text{ m/s}}{5 \text{ s}} = 8 \text{ m/s}^2$

(ii) Force causing acceleration, $F = m \times a$

$= 1000 \times 8$ (kg) (m/s²)

$= 8000 \text{ N}$

(iii) From: speed $= \dfrac{\text{distance}}{\text{time}}$

distance travelled $=$ average speed $\times$ time

$= \dfrac{40 \text{ m/s}}{2} \times 5 \text{ s} = 100 \text{ m}$

22.11 Forces on falling objects

When an object is dropped, it falls to the ground. Its weight acts on it as an unbalanced force and it accelerates downwards due to the force of gravity. In the air, a penny falls faster than a tiny piece of paper. But in a vacuum, the penny and the paper fall at the same rate (Figure 22.10).

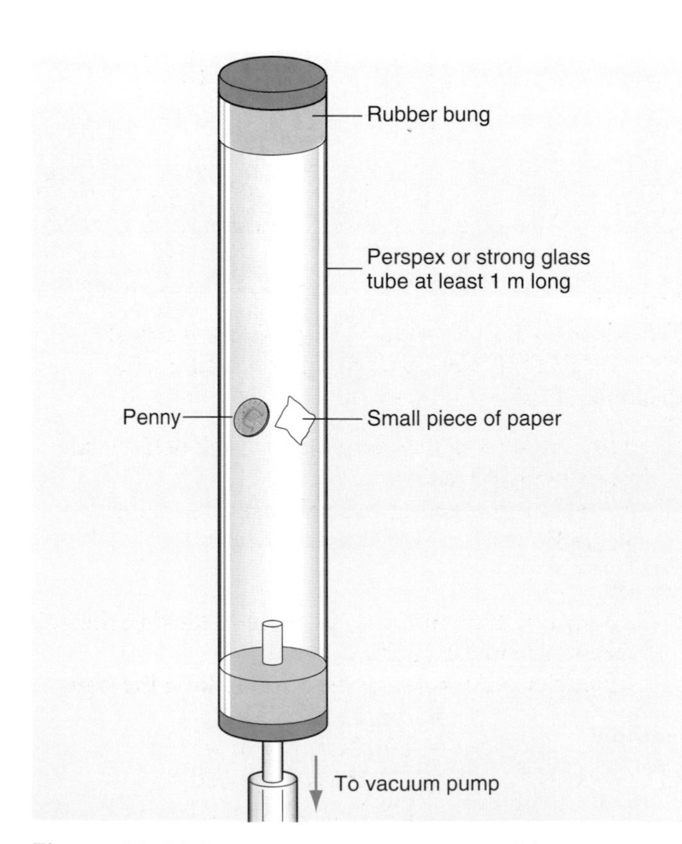

Figure 22.10 When the tube is evacuated and then quickly inverted, the penny and the piece of paper fall at the same rate. If the tube is *not* evacuated, the penny falls faster

The difference in the rate at which the penny and the paper fall in air is due to air resistance (i.e. friction or drag). This affects the fall of less dense objects like paper more than denser objects like the penny. Accurate experiments show that the **acceleration due to gravity** (symbol g) is 9.8 m/s². For most calculations, its value is taken to be **10 m/s²**.

This means that the speed of a falling object increases by 10 m/s every second. Acceleration due to gravity also means that if something is thrown upwards, its velocity will *decrease* by 10 m/s every second until it reaches its highest point. From there, its velocity will increase downwards at the rate of 10 m/s every second.

Example

A boy throws a stone vertically upwards at a speed of 30 m/s.

(i) What is the acceleration of the stone,
(ii) How long does it take to reach its highest point,
(iii) How high will it go?

Solution

(i) Acceleration of stone $=$ acceleration due to gravity

$= 10 \text{ m/s}^2$ downwards

i.e. -10 m/s^2

(ii) Now, acceleration $= \dfrac{\text{change in speed}}{\text{time taken}}$

At its highest point, the speed of the stone is zero,

$\therefore -10 = \dfrac{0 - 30}{\text{time taken}}$

$\therefore$ time to reach highest point $= \dfrac{-30}{-10} = 3$ seconds

(iii) Distance to highest point $=$ average speed $\times$ time

$= \dfrac{30 \text{ m/s}}{2} \times 3 \text{ s}$

$= 45 \text{ m}$

When an object falls, air resistance (drag) increases as its speed increases. So when an object falls:

1 it accelerates initially due to the force of gravity;

2 frictional forces due to air resistance increase as it travels faster until eventually they balance the gravitational forces due to the object's weight;

3 the forces acting on the falling object are now *balanced*, the overall force is zero and so the object falls with constant velocity. This is called its **terminal velocity**. The terminal velocity of any object will depend on its weight, its size and its shape.

Weight and gravity

Weight is the force of gravity on an object.

But force = mass × acceleration

$$\therefore \text{ weight} = \text{mass of object} \times \text{acceleration due to gravity}$$

i.e. $w = m \times g$

So, the weight of a person whose mass is 55 kg is:

$$\text{weight} = m \times g = 55 \times 10 = 550 \text{ N}$$
(kg) (m/s²)

A sky diver falling at terminal velocity during a parachute jump

22.12 Work and energy

When a force moves an object, we can say that

work done (joules) = **force** (newtons) × **distance** (metres)

So, if Franco, a weight lifter, raises 80 kg through 2.5 m;

mass lifted = 80 kg

∴ weight lifted = mass lifted × gravitational field strength
= 80 kg × 10 N/kg
= 800 N

Distance lifted = 2.5 m

So, work done = force × distance
= 800 N × 2.5 m
= 2000 J

When Franco lifts the weights, he is using up energy. Chemical energy in the chemicals in his muscles is being converted into work. So energy can be measured by the amount of work done. This means that the units of both energy and work are joules.

When we do work, we use energy. When a machine does work, it uses energy. **Energy enables us and our machines to do work.**

Potential energy and kinetic energy

When weights or bricks or water are raised to a higher level, they gain energy because, in falling, they are able to do some work. The energy which objects possess in being raised to a higher level is called **gravitational potential energy** or potential energy for short. Their extra height gives them the *potential* to do work when they fall.

Now,

potential energy = work done in being raised
= force × distance
= weight × height raised
= mass × gravitational field strength × height lifted

i.e.

$$\text{P.E.} = m \times g \times h$$

So, Franco transfers 2000 J to his weights when he raises them 2.5 m.

Moving objects – a hammer, falling water, a moving car, can all do work if they hit something. So, moving objects possess energy. This energy is called **kinetic energy**. (The word 'kinetic' comes from a Greek word which means 'moving'.)

$$\text{Kinetic energy} = \tfrac{1}{2} \times \text{mass} \times (\text{speed})^2$$
(kg) (m/s)

i.e.

$$\text{K.E.} = \tfrac{1}{2} \times m \times v^2$$

Example

Eric kicks a football (mass 250 g) with a force of 500 N. His foot is in contact with the ball over a distance of 0.1 m.

(i) How much kinetic energy does the ball have just after being kicked?
(ii) What is the speed of the ball just after being kicked?

Solution

(i) Assuming that all the work done in kicking the ball becomes kinetic energy;
kinetic energy = work done in kicking the ball
= force × distance
= 500 N × 0.1 m = 50 J

(ii) Now, K.E. $= \tfrac{1}{2} \times \text{mass} \times (\text{speed})^2$
∴ 50 $= \tfrac{1}{2} \times 0.25 \times (\text{speed})^2$
(kg) (m/s)
So, $(\text{speed})^2 = 400$
speed = 20 m/s

1 The graph shows distance plotted against time for a short car journey.

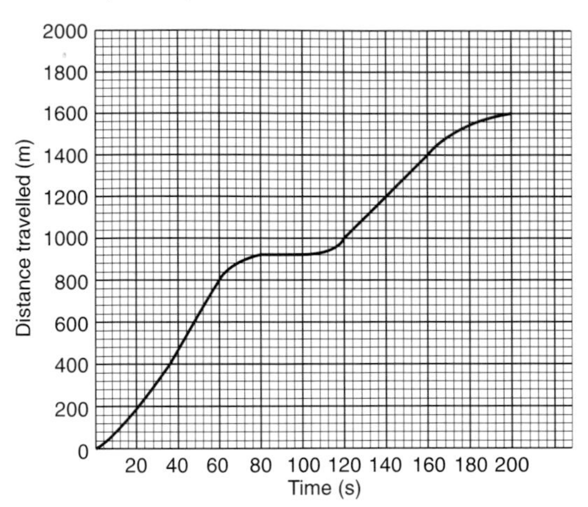

a) Use information from the graph to answer the questions
i) What does the graph tell us about the speed of the car between 20 and 60 seconds after starting the journey?
ii) How far did the car travel between 20 and 60 seconds?
iii) Calculate the speed of the car between 20 and 60 seconds. Show your working.

b) What happened to the car between 80 and 100 seconds after starting the journey?

c) Below is a velocity-time graph for a toy car travelling in a straight line.

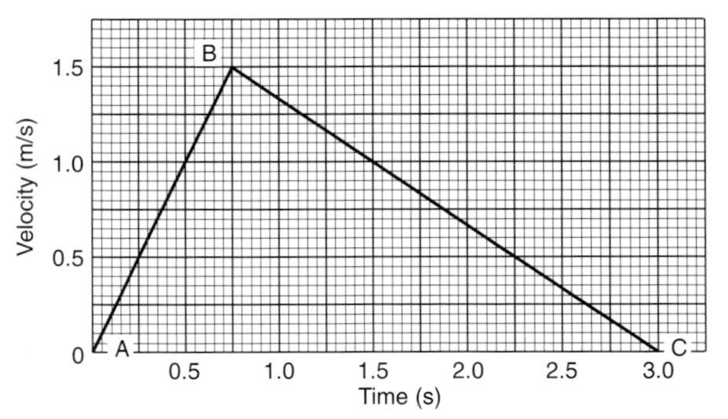

i) Calculate the acceleration of the car from A to B.
ii) Calculate the **total** distance travelled by the car from A to C.

NEAB

2

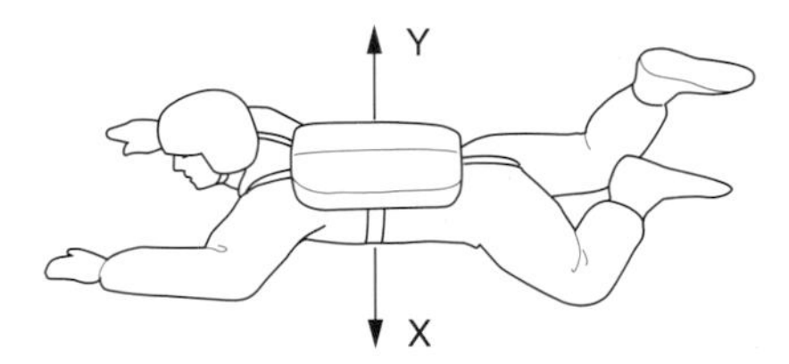

A sky diver is falling from an aeroplane.

a) Name:
i) force X,
ii) force Y.

b) State how **each** force changes as the sky diver speeds up.
i) Force X,
ii) Force Y.

c) Why does the sky diver reach a steady speed (terminal velocity)?

d) Describe and explain what happens when the sky diver deploys (opens) the parachute

WJEC

3 A jack is a machine which may be used to lift one side of a car so that a wheel may be changed.

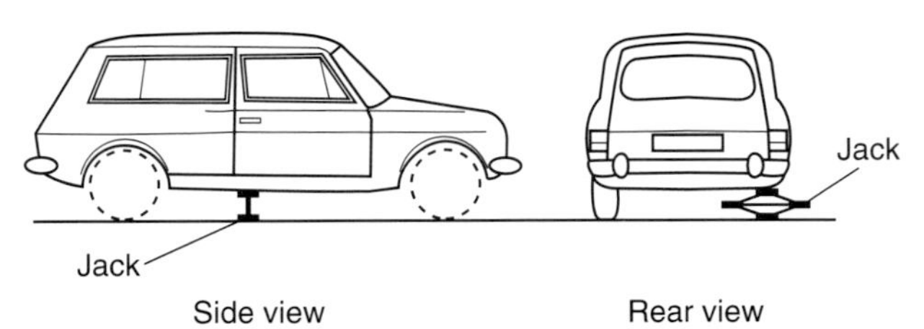

The car shown in the diagram has had both wheels on the right hand side removed and it is now supported, on this side, only by the jack which exerts an upward force of 2800 N.

a) If the weight of the car is now 6000 N, what is the total force exerted upwards by the remaining wheels?

b) Calculate the work that would be done, by the jack, in raising the right hand side of the car a further 3 cm. Show clearly how you obtain your answer.

NICCEA

4

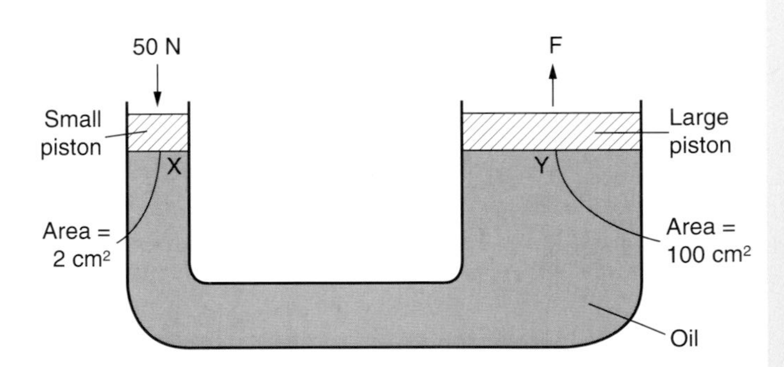

The diagram above shows the principle of the hydraulic car jack. A small force applied to the small piston enables a large load (the car) to be lifted by the large piston.

a) i) If a force of 50 N is applied to the small piston, calculate the pressure measured in N/cm^2 produced in the oil at X.
ii) What is the pressure exerted by the oil at Y?
iii) Calculate the upward force, F, acting on the large piston.

b) Why does having air trapped in a hydraulic system make it less effective?

WJEC

CHAPTER

23

Energy transfers

Energy transfer by conduction and convection	Reducing the energy loss from our homes
Energy transfer by radiation	Energy efficiency

Which parts of this mountaineer are insulated best from the cold? Which parts of his body will lose the most heat?

23.1 Energy transfer by conduction and convection

When different materials or different parts of the same substance are at different temperatures, energy is transferred from places where the temperature is higher to places where the temperature is lower.

When you walk around barefooted, you will notice that tiles feel much colder to your feet than carpet. Your feet feel the cold because they are losing energy to the tiles in the form of heat. The transfer of heat from your feet to the tiles is an example of **conduction**.

> **Conduction** is the transfer of energy (heat) between materials in contact, or between different parts of the same substance, without the materials or the substance moving.

All metals are very good conductors, but materials like plastic, wood and carpet are poor conductors. They are **insulators**. Gases are very poor conductors.

When heat is transferred by conduction (Figure 23.1), hot atoms (i.e. those with more kinetic energy) pass some of their kinetic energy to neighbouring atoms.

a) In a metal, heat is conducted quickly by fast-moving, mobile electrons

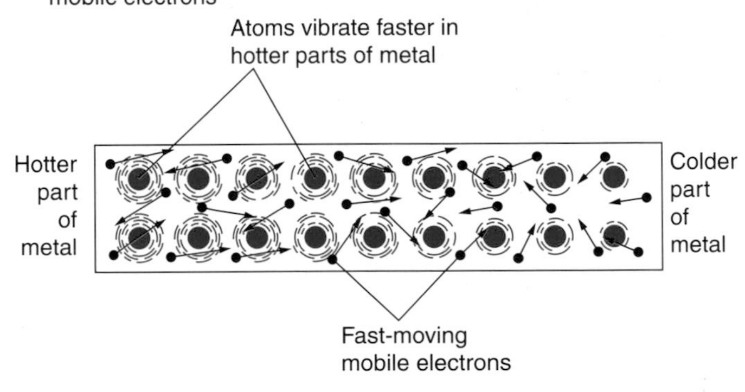

b) In an insulator, there are no mobile electrons, so heat is conducted slowly by atoms colliding as they vibrate.

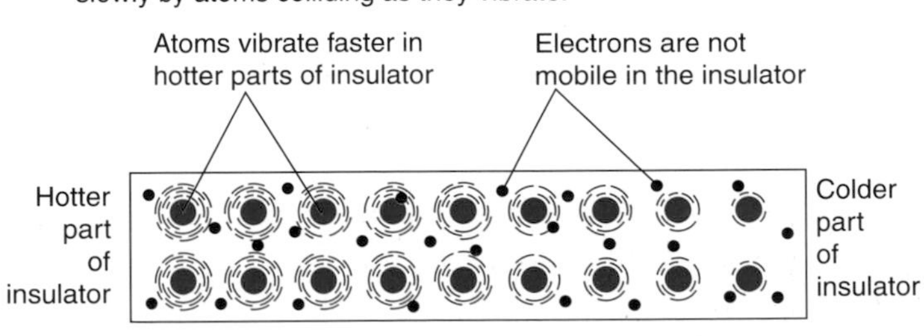

Figure 23.1 Conduction in metals and insulators

Metals contain electrons that are free to move (mobile). The hotter the metal, the more kinetic energy its mobile electrons have and the faster they diffuse through the metal (Figure 23.1a on page 225). The vigorous movements of electrons in the hotter parts of the metal are also transferred, via collisions, to adjacent electrons. These, in turn, transfer energy to other electrons and heat is conducted through the metal.

In thermal insulators, there are no free electrons and so heat is transferred much more slowly by hot atoms bumping into colder atoms (Figure 23.1b).

A second way in which heat (energy) can be transferred from one place to another is illustrated in Figure 23.2. When the electric heater is working, air near the heater is warmed. The warmed air expands, becomes less dense and rises. As this air rises, cooler air falls and moves in to replace it near the heater. This movement of air is an example of **convection**.

Convection currents in the air, called thermals, are essential for hang-gliding

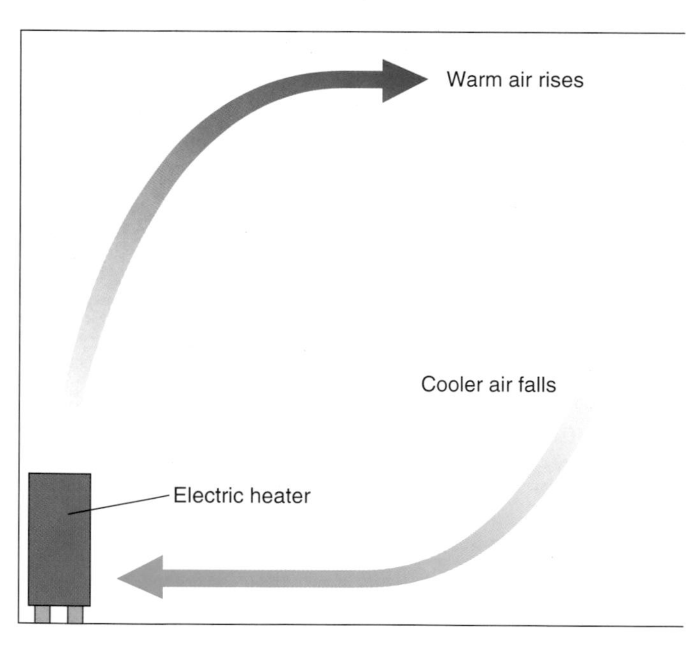

Figure 23.2 Energy is transferred by convection currents when an electric heater is working

> **Convection** is the transfer of energy (heat) by the movement of a liquid or a gas due to differences in density.

Convection, like conduction, involves energy being transferred through the movement of particles. Convection currents occur in liquids and gases because their particles move faster when they are hot, causing the liquid to expand. Warmer parts are then less dense than colder parts. The warmer parts rise and cooler parts fall to replace them.

23.2 Energy transfer by radiation

Although conduction and convection result in some essential energy transfers, the most important method of energy (heat) transfer is by **radiation**. Indeed, radiation is responsible for our survival. Energy from the Sun reaches the Earth by radiation as it travels millions of miles as electromagnetic waves through empty space (section 28.2).

> **Radiation** is the transfer of energy (heat) from one place to another by means of electromagnetic waves.

Energy is transferred to and from all objects by radiation. All objects emit radiation and the hotter they are, the more radiation they emit. This radiation is composed of electromagnetic waves mainly from the infra-red region of the electromagnetic spectrum. Unlike conduction and convection, radiation does *not* require particles of matter for the transfer of energy. This explains why radiation can travel through empty space (a vacuum).

Absorbing and emitting radiation

In some countries, infra-red radiation from the Sun can be used for cooking and heating. Radiation also behaves like light. It can be reflected and focused using a mirror. This is the idea behind a solar furnace (Figure 23.3).

Figure 23.3 The huge reflector at Odeillo power station in the East Pyrenees, France. Positioned in front of the reflector (out of view in the photograph) are other mirrors that automatically track the sun and reflect incident radiation onto the huge reflector. This then concentrates the Sun's rays onto the furnace at its focus in the central tower producing electricity at the rate of 1000 kilowatts

The light, shiny surface of the huge mirror is a poor absorber of radiation but it is a good reflector. In contrast, the dark, matt surface at the focus of the mirror inside the control tower is a very good absorber of radiation.

Figure 23.4 shows the equipment for an experiment to investigate the emission of radiation by different surfaces and to see which teapot would keep your tea the hottest. The two teapots are identical except one has a shiny white surface and the other is dull black.

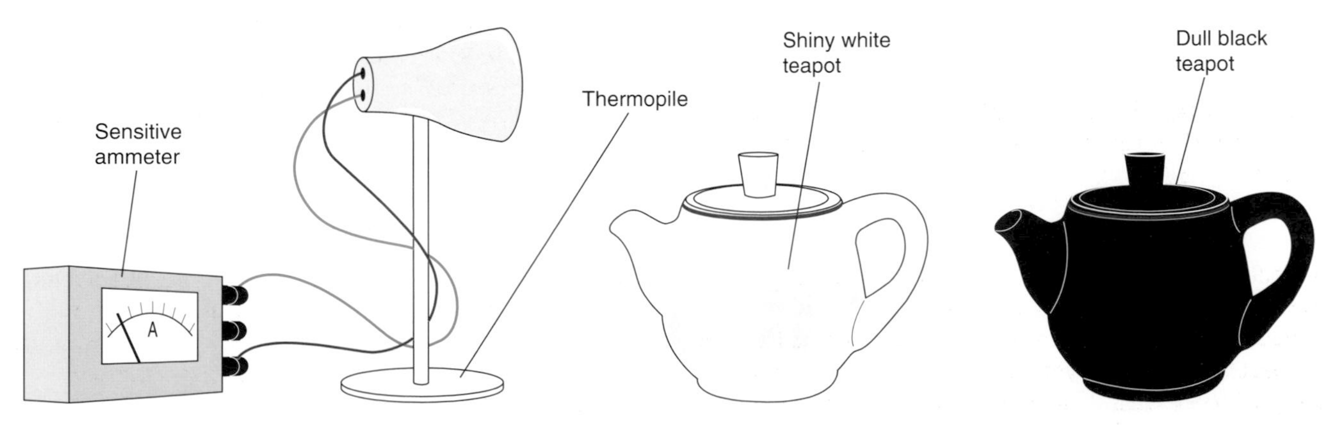

Figure 23.4 An experiment to investigate the emission of radiation and find out which teapot is 'best'

Radiation emitted from the teapots can be detected by a thermopile connected to a sensitive ammeter. Experiments show that the dull black teapot emits more radiation than the shiny white one. So, the shiny teapot would keep your tea warmer.

- Dark, dull surfaces are good emitters and absorbers of radiation.
- Light, shiny surfaces are poor emitters and absorbers of radiation.

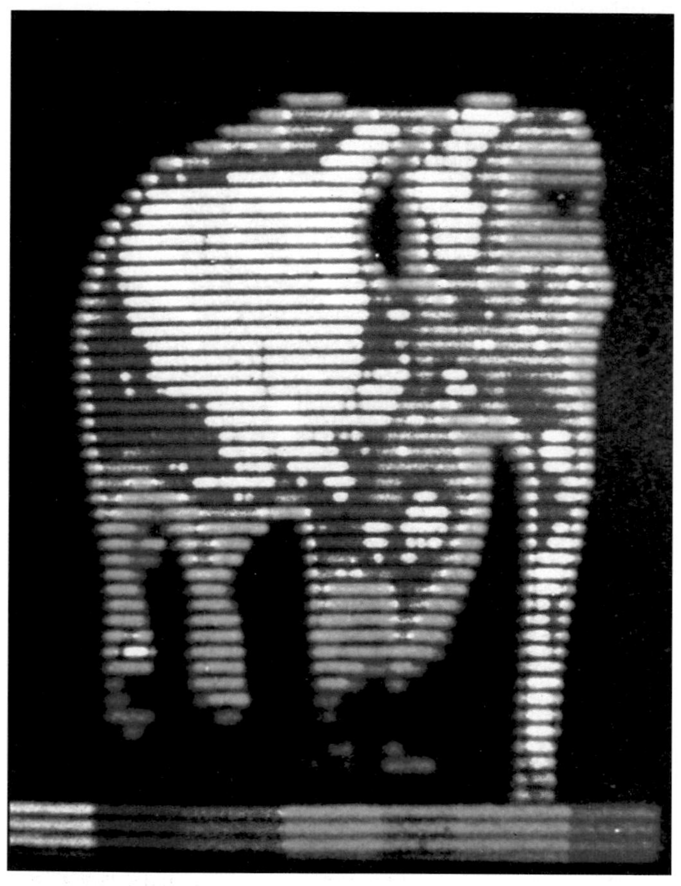

This thermal photograph (thermograph) of an elephant was taken using an infrared camera which detects heat radiation. Ordinary cameras detect only light. The colour represents the temperature of the elephants skin ranging from white (hottest) through yellow, then red to blue (coldest)

23.3 Reducing the energy loss from our homes

Heat escapes through the walls, the roof, the floor, the windows and the doors of your home. Figure 23.5 shows the percentages of heat loss from different parts of a house.

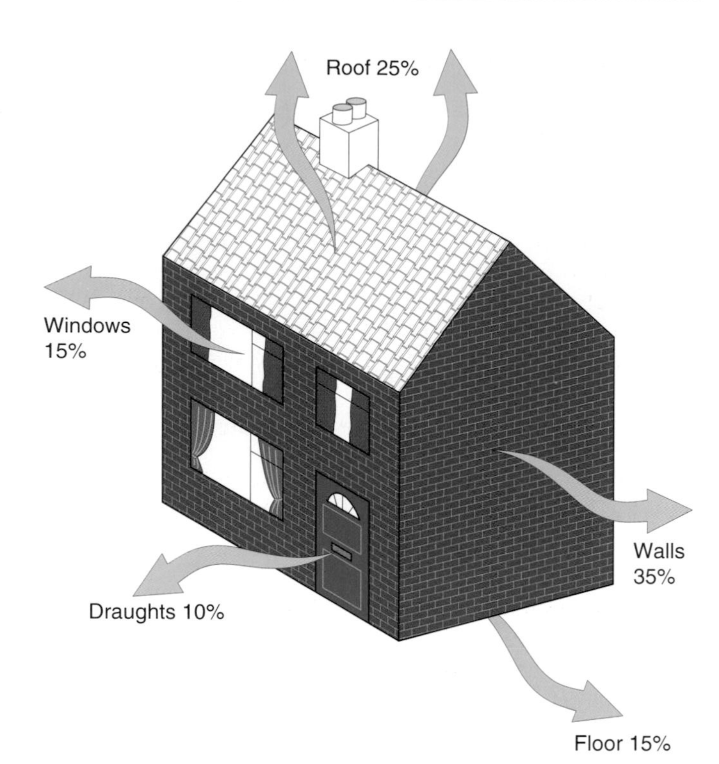

Figure 23.5 Heat loss from the different parts of a house

Heat is lost from the house by conduction, convection and radiation. By insulating a house, we can keep the heat in it for longer. This reduces the energy needed to warm the house in the winter months and will reduce the fuel bills.

Look closely at Table 23.1. It shows the relative conductivities of various materials used in buildings and in their insulation.

Table 23.1 The relative conductivities of different materials

Material	Relative conductivity
aluminium	8800
steel	3100
concrete	175
glass	35
water	25
brick	23
breeze block	9
wood	6
felt	1.7
wool	1.2
fibreglass	1.2
air	1.0

Notice the following points from Table 23.1.

- Metals are far better conductors than other building materials. This has led to their use in radiators, electrical wires and cables.
- Poor conductors (insulators) such as fibreglass, felt, brick and glass are used to prevent heat transfer from rooms through the cavity walls to the outside.
- Air is the poorest conductor of all the materials in Table 23.1. Materials which trap air such as fibreglass, wool, felt, feathers and hair are good insulators. They are used in cavity wall insulation, carpets and lagging for pipes and water tanks.

Every year, families spend large amounts of their income in keeping homes warm during the cold winter months. We have to use fuels and energy to heat our homes, but we cannot avoid losing some heat to the air outside so it is important to use effective insulation measures to keep these losses to a minimum. This saves energy resources such as coal and oil and keeps the cost of heating as low as possible.

Figure 23.6 shows the five most common methods of preventing heat loss from our homes. Each method involves trapping pockets or layers of air. This significantly reduces the transfer of heat by conduction and convection.

Example

The Wadsworth family spends £600 each year on heating bills. At present, their house has no loft insulation, no cavity wall insulation and no double glazing. There are no draught excluders fitted to the doors or windows and the hot water tank is not lagged. The Wadsworths approached various firms which provided them with the following estimates.

Table 23.2 The costs and savings from various methods of insulation

Type of insulation	Cost (£)	Saving in one year (£)
loft insulation	400	200
cavity wall insulation	1000	100
double glazing	3000	150
draught excluders	50	10
blanket for hot water tank	25	10

If the Wadsworths fit loft insulation, it will cost £400, but they would then save £200 each year on their fuel bills. So, it would take only 400/200 = 2 years to save the cost of loft insulation.

Figure 23.6 Preventing heat loss from our homes

Joists
Fibreglass
Loft insulation
Thick rolls of fibreglass are laid between the joists in the loft. The fibreglass contains pockets of trapped air

Lagged hot water tank
A thick blanket is fitted around the hot water tank to keep the water hot. The blanket contains pockets of trapped air

Draught excluders
on windows and doors trap warm air inside the house
Metal draught excluder

Cavity wall insulation
The space between the inner and outer walls is filled with foam or fibreglass. Both of these contain pockets of air. The foam can be pumped in and allowed to set preventing convection currents
Inner wall
Outer brick wall
Cavity

Glass
Double glazing
A layer of air is trapped between two sheets of glass. The layer must be thin to prevent convection currents

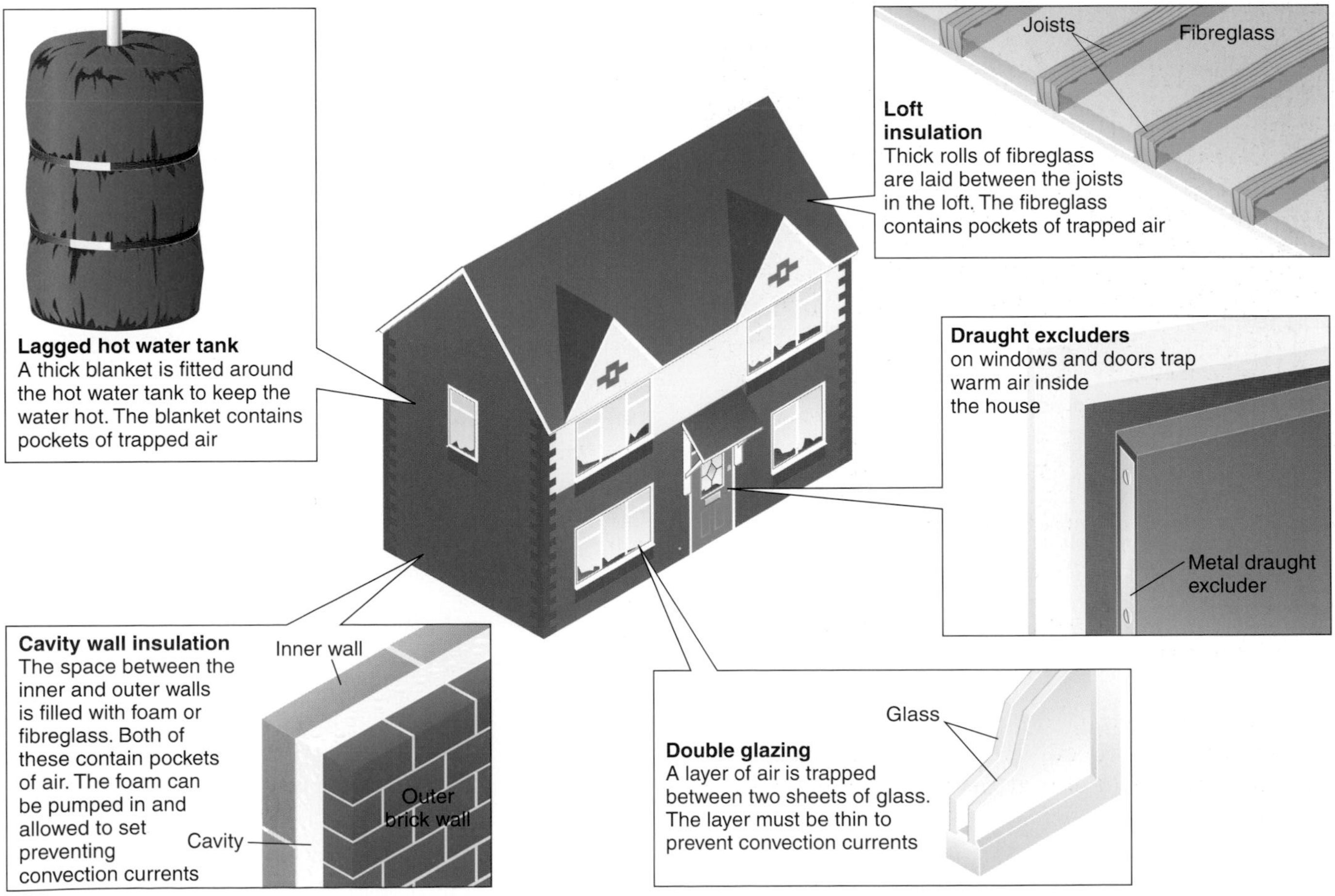

On the other hand, if they decide to have cavity wall insulation, it would cost £1000 with a saving of £100 a year. Therefore, it would take 1000/100 = 10 years to save the cost of cavity wall insulation.

(i) How long would it take to save the cost of a) double glazing, b) draught excluders and c) lagging the hot water tank?
(ii) How would you advise the Wadsworths to insulate their house?

Why is bubble sheeting better than ordinary plastic sheeting for retaining the heat in the water in this swimming pool?

23.4 Energy efficiency

Machines can make jobs easier, but no machine is perfect. Look at Figure 23.7 which shows Julie hoisting her sail by using a pulley.

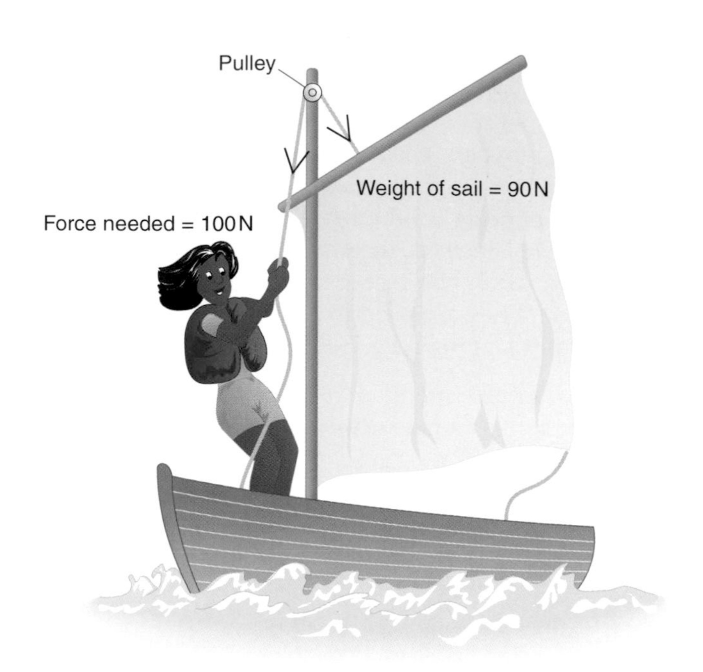

Figure 23.7

The weight of the sail is only 90 N, but Julie has to pull with a force of 100 N. This is needed to overcome the weight of the sail as well as the friction between the rope and pulley. If Julie pulls in 1 m of rope, the sail goes up 1 m.

So, the energy (work) she puts in
= force × distance
= 100 N × 1 m = 100 J
and the useful energy (work) output in raising the sail
= 90 N × 1 m = 90 J.

Notice that 90 J of energy have been transferred from Julie to the sail, but 10 J have been wasted. In practise, it is impossible to transfer all the energy that you want to put into a machine. Whenever energy is being transferred, there is always some wasted energy which is not transferred in a useful way. The main causes for wasted energy in machines are:

- friction producing heat between the moving parts of the machine,
- energy used in moving or lifting the machine itself.

The energy which is 'wasted' during energy transfers usually ends up being transferred to parts of the machine or the surroundings, which become warmer. As a result of this, the energy is dissipated (increasingly spread out) and becomes more difficult to use.

Different machines waste different amounts of energy. The fraction of the energy supplied to a machine or device which is usefully transferred is called its **efficiency**.

$$\text{Efficiency} = \frac{\text{useful energy (work) output}}{\text{total energy input}}$$

So, the efficiency of Julie's pulley in Figure 23.7

$$= \frac{90\text{ J}}{100\text{ J}} = 0.9 \text{ or } 90\%.$$

We can write the efficiency as a decimal (0.9) or as a percentage (90%).

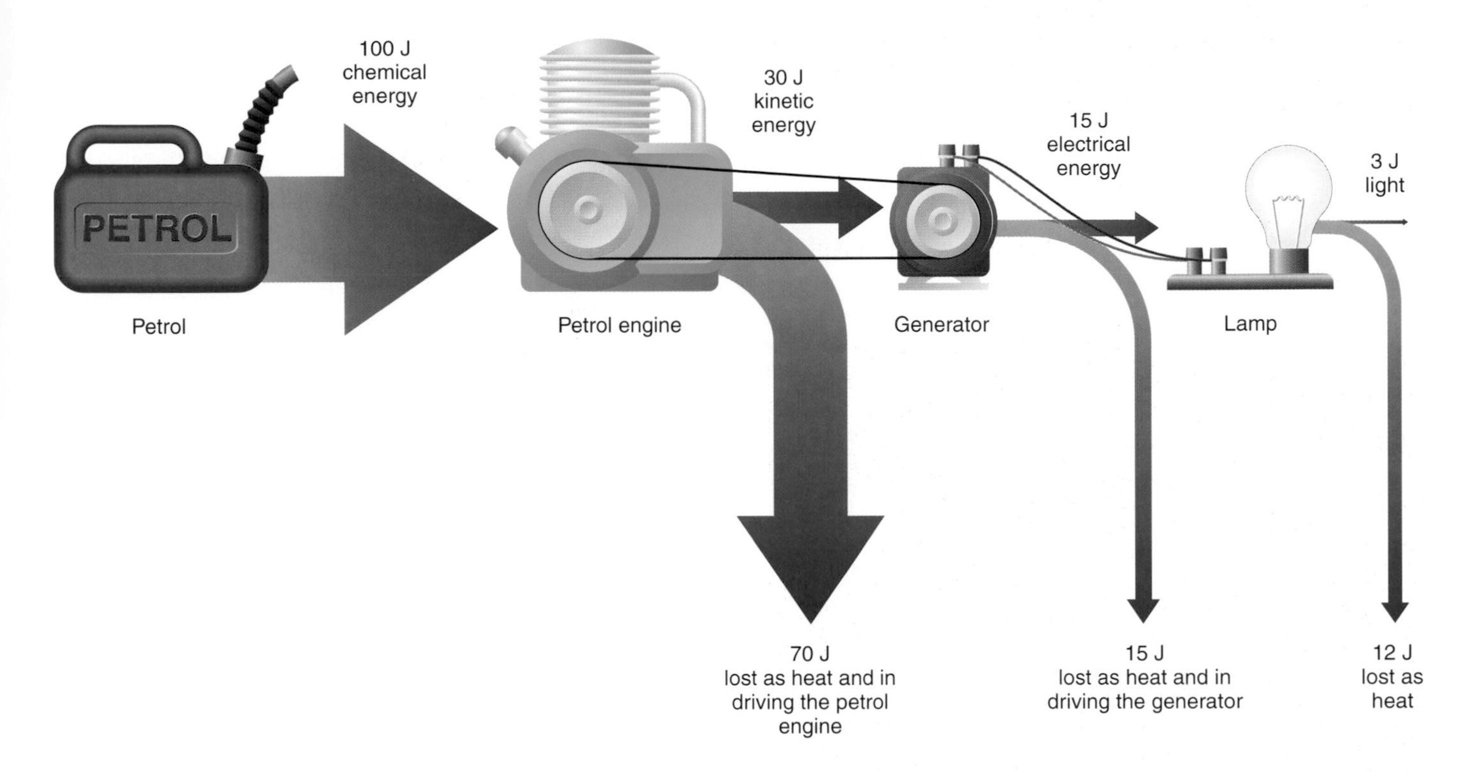

Figure 23.8 A flow diagram of the energy conversions when a petrol engine and generator are used to light a lamp

How efficient are energy conversions?

Figure 23.8 shows the energy conversions which occur when a small petrol engine is used to drive a generator which produces electricity to light a lamp. The energy changes are shown in Figure 23.8 as a flow diagram starting with 100 J of chemical energy in petrol.

Some energy is transferred usefully from one stage to the next, but in each case some energy is wasted as heat or in moving the machine itself.

From the energy transfers in Figure 23.8, we can calculate the efficiency of each machine.

$\therefore$ Efficiency of generator =

$$\frac{\text{useful energy output}}{\text{total energy input}} = \frac{15\text{ J}}{30\text{ J}} = 0.5 \text{ or } 50\%.$$

What is the efficiency of (i) the petrol engine, (ii) the lamp?

How does this Olympic cyclist and his bike avoid wasted energy?

1 When a new small town was built, two different schemes to supply energy were considered.

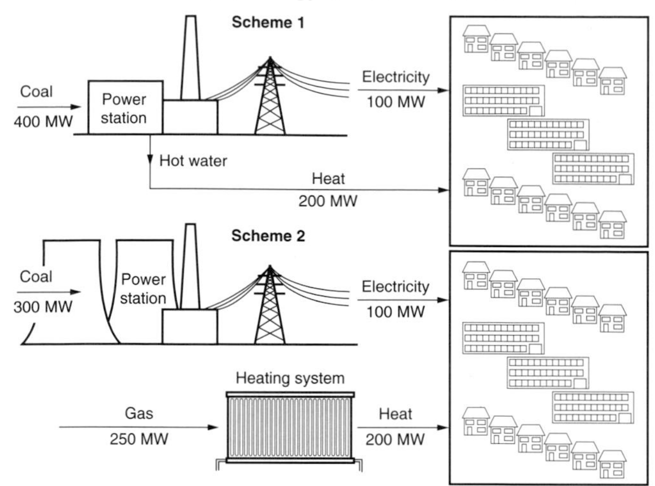

a) Which fossil fuel provides the energy for heating in:
i) scheme 1,
ii) scheme 2?

b) Copy and complete the table below.

	Total power needed by the town in MW	Total input power in MW
Scheme 1	300	
Scheme 2	300	

c) The efficiency of a power station is measured using the formula

$$\text{efficiency} = \frac{\text{useful output (in MW)}}{\text{input (in MW)}} \times 100$$

The answer is in the form of a percentage (%). Use this formula to calculate the efficiency of the power station in scheme 2. Your answer should be correct to the nearest whole number.

d) Suggest a reason why power stations are **not** 100% efficient. **MEG**

2 Electricity can be generated in several different ways.

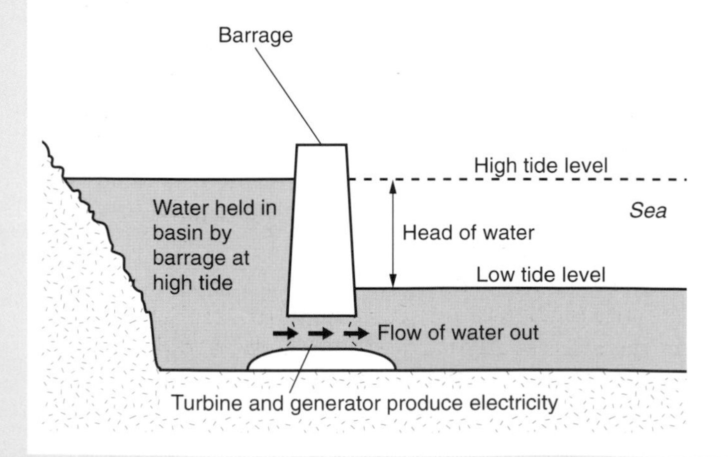

a) Describe how the energy in coal is transferred to drive generators in coal-fired power stations.

b) The diagram at the bottom left of the page shows a section through a tidal power generating system. Describe fully how this method of driving turbines to generate electricity is **different** from using coal.

c) Compare in detail the advantages and disadvantages of these two methods of generating electricity. **NEAB**

3 When an electric kettle is heating water, energy is being transferred to the water and to the surroundings. The diagram below illustrates three ways in which energy is transferred to the surroundings.

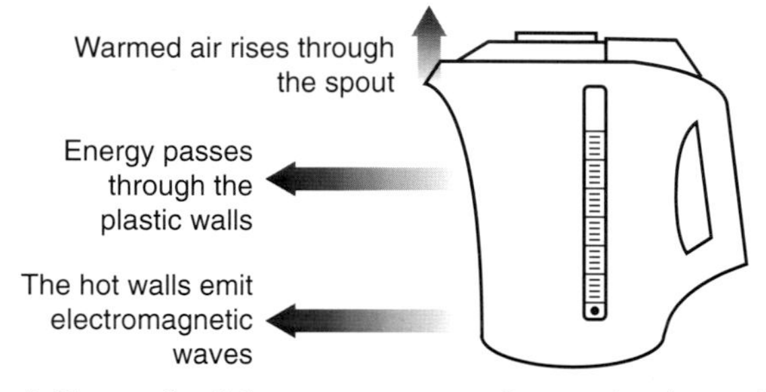

a) For each of these energy transfers, write down the word from the list which describes it.

conduction convection radiation

You may use each word once, more than once or not at all.
i) Warmed air rises through the spout.
ii) Energy passes through the plastic walls.

b) i) Describe, in terms of the motion of particles, how energy is transferred through the kettle walls.
ii) Some kettles are double-walled. The diagram below shows how a layer of air is trapped between the inner wall and the outer wall.

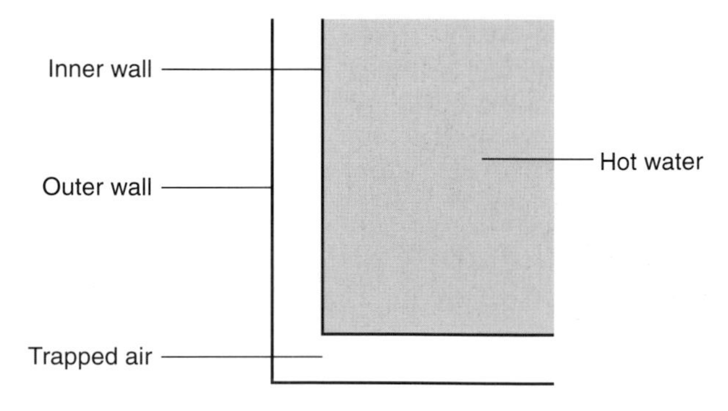

Suggest **one** advantage of using a double-walled kettle.

c) i) 4200 J of energy is needed to raise the temperature of 1 kg of water by 1 °C. The kettle is filled with 1.5 kg of water at 10 °C. Calculate the quantity of energy which needs to be transferred to the water to bring it to the boil.
ii) Suggest **two** reasons why the energy transferred to the kettle needs to be more than the answer to c)i). **MEG**

24

Currents and circuits

Circuit diagrams	Voltage–current graphs – Ohm's law
Series and parallel circuits	The variation of current with voltage in different circuit components
Energy transfer in electric circuits	

Blackpool Illuminations would be impossible without electric currents and circuits

24.1 Circuit diagrams

Figure 24.1 shows an electrical circuit. This type of drawing is called a **circuit diagram**. The circuit contains a cell connected to a lamp (bulb), a motor, an ammeter and a switch. Notice that there are standard symbols for the different pieces of equipment, usually called **components**, in the circuit.

A full list of the standard symbols that you should know are shown in Table 24.1, over the page.

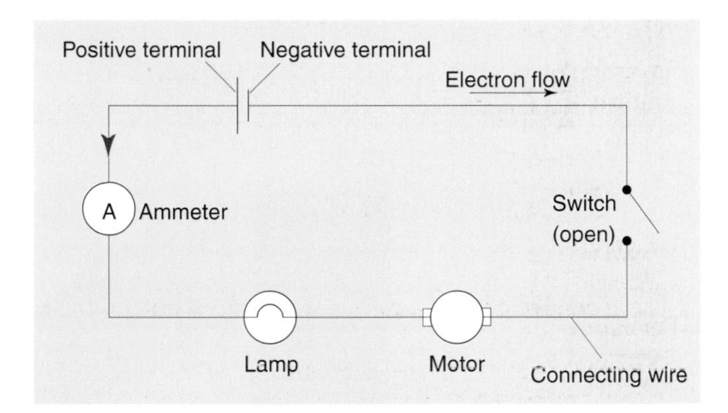

Figure 24.1 A circuit diagram

At present, the switch is open in Figure 24.1 and this is called an **open circuit**. When the switch is closed, there is a complete **closed circuit** of conductors. An electric current therefore flows around the circuit, the bulb lights up and the motor starts. The electric current is measured in **amperes** on the **ammeter**.

The electric current in the circuit comprises of a flow of negatively-charged electrons from the negative terminal of the cell through the circuit to the positive terminal. This can be shown in a circuit diagram *using a labelled arrow at the side of the circuit wires* (see the black arrow in Figure 24.1).

In the 19th century, however, scientists agreed to show current flowing from the positive terminal to the negative terminal *by an arrow on the connecting wires in the circuit diagram* (see red arrow head in Figure 24.1). In order to distinguish it from the electron flow, this is called the **conventional current**.

Table 24.1 Standard symbols for circuit diagrams

Component	Standard symbol	Component	Standard symbol
switch (open)		light emitting diode (LED)	
switch (closed)		ammeter	A
cell		voltmeter	V
battery of cells		motor	M or
lamp	or	light dependent resistor (LDR)	
resistor		fuse	
variable resistor		thermistor	
diode		variable power supply	0–12V + – d.c.

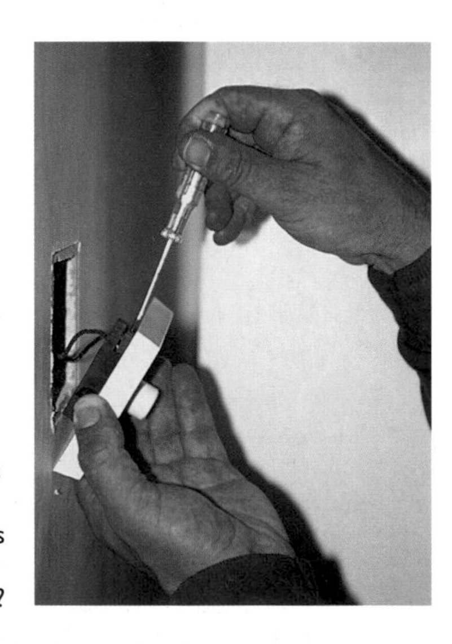

The handles of tools used by electricians, such as screwdrivers must have insulated handles. Why is this?

24.2 Series and parallel circuits

The circuit in Figure 24.1 can be described as a **series circuit**. In a series circuit, there is only *one path* for the current around the circuit. Another series circuit is shown in Figure 24.2 containing a battery of two cells, three identical bulbs and three ammeters. A **battery** is two or more cells connected together.

When the circuit in Figure 24.2 is complete, all the bulbs light and all three ammeters show the same current of 0.5 amperes (0.5 A). If more bulbs are added to the circuit in Figure 24.2, the current is smaller.

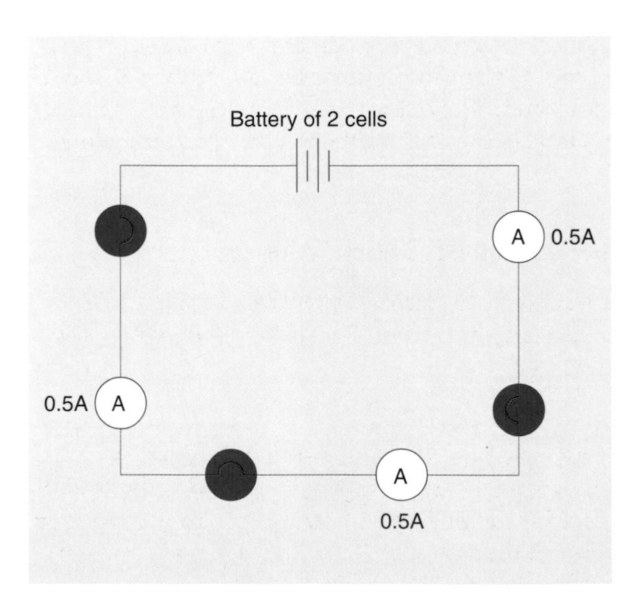

Figure 24.2 Components in a series circuit

For example, if three more bulbs are added to the circuit, the current falls to 0.25 A. The bulb (and other components in circuits) are said to cause a **resistance** to the current. The very thin nichrome or tungsten wire in the bulbs impedes the flow of electrons and acts as a resistance. The bigger the resistance of the components in a circuit, the smaller the current. The wire connecting the components in the circuit is made of much thicker copper wire. This causes negligible resistance to the current.

Now look at Figure 24.3 which also contains three identical bulbs, X, Y and Z. But, in this circuit there is a branch, and when the current reaches point P, it can either go through bulbs X and Y or it can go through bulb Z. In Figure 24.3, bulbs X and Y are in series with each other, but they are **in parallel** with bulb Z.

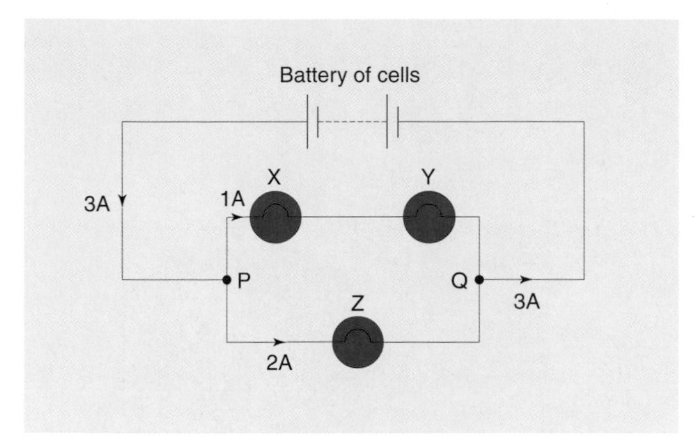

Figure 24.3 Components in a parallel circuit

In parallel circuits, the *current divides equally* at a junction if the two parallel sections are the same. The current will not, however, divide equally if one path is easier than the other. In Figure 24.3, for example, at point P it is easier for the current to go through just one bulb (Z) than to go through two bulbs (X and Y). So, the current divides at point P, with twice as much going via Z (2 A) and only one ampere going via X and Y.

Rules about electric currents in circuits

Discussion of the circuits in Figures 24.2 and 24.3 illustrate some important rules about **electric currents**.

- The current flowing through a component in a circuit is measured in amperes (A) using an ammeter connected in series with the component.
- The current is not used up as it passes through bulbs and other components in a circuit.
- In a series circuit, the current is the same at all points.
- In a parallel circuit, more current takes the easier path.
- In a parallel circuit, the total currents into a junction equals the total currents out of a junction.
- As the number of cells in a circuit increases, the current gets larger.
- As the number of bulbs and other components in a circuit increases, the resistance increases and the current gets smaller.

24.3 Energy transfer in electric circuits

The chemicals in a cell or a battery contain chemical energy. When the battery is used in a closed circuit, this chemical energy is transferred from the materials of the cell to electrons in the electric current.

The electrons move through the circuit as the electric current from the negative terminal, where they have high electrical potential energy, towards the positive terminal, where they have zero electrical potential energy. As the electrons pass through other components in the circuit (lamps, resistors, motors, LEDs, etc.), they lose some of their electrical potential energy. This 'lost' energy is turned into heat in resistors and motors, heat and light in lamps and LEDs and heat and sound in buzzers (Figure 24.4).

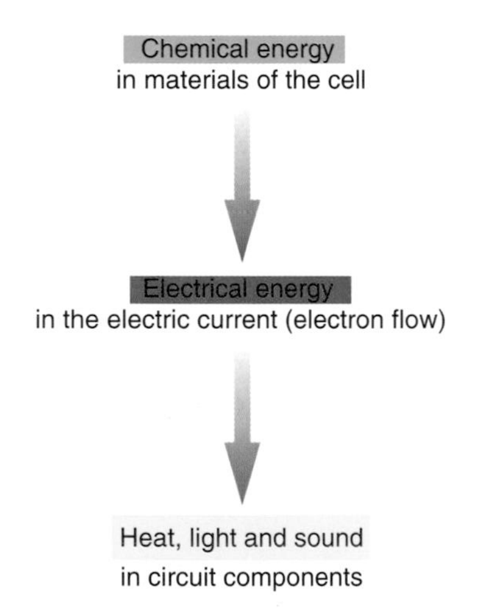

Figure 24.4 Energy transfers in an electric circuit

The difference in electrical potential energy between the terminals of the cell or battery is indicated by its **voltage** or **potential difference (p.d.)**. The greater the voltage of a cell or battery, the more energy it can provide for the circuit components. For example, the cells used in a small calculator might provide 1 or 2 volts, car batteries usually give 12 volts and the voltage of the electrical supply to our homes is 240 volts.

Voltages (potential differences) are measured in **volts (V)** using a **voltmeter**. The voltmeter is connected across (i.e. *in parallel with*) the portion of the circuit where the voltage is being measured.

Special cells down the backbone of this electric eel can produce a large voltage between its nose and tail. The eel uses this voltage to kill its prey

Figure 24.5 shows how the lamps are arranged in a car's headlights and in Christmas tree lights.

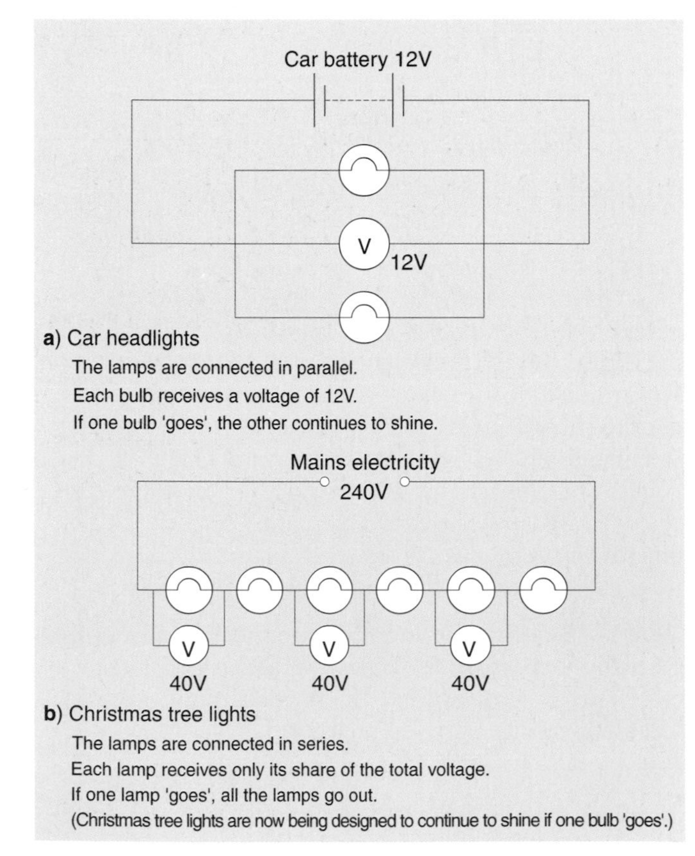

Figure 24.5 Voltages across lamps in series and in parallel

Often the lights on Christmas trees are all connected in series. This means that if one light bulb fails, all of them go out. Today, some fairy lights are designed so that this does not happen

The arrangements in Figure 24.5 illustrate important rules regarding **voltages in circuits**.

- The voltage (p.d.) across a component in a circuit is measured in volts (V).
- A current will flow through an electrical component only if there is a voltage (p.d.) across its ends. The bigger the voltage across a component, the larger the current flowing through it.
- Bulbs or circuit components in parallel have the same voltage across them.
- Bulbs or components in series share the total voltage.
- As the number of cells in a circuit increases, the voltage increases and the current is larger.

Example
Look at the circuits below in which all the cells are identical and all the lamps are identical.

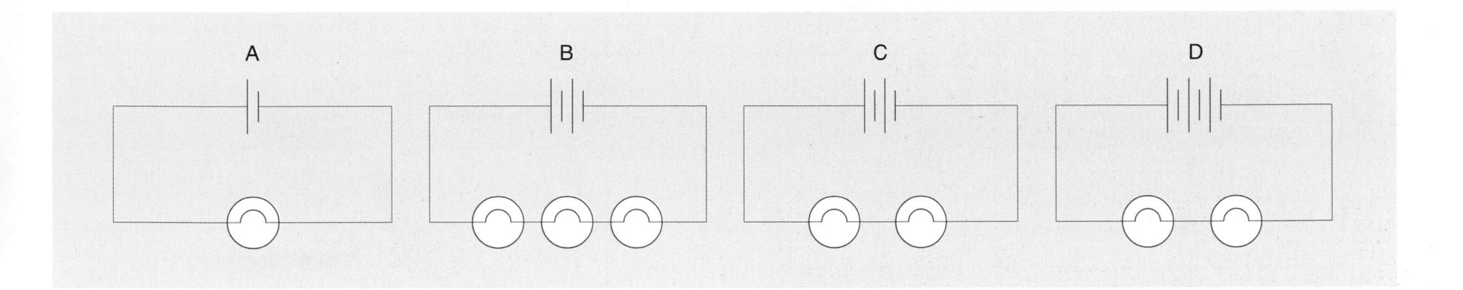

(i) Which circuit has the smallest current?
(ii) Which circuit has the largest current?
(iii) Which two circuits have the same current?

Most of the meters in the cockpit of this aircraft are modified ammeters and voltmeters

24.4 Voltage – current graphs – Ohm's law

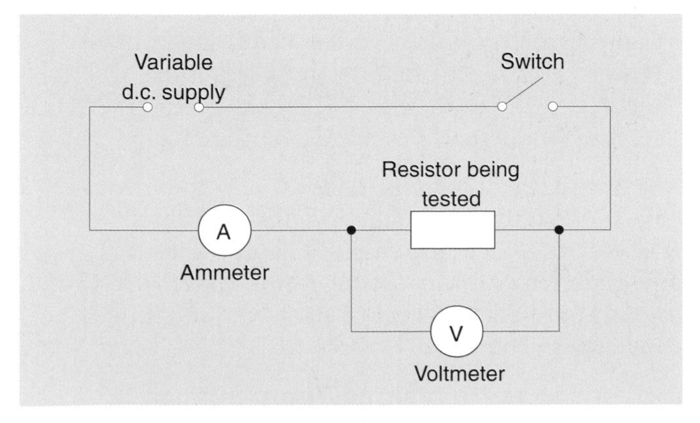

Figure 24.6 A circuit used to investigate voltage–current relationships

Voltage–current graphs are used for two main reasons:

- to show how the current through a component (device) varies with the voltage across it,
- to measure the resistance of a device and understand how the resistance may vary with current and voltage.

During the 1820s, the German physicist, Georg Ohm, investigated the resistance of various metals. The unit which we now use for resistance is the **ohm** in honour of him. The symbol for the ohm is Ω, so twenty ohms is written as 20 Ω.

Figure 24.6 shows a circuit which can be used to investigate voltage–current relationships. Using this circuit, it is possible to measure the voltage across the resistor for different values of the current through it. The current can be altered using the variable direct current (d.c.) supply.

Figure 24.7 shows a graph of the results obtained when the resistor is a metal.

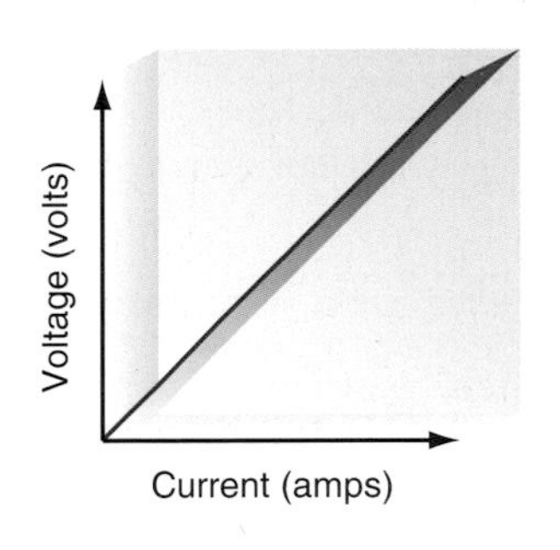

Figure 24.7 A graph of voltage against current for a metal resistor

The results produce a straight line through the origin (O) which means that:

The voltage across a metal resistor is proportional to the current through it, provided its temperature is constant.
i.e. $V \propto I$ (at constant temperature)

This statement is sometimes called **Ohm's law** because it was first discovered by Georg Ohm.

Because V is proportional to I, we can say that:

$$\frac{V}{I} = \text{a constant}$$

This means that doubling the voltage, doubles the current. Trebling the voltage, trebles the current and so on. The larger the resistance, the greater is the voltage needed to push each ampere of current through it. This led scientists to a clear definition of resistance.

A resistor has a resistance of one ohm (1 Ω) if a voltage of one volt (1 V) drives a current of one ampere (1 A) through it.

If a voltage of 2 V is needed for a current of 1 A, the resistance of the resistor is 2 Ω. If 10 V are needed to drive a current of 1 A through the resistor, its resistance is 10 Ω. Notice that **resistance is the voltage per unit current**. i.e.;

$$\text{Resistance (ohms)} = \frac{\text{Voltage (volts)}}{\text{Current (amps)}}$$

$$R = \frac{V}{I}$$

Example

The element in an electric kettle has a resistance of 24 Ω. What is the current in the element of the kettle when it is connected to the 240 V mains supply?

Solution

Using the formula $R = \frac{V}{I}$, $I = \frac{V}{R} = \frac{240}{24} = 10\text{ A}$

As resistance equals voltage divided by current, you should also appreciate that:

gradient of a voltage–current graph = **resistance**

provided voltage is plotted vertically and current horizontally.

24.5 The variation of current with voltage in different circuit components

Figure 24.8 shows voltage–current graphs for three different circuit components – a resistor at constant temperature, a filament lamp and a thermistor (LDR).

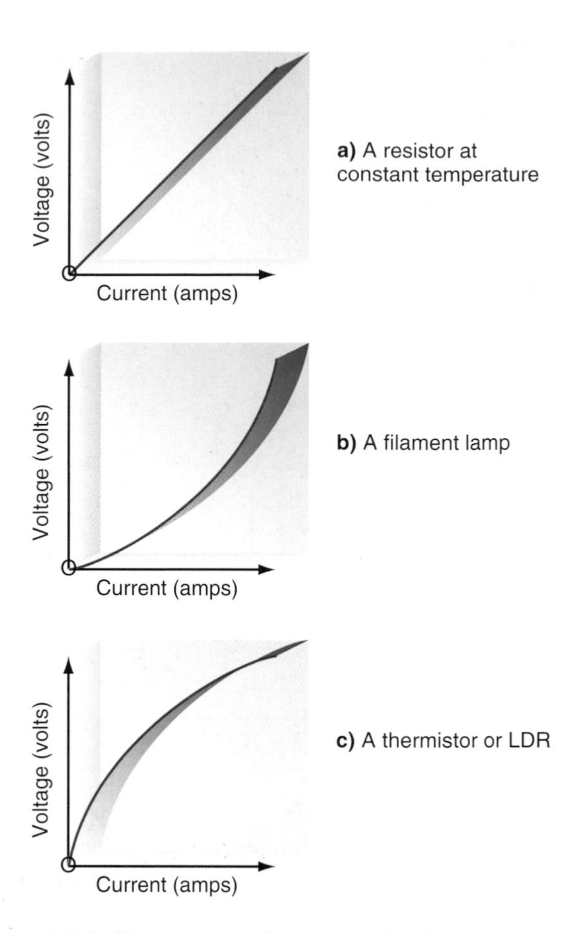

Figure 24.8 The variation of current with voltage in three different circuit components

Notice that the filament lamp and the thermistor do not obey Ohm's law (i.e. voltage is *not* proportional to the current). These are known as **non-ohmic conductors** to differentiate them from **ohmic conductors** – resistors at constant temperature.

The current through a resistor (at constant temperature) is proportional to the voltage (Figure 24.8a). The voltage–current graph is a straight line, V/I is constant and therefore the resistance is constant.

For the filament lamp in Figure 24.8b, the voltage–current graph curves upwards as current increases. This means that the gradient of the voltage–current graph rises, V/I increases and therefore the resistance increases as the current increases.

Now, the filament lamp is made of very thin metal wire. As the current through the filament wire increases, its temperature rises sharply. With this increase in temperature, metal atoms in the wire vibrate faster and impede the flow of electrons which make up the electric current.

So, the resistance of a **filament lamp** *increases as the current increases due to the increase in filament temperature.*

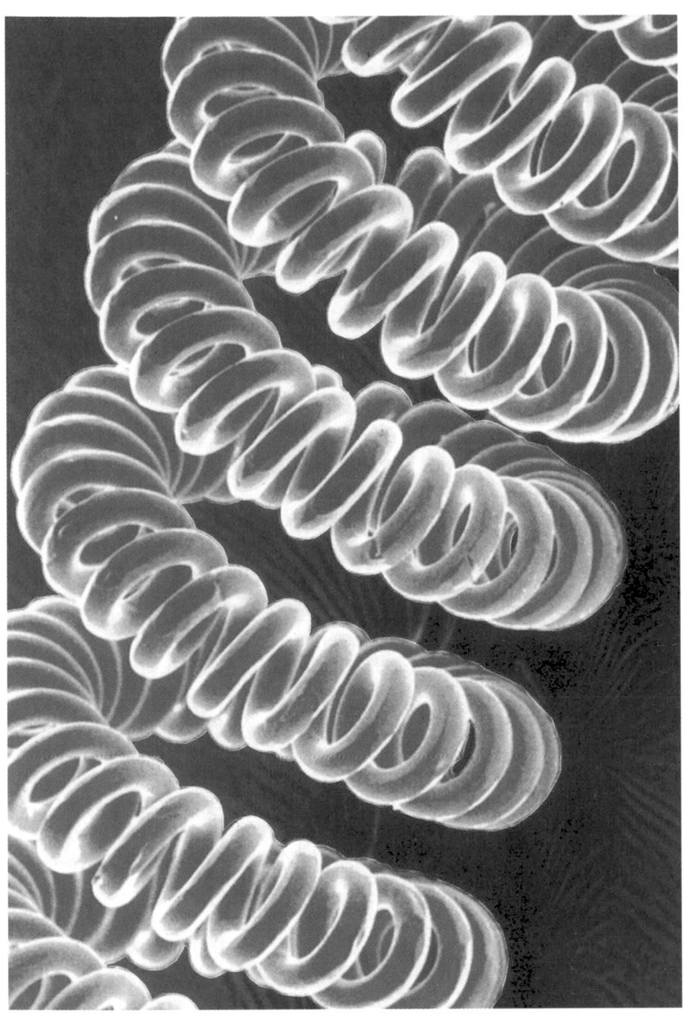

The tightly coiled hot tungsten filaments in a light bulb

At very low temperatures, the atoms in a metal vibrate very little. At extremely low temperatures, the electrons in some metals are not really impeded in any way and they become **superconductors**.

Figure 24.8c shows the voltage–current graph for a thermistor or a light dependent resistor. In this case, the graph curves over as current increases. This means that V/I and therefore resistance, decreases as the current increases.

Unlike resistors and lamps, which are made from metals and alloys, thermistors and LDRs are made of **semiconductors**, such as silicon and germanium.

In semiconductors, electrons are held strongly by the atoms in covalent bonds (section 17.9). As the current increases, the temperature rises, more electrons have sufficient energy to escape from their atoms and the resistance appears to fall. This results in the decreasing gradient of the voltage–current graph.

In **thermistors**, *the resistance usually decreases sharply as the temperature increases.* **LDRs** also have a resistance which decreases with temperature. LDRs are, however, as their name implies, more easily influenced by light and their *resistance decreases sharply as light intensity increases.*

Diodes are also made from semiconductors. In this case, the semiconductor materials are carefully chosen and combined so that current will flow through the diode in one direction only. In the reverse direction, the diode has a very high resistance and the current is negligible even at relatively high voltages.

QUESTIONS

1 The circuit below was used to measure the resistance of a lamp bulb.

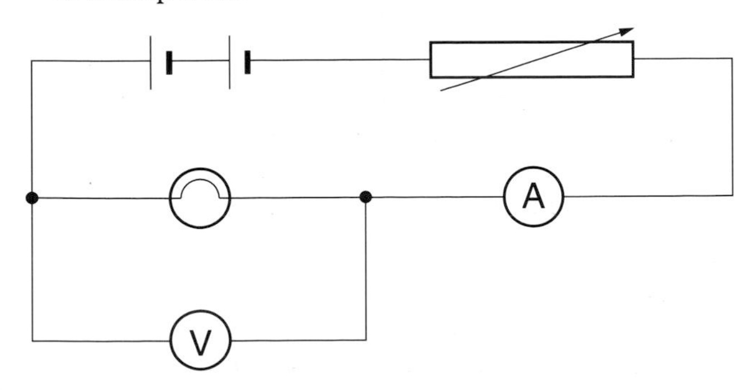

a) The first readings were:
voltmeter 4.0 V.
ammeter 1.25 A.

i) Calculate the resistance of the lamp at this current.

ii) If the resistance of the lamp did **not** change, what current would a voltage of 12 V drive through this bulb?

b) Further sets of readings were taken. The resistance of the lamp was calculated for each set of readings. The graph shows how the resistance of the lamp changes with the current passing through it.

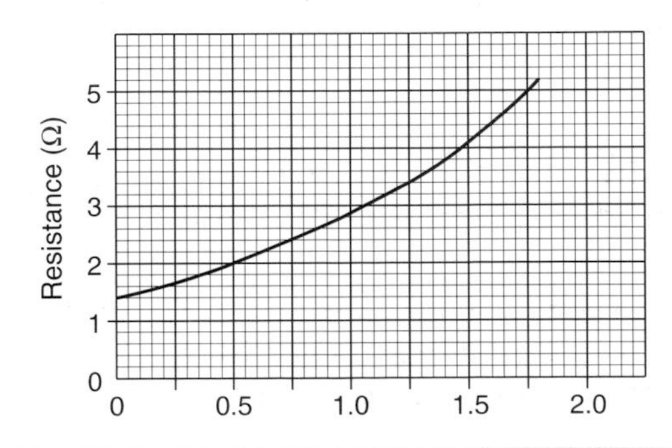

2 Study the diagram below and read the following passage carefully before answering the questions.

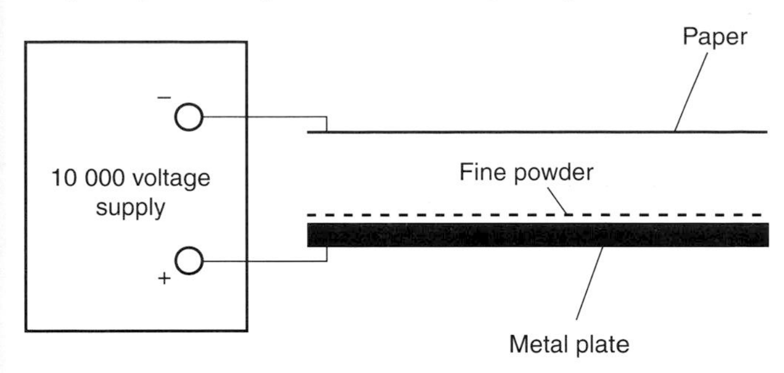

A fingerprint on the surface of a piece of paper can be revealed using electrostatically charged paper.

The piece of paper is connected to the negative terminal of a 10 000 V supply.

The metal plate coated with a fine powder is given a charge from this supply.

The powder particles are repelled from the metal plate towards the paper. When they strike the paper they lose their charge before picking up another charge which causes them to be repelled back towards the plate. However, particles which strike the ridges of the fingerprint stick to them. This enables the fingerprint to be clearly seen.

a) What is the charge on the powder when it leaves the plate?

b) Explain why the powder is repelled from the plate.

c) Explain in terms of *electrons* what happens when the particles come into contact with the paper. **WJEC**

3 a) State **two** disadvantages of transporting oil in large tankers from the Middle East to electric power stations in Northern Ireland.

b) In Northern Ireland, electric power stations use fossil fuels and have an efficiency of about 0.25 (25%). Most of the energy lost occurs as heat loss. This heat is carried away by a large volume of cooling water.
i) Write down an equation which explains the meaning of efficiency.
ii) How much fuel energy has to be used to produce 1 kWh of electrical energy? Give your answer in megajoules. Show clearly how you obtain your answer.
iii) Very large volumes of waste hot water are produced in our power stations. Suggest why this hot water is not usually piped directly to homes as it is in Iceland.

c) There is a hydroelectric power station at Foyers, near Loch Ness, in Scotland. When the station is generating electricity, water flows from Loch Mhor to Loch Ness via the power station as shown in the diagram.

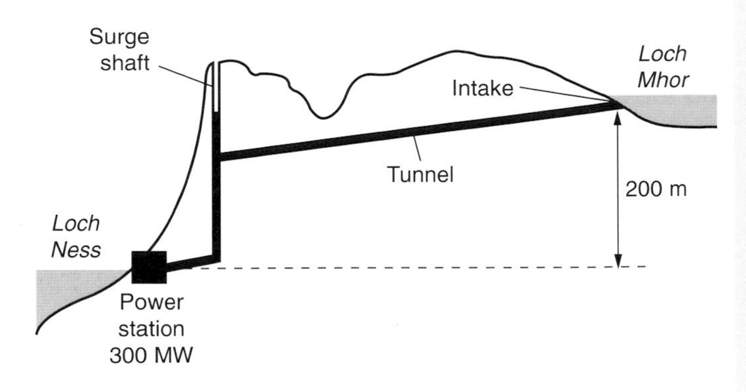

(Not to scale)

i) Name the **two** major types of energy possessed by the water in the pipe line (tunnel).
ii) Explain why a hydroelectric power station is much more efficient than one burning fossil fuels.
NICCEA

4

You may find these formulae useful when answering this question energy transferred (kWh) = power (kW) × time (hr) total cost = number of Units × cost per Unit

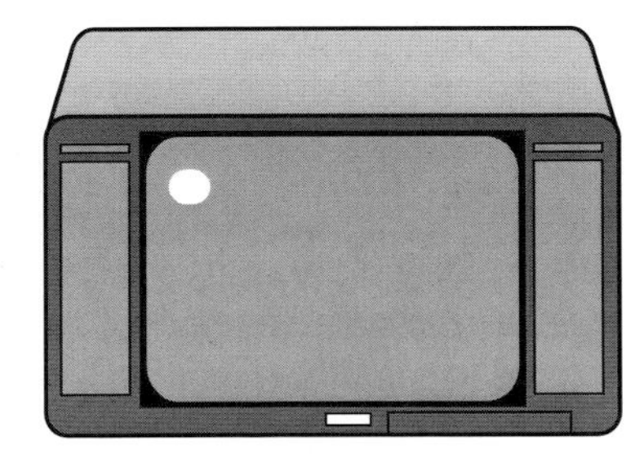

a) Copy and complete these sentences:
The television set transfers electrical energy into ______________ energy, ______________ energy and ______________ energy.
The unit of energy is the ______________ .

b) The power of the television set is 200 W. Electricity costs 9p per unit. Calculate how much it costs to watch the television for 5 hours. Show your working.
NEAB

CHAPTER

26

Electromagnetism

26.1 Magnets and magnetic poles

More than 4000 years ago, Chinese travellers used magnetic iron ore as a crude compass. Since then, magnets have been used in scores of other machines like motors, generators, transformers, electric bells and televisions.

Sailors have used magnetic compasses to navigate their ships for more than 4000 years

When a magnet is dipped into a box of pins, they cling to it (Figure 26.1). A similar thing happens with iron filings or steel paper clips, but not with brass screws. The ends of the magnet where most of the pins cling are called **poles**.

Magnets will only attract unmagnetised metals and alloys which contain **iron**, **cobalt** or **nickel**. So, steel which is an alloy of iron is attracted by magnets, but brass (an alloy of copper and zinc) is unaffected.

Figure 26.1 Pins cling to a bar magnet

North poles and south poles

If a bar magnet is hung horizontally so that it can turn freely, it always comes to rest in a north–south direction. The end (pole) of the magnet which always points north is called the **north-seeking pole** (or just **north pole** for short). The pole of the magnet which points south is called the **south-seeking pole** or **south pole**. Because of this, a small magnet can be used as a compass (Figure 26.2).

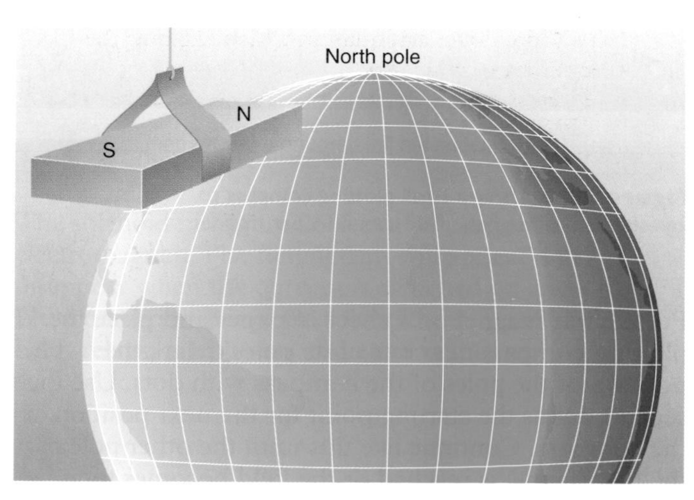

Figure 26.2 A small magnet can be used as a compass

CHAPTER

27

The properties of waves

27.1 Reflecting sound – echoes

If you speak in a completely empty room, you will hear your voice make a 'ringing' sound. Without furniture or fabrics in the room, sounds are reflected from the walls like balls bouncing off the sides of a snooker table. The most obvious examples of sound being reflected are **echoes**.

Using echoes, it is possible to measure the speed of sound in air (Figure 27.1). Stand 50 to 100 metres from a large flat-sided building. Now clap your hands and listen for the echo. Try to clap your hands continuously *so that each clap coincides with the echo from the previous clap*. If you can do this, the sound has travelled to the building and back in the time between one clap and the next. While you are clapping, your partner should find the time for 20 of your claps.

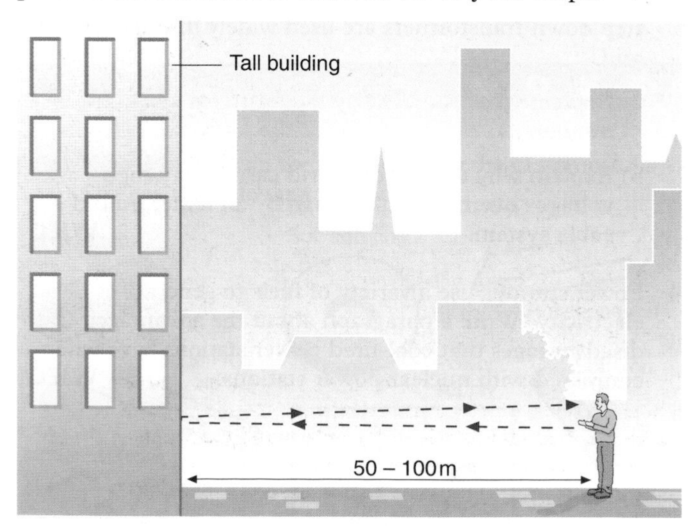

Figure 27.1 Measuring the speed of sound in air

Example
Suppose you stand 60 metres from the building and 20 of your claps take 7 seconds. What is the speed of sound?

Solution
Distance travelled by sound between claps

$$= 2 \times 60 \text{ m} = 120 \text{ m}$$

$$\text{Time between claps} = \frac{7 \text{ s}}{20} = 0.35 \text{ s}$$

$$\therefore \text{ speed of sound} = \frac{\text{distance travelled}}{\text{time taken}}$$

$$= \frac{120 \text{ m}}{0.35 \text{ s}} = 343 \text{ m/s}$$

27.2 Echoes and ultrasounds

Echoes can make speech and music very unclear because each sound is mixed with echoes of sounds made a fraction earlier. Because of this, echoes must be carefully controlled in halls and theatres, particularly where music is played.

Echoes have some very important uses:

- At sea, to measure the depth of the sea bed, to locate ship wrecks and to detect shoals of fish (Figure 27.2).
- By geologists and mining engineers, to study the structure of the Earth and search for minerals including oil.
- By doctors, to examine a pregnant woman's baby and body tissues. Unlike X-rays, low power ultrasound will not damage body cells or tissues.

The acoustics (sound qualities) in the music studio above have been improved by fixing large baffles to the wall. These absorb the sounds and reflect them in different directions, thus reducing echoes

These uses involve short bursts of sound waves caused by vibrations which are so rapid that we cannot hear them. The name **ultrasound** is used to describe these waves with high frequencies that cannot be detected (heard) by the human ear. Ultrasounds are used because narrow beams of sound can be sent from a small source with very little spreading of the sound.

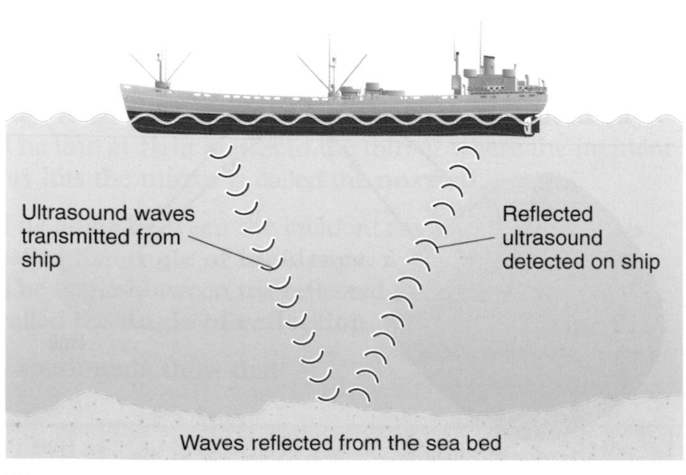

Figure 27.2 Using ultrasound to measure the depth of the sea

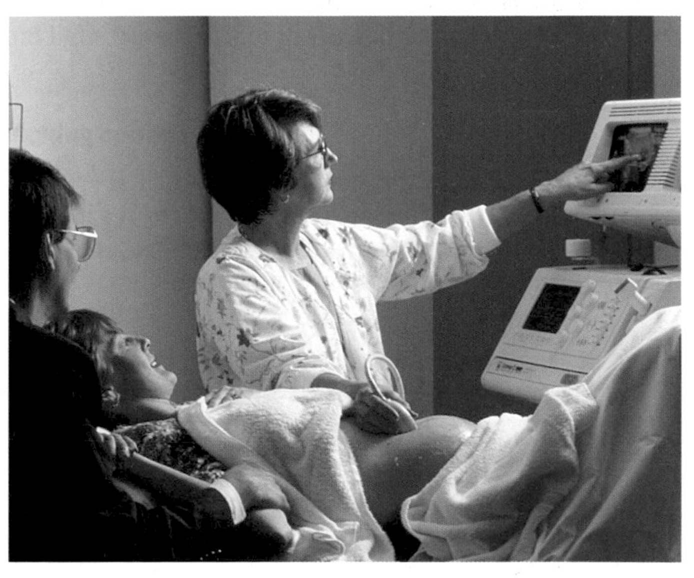

Ultrasound scans are routine for pregnant women. The scan (picture) provides a check on the size and development of the foetus. The scan is built up by reflections of ultrasound from any boundary between body tissues. The foetus reflects more of the sound than the surrounding amniotic fluid. This means that the foetus shows up lighter and can be defined

When food is heated in a conventional oven, the oven itself is heated and heat is lost from the oven. Heat is transferred through the food to be cooked by conduction. This is a relatively slow process. In contrast, microwaves are reflected inside the oven and they do not heat the oven itself. Microwaves heat the food more quickly by penetrating several centimetres before being absorbed. Using a microwave oven, cooking times can be reduced to about a quarter.

Infrared radiation for cooking and detection

Infrared waves fall between visible light and microwaves in the electromagnetic spectrum. Infrared radiation is absorbed by all materials. It is used in our homes for cooking, for radiant heating and for remote control of televisions, videos and other hi-fi equipment.

All objects emit infrared rays, the hotter the object, the more penetrating the infrared rays. Infrared detectors can therefore be used to track and observe warm objects such as rockets, nocturnal animals and parts of our bodies.

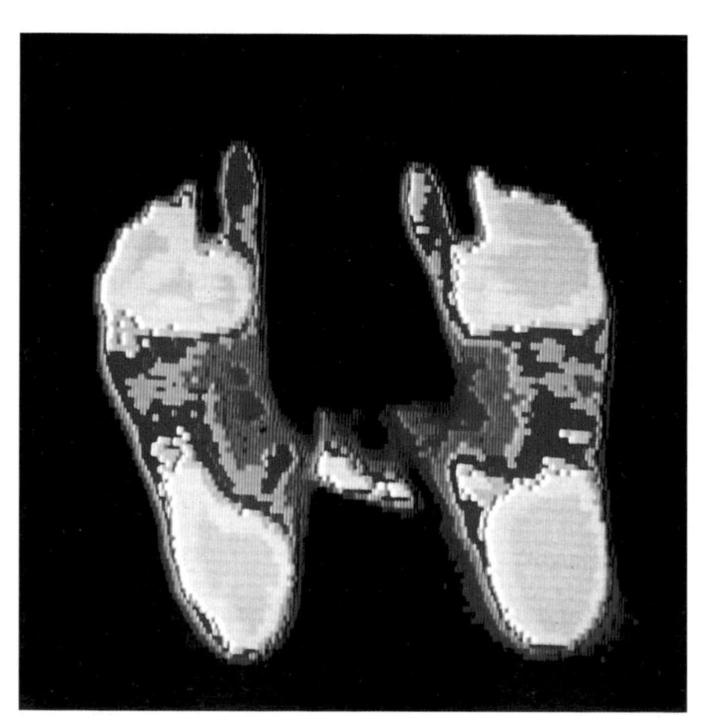

This thermal photograph (thermograph) shows the soles of someone's feet. The thermograph detects thermal (infrared) radiation rather than light. The colour varies from white (hottest) through yellow, to red and blue (coldest)

Ultraviolet radiation – sunbathing and energy-efficient lamps

Ultraviolet (UV) radiation is emitted by the Sun and by other white hot objects. It will pass through air and other gases, but not through solids or liquids.

UV rays increase the formation of vitamin D and also melanin, the brown pigment that gives us a suntan. Unfortunately, UV rays can also damage cells and cause skin cancer. The Sun causes sunburn from the infrared radiation and skin cancer from the ultraviolet radiation. There is less risk for people with dark skins than people with fairer skins. The darker the skin, the more UV radiation is absorbed by the epidermis and less reaches deeper tissues where there may be permanent damage. However, everyone should limit their exposure to the Sun and use appropriate cream for extra protection. The use of sunbeds is not advisable.

Sunbathing can be very pleasant but people should take care to limit their exposure to the Sun

Ultraviolet radiation is also produced in fluorescent lights. When electricity is passed through a fluorescent tube, reactions occur and ultraviolet radiation is emitted. This UV radiation is absorbed by the coating on the inside of the glass tube. The coating fluoresces and re-emits the energy as light.

Fluorescent lamps are more efficient that ordinary electric lights. They rely on the emission of UV radiation rather than heating a metal coil to white heat. Therefore, there is less heat energy loss when the lamps are operating.

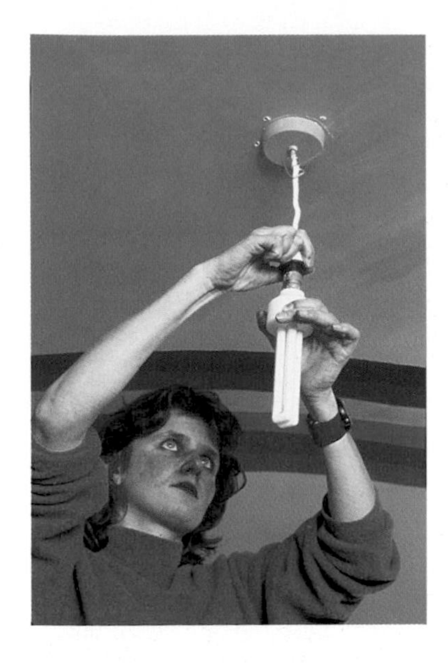

Energy-efficient lamps like this one are more efficient than ordinary lamps because they do not rely on heating a metal coil to white heat

28.5 Using waves in communications

From section 27.10 you will realise that information and messages can be transmitted using electromagnetic radiation. In fact, electromagnetic radiations as radiowaves, microwaves, infrared rays and visible light are crucial for our communications. Increasingly, telephone links are using narrow beams of microwaves to transmit information and messages. Radiowaves and microwaves are also important in radio and television broadcasting. These waves have wavelengths between 1 metre and 1 kilometre. Remember that radiowaves are electromagnetic transverse waves. They are not the same as sound waves which are longitudinal and *not* electromagnetic.

The way in which radiowaves and microwaves travel depends on their wavelength.

- Long wavelength radiowaves ($\lambda > 10$ m) are transmitted around the Earth by reflections from the ionosphere (layers of ionised gas 100 to 400 km above the Earth). Radio stations which use radiowaves with wavelengths of 10 m or more can therefore broadcast from one transmitter to the whole of the U.K. Another advantage of long wavelength radiowaves for broadcasting is that their long wavelengths allow diffraction around landmasses.
- More information can, however, be transmitted by shorter wavelength, higher frequency waves. Stereo radio and TV channels therefore broadcast using shorter radiowaves and microwaves with a wavelength of about 1 m. Unfortunately, these waves are not reflected by the ionosphere, so their range is limited by the Earth's curvature. National broadcasts at this wavelength are transmitted from the British Telecom Tower in London.

Stereo radio and TV programmes are transmitted from the British Telecom Tower in London by narrow beams of radiowaves. These narrow beams are passed through a network of repeater stations throughout the country. At each local transmitting station, the signal is transmitted in all directions to our homes

28.6 Using X-rays and gamma rays in medicine

X-rays and gamma rays are at the short wavelength end of the electromagnetic spectrum. They have wavelengths less than 10^{-8} m and are the most penetrating. The penetration of matter by these rays depends very much on the relative atomic mass of the constituent atoms. The smaller the relative atomic mass, the more penetrating is the radiation. X-rays are produced by accelerating electrons onto a metal target. Gamma rays are emitted by unstable radioactive materials.

In medicine, X-rays are used for diagnostic purposes, while gamma rays are used for treating cancer. They are used with great care because their short wavelength radiations can destroy and damage cells and human tissue causing mutations and cancer.

X-ray photography

X-rays are absorbed to some extent by all body tissue, but bones and teeth containing calcium (Ca = 40) and phosphorus (P = 31) absorb more effectively than flesh which contains mainly carbon (C = 12), hydrogen (H = 1) and oxygen (O = 16). X-rays are passed through the body, forming an image on a photographic film (Figure 28.6).

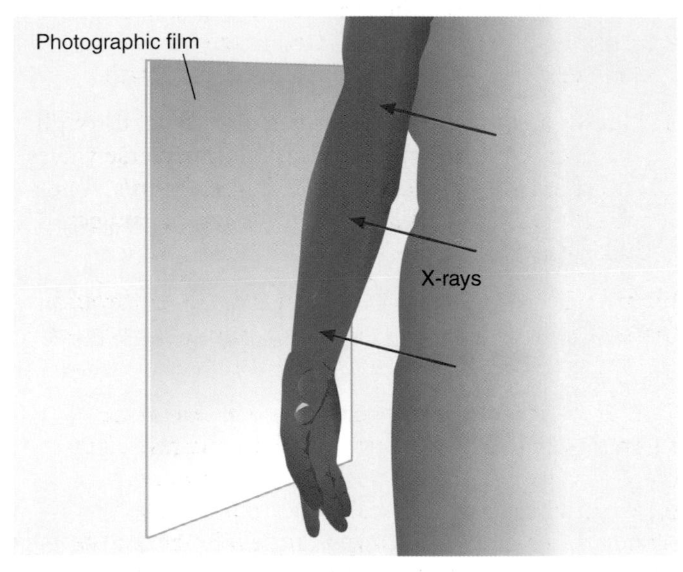

Figure 28.6 Taking an X-ray photograph of an arm

Figure 28.7 shows an X-ray photograph of a patient's arm. Notice how the bone shows up white because it has absorbed the X-rays and prevented them darkening the photographic film.

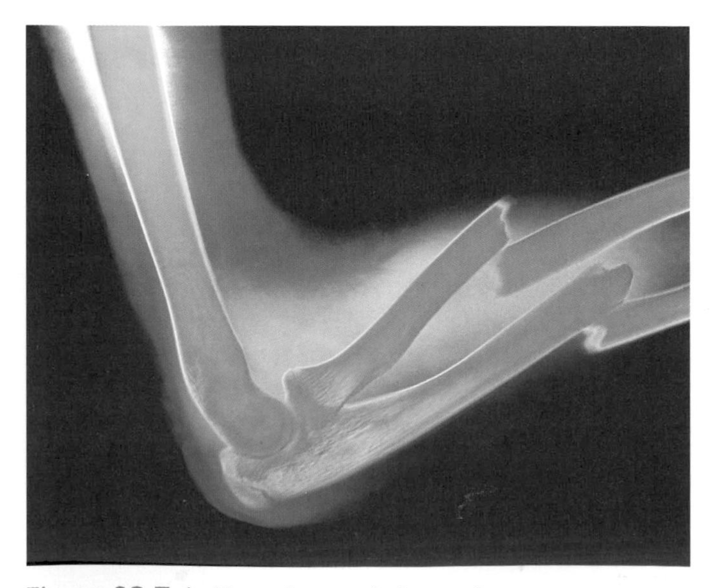

Figure 28.7 An X-ray photograph showing fractures to the radius and ulna bones of the lower arm

When X-rays and gamma rays are absorbed by body tissue, their energy can cause ionization by removing electrons from atoms. This can change the molecules that control the way in which a cell operates. This can result in damage to the central nervous system, mutation of genes or even cancer.

The dangers from X-rays and gamma rays mean that exposure to such ionizing radiations must be kept to a minimum. Radiographers, who use X-ray equipment, work from behind a lead screen. Lead is also used to protect the parts of a patient's body which are not being X-rayed. Lead, with a high relative atomic mass (Pb = 207) is particularly effective in absorbing X-rays and gamma rays.

Treatment of cancer using X-rays and gamma rays

Although X-rays and gamma rays can *cause* cancer, they are used to *treat* cancer. Cancer cells are more vulnerable to X-rays and gamma rays than normal, healthy cells. This is because cancer cells are growing and dividing more rapidly. Figure 28.8 shows how a beam of X-rays or gamma rays from cobalt-60 can be used to treat a cancerous tumour. Directing the radiation at the tumour from several directions ensures that the cancer receives a much higher dose of radiation than the surrounding tissues.

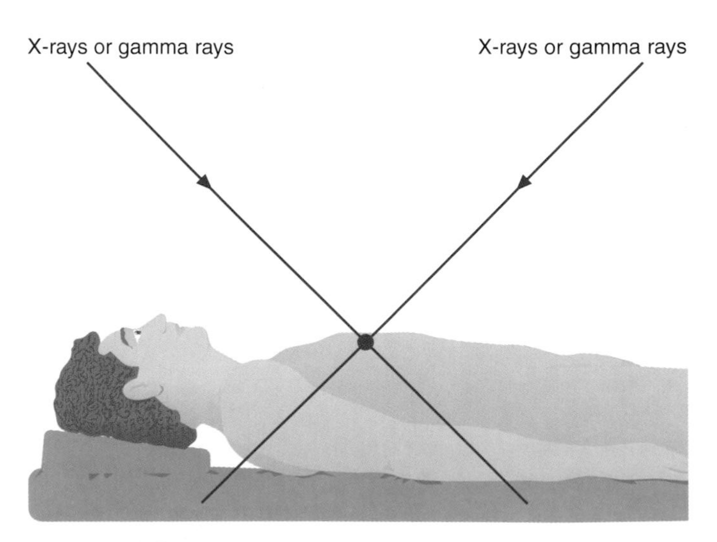

Figure 28.8 Treating a cancerous tumour using X-rays or gamma rays

28.7 Earthquake waves – evidence for the Earth's structure

Earthquakes provide good evidence for the layered structure of the Earth (see section 19.8). When an earthquake occurs, three kinds of shock waves (**seismic waves**) travel outwards from the centre of the quake (**epicentre**). These seismic waves are continually monitored by geologists using special instruments called **seismometers**. Seismometers are designed so that even the tiniest movement of the Earth's surface can be detected and charted using a pen recorder.

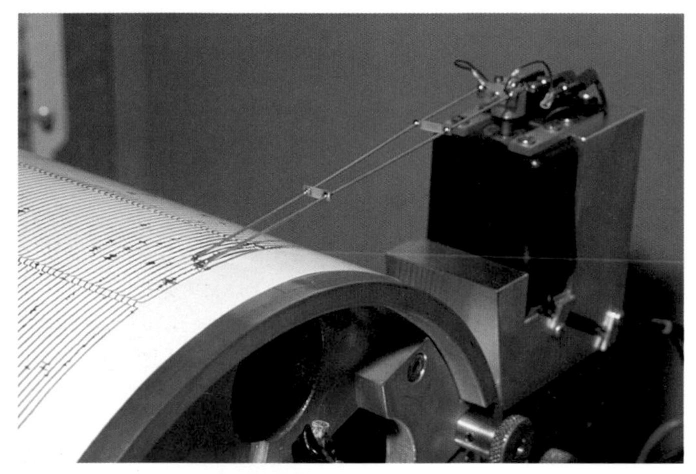

A seismometer pen recorder recording Earth movements

Seismometers detect the three types of shock waves.

1 **Surface waves** travel out from the epicentre in the Earth's crust. Surface waves are transverse waves. They are the slowest moving earthquake waves (2 to 4 km/s), but they do the most damage to buildings because they cause the ground to move.

2 **Primary waves** or **P waves** travel through the body of the Earth. P waves are longitudinal waves. They are the fastest moving earthquake waves travelling through the mantle and the core at 8 to 13 km/s. P waves are refracted at the boundary between the mantle and the core of the Earth.

3 **Secondary waves** or **S waves** also travel through the body of the Earth. S waves are transverse waves travelling slower than P waves at 4 to 7 km/s. They travel through the Earth's mantle, but are reflected when they hit the core.

Figure 28.9 shows how S and P waves travel through the body of the Earth and how they provide evidence for its layered structure.

The reflection of S waves causes a 'shadow zone' where S waves are not detected. This suggests that there is a boundary (the mantle/core boundary) through which the S waves cannot pass so are reflected. This provides important evidence for the layered structure of the Earth as the S waves pass freely through one layer but cannot pass through the core.

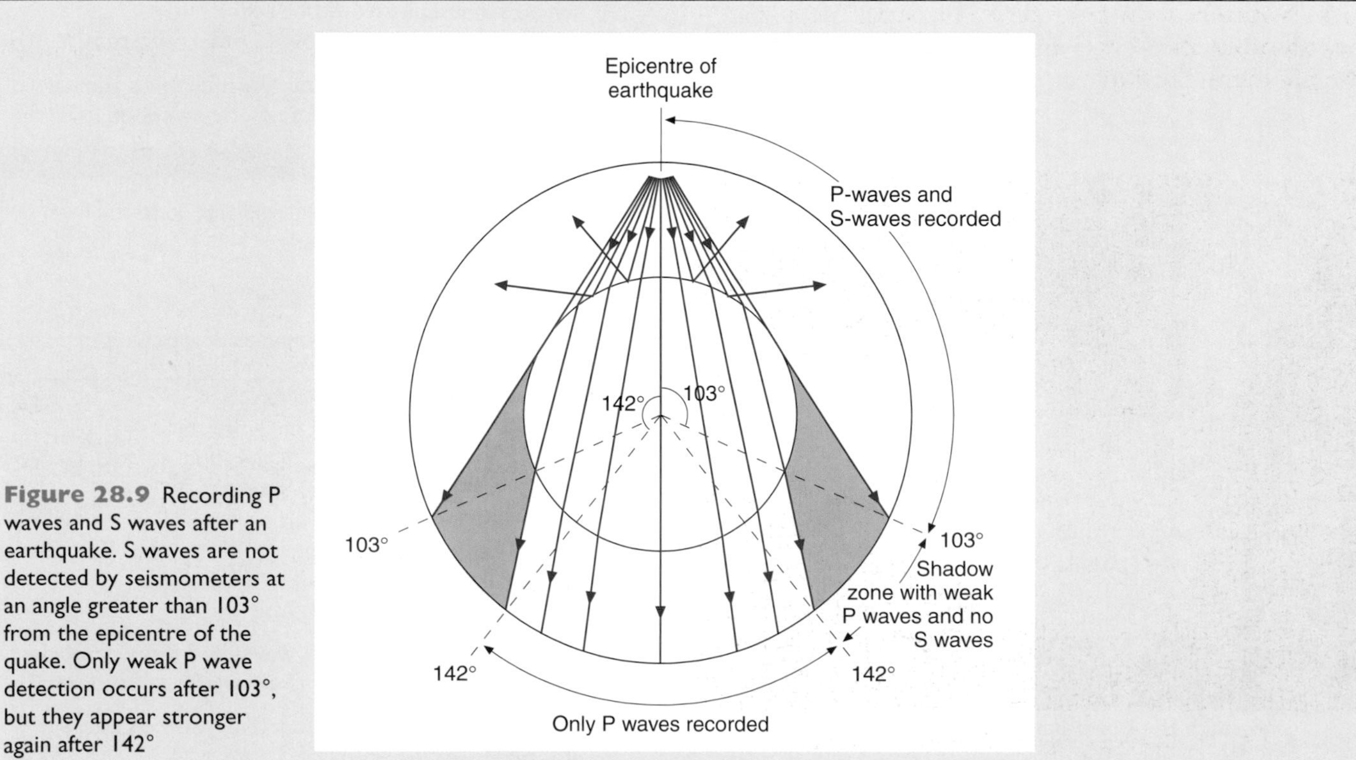

Figure 28.9 Recording P waves and S waves after an earthquake. S waves are not detected by seismometers at an angle greater than 103° from the epicentre of the quake. Only weak P wave detection occurs after 103°, but they appear stronger again after 142°

1 Diagrams A, B, C and D show oscilloscope traces of four different sounds.

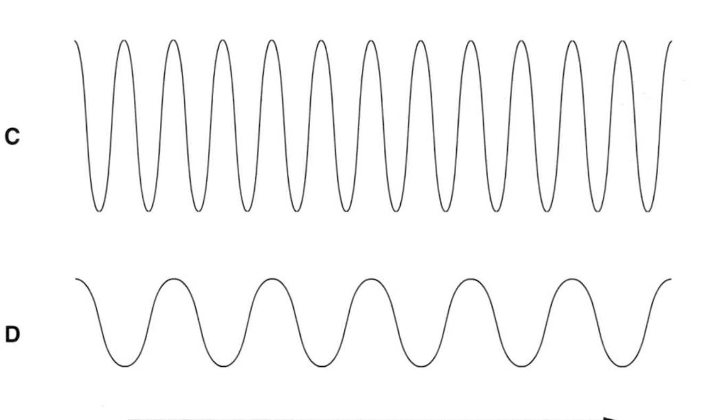

a) Which trace A, B, C or D, represents:
 i) the loudest sound?
 ii) the sound with the highest pitch?

b) The sound labelled C has a frequency of 500 Hz and a speed of 340 m/s. Calculate the wavelength of this sound. Show your working.

c) The drawing shows how the speed of P waves change as they travel through the Earth.

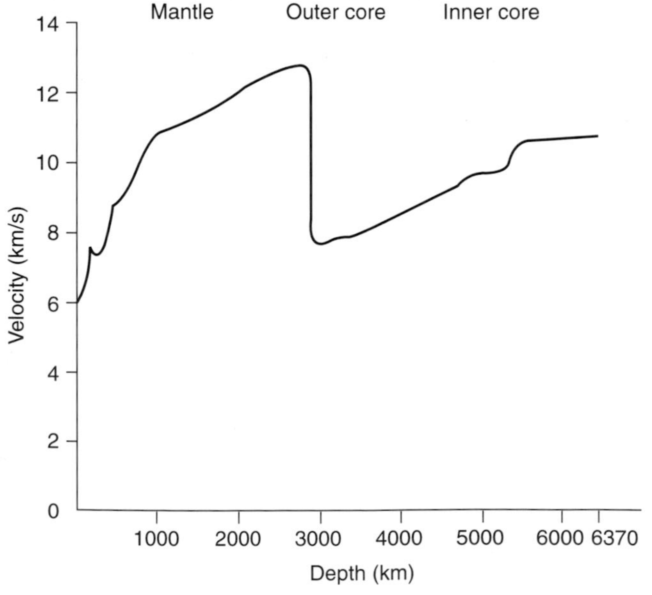

i) Explain why the speed of the waves falls at a depth of 3000 m.
ii) Explain what evidence the speed of the P waves through the Earth has given scientists about its internal structure. **NEAB**

2 The table below shows part of the electromagnetic spectrum.

Radio waves	Microwaves	p	Visible light	Ultraviolet light	q	Gamma rays

Name

a) the missing radiation p,

b) the missing radiation q,

c) the radiation with the longest wavelength,

d) the radiation with the lowest frequency,

e) the radiation used to send information to and from satellites. **WJEC**

3 A prism can be used to produce a spectrum on a screen.

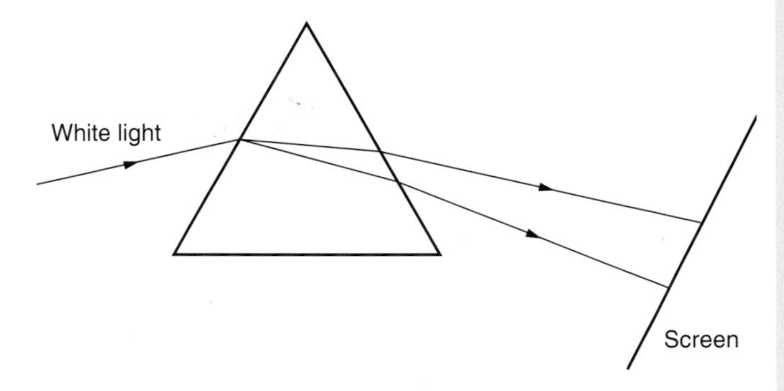

a) i) Copy the diagram and write the letter R on the screen at the place where red light would be seen.
 ii) Write the letter V on the screen at the place where violet light would be seen.
 iii) Name two other colours of this spectrum.

b) Copy and complete the table which gives the uses of different types of electromagnetic radiation.

Type of radiation	Use
	Sending information to and from satellites
	Making toast
	Producing shadow pictures of bones

NEAB

4 a) The two main earthquake waves are the primary (P) waves and the secondary (S) waves.
State which wave:
 i) is a longitudinal wave;
 ii) travels at the slower speed;
 iii) travels through solids only.

b) The average speed of (P) waves is 6 km/s. Calculate the distance travelled by these waves in $\frac{1}{2}$ minute.

c) Explain what can be deduced about the earth's interior from the study of S waves. **WJEC**

CHAPTER

29

Radioactivity

29.1 Radioactive materials

Most radioactive materials that we use today are man-made, but some are naturally occurring. Large amounts of radioactive uranium and plutonium are used to generate electricity in power stations. Yet tiny amounts of these and other elements can be used to generate electricity in heart pacemakers. Radioactive materials can have both beneficial effects and harmful effects. They can cause cancer, but they can also cure it.

Marie Curie began her studies of radioactivity with uranium compounds. Between 1898 and 1902, she spent almost 4 years in isolating a new element, radium, which was 2 million times more radioactive than uranium. In 1903, Marie Curie shared the Nobel Prize for Physics. Then in 1911, she was awarded the Nobel Prize for Chemistry. She was the first person to win two Nobel Prizes

The first investigations of radioactivity were carried out by a Frenchman, Henri Becquerel in 1896 and by his assistant Marie Curie in the years following. They discovered that all uranium compounds emitted **radiation** which could pass through paper and affect a photographic plate.

Becquerel called this process **radioactivity** and he described the uranium compounds as **radioactive**.

29.2 Detecting radioactivity – unstable nuclei

Nowadays, the best method to detect the emissions from radioactive materials is to use a Geiger Müller tube (Figure 29.1).

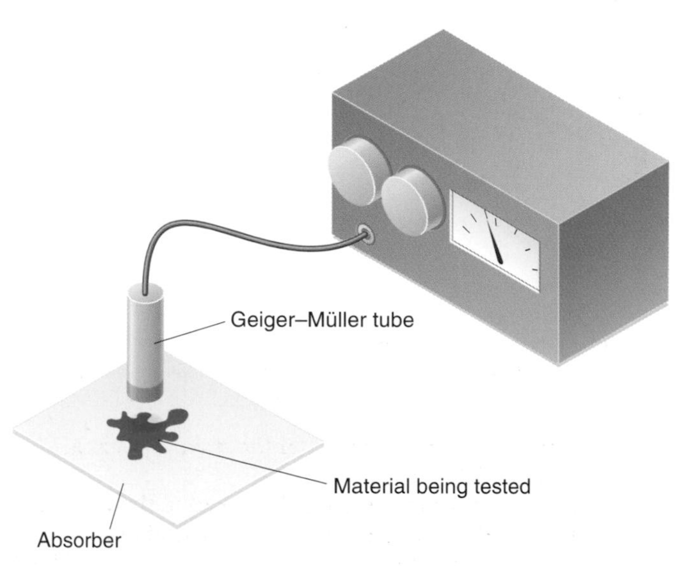

Figure 29.1 Detecting radioactivity

> Radioactivity results from the breakdown of isotopes (nuclei) that are unstable.

These unstable isotopes (nuclei) are known as **radio isotopes** (**radionuclides**). For example, most carbon and carbon compounds are composed of the isotope $^{12}_{6}C$ (carbon-12) which is completely stable. However, all these samples also contain a tiny percentage of $^{14}_{6}C$ (carbon-14) which is unstable. These unstable atoms of $^{14}_{6}C$ can break down forming stable atoms of nitrogen.

When the nuclei of radioactive atoms break up, they emit energy and lose three kinds of radiation. These radiations or emissions are called **alpha rays** (α rays), **beta rays** (β rays) and **gamma rays** (γ rays).

α rays and β rays contain *particles*, so they are usually called alpha particles and beta particles. Gamma rays are very penetrating *electromagnetic waves*.

The nature and properties of the three kinds of radioactive emissions are summarised in Table 29.1. Notice the different penetrating power of the different radiations in Table 29.1. γ rays are much more penetrating than β particles which are themselves more penetrating than α particles. α particles can be absorbed by a sheet of paper. β particles pass through paper but are absorbed by aluminium foil, but γ rays can only be absorbed by thick lead. Because of this, radioactive substances which emit γ rays are usually stored in thick lead containers.

The other important point to note from Table 29.1 is the relative ionizing power of the radiations. α radiation is the most intensely ionizing radiation even though it is not very penetrating. Exposure to this ionizing radiation can be very harmful because it damages the cells in our skin and other tissues.

29.3 Writing nuclear decay equations

When unstable nuclei break down, they may lose α particles or β particles. The loss of an α particle results in the loss of 2 protons and 2 neutrons from a nucleus. This means that the mass number (nucleon number) of the original nucleus falls by 4 and its atomic number (proton number) falls by 2.

So, when an atom of radium-226 ($^{226}_{88}Ra$) loses an α particle ($^{4}_{2}He$), the fragment left behind will have a mass number of 222 and an atomic number of 86. All atoms of atomic number 86 are those of radon, Rn. So the radioactive decay for radium-226 can be summarised in a nuclear equation as:

$$\begin{matrix}\text{Mass} \rightarrow \\ \text{Atomic number} \rightarrow\end{matrix} \quad ^{226}_{88}Ra \longrightarrow \, ^{222}_{86}Rn + \, ^{4}_{2}He$$

Table 29.1 The nature and properties of alpha, beta and gamma radiations

Radiation	Nature	Symbol	Penetrating power	Effect of electric and magnetic fields	Ionizing power
α particles	helium nuclei containing 2 protons and 2 neutrons	$^{4}_{2}He^{2+}$	a few centimetres of air, absorbed by thin paper	very small deflection	strong
β particles	electrons	$^{0}_{-1}e^{-}$	a few metres of air, pass through paper, but absorbed by 3 mm of aluminium foil	large deflection	weak
γ rays	electromagnetic waves		a few kilometres of air, pass through paper and Al foil, but absorbed by very thick lead	none	very weak

Radioactive decay with the loss of an α particle is common for large isotopes with an atomic number greater than 83. These include uranium-238, plutonium-238 and, of course, radium-226. These isotopes decay partly because they are too heavy. They lose mass and try to become stable by emitting an α particle.

Radioactive isotopes with atomic numbers below 83 usually emit β particles when they decay. β particles are electrons. Their mass is effectively zero compared to a proton or neutron and their charge is negative, i.e. ${}^{0}_{-1}e$.

When carbon-14 (${}^{14}_{6}C$) undergoes β decay, the process can be represented as

$$\begin{array}{l} \text{Mass} \rightarrow \\ \text{Atomic number} \rightarrow \end{array} {}^{14}_{6}C \longrightarrow {}^{14}_{7}N + {}^{0}_{-1}e$$

Notice that the *total* mass and the *total* charge are the same before and after β decay. During β decay, a neutron in the nucleus of the radioactive atom splits up into a proton and an electron. The proton stays in the nucleus, but the electron is ejected as a β particle. Thus, the mass number of the remaining fragment stays the same, but its atomic number *increases* by one.

From these **nuclear equations** you should appreciate that radioactivity involves nuclear reactions with changes to **nuclei** in the **central parts of atoms**.

On the other hand, **chemical reactions** involve **electrons** in the **outer parts of atoms**.

29.4 Background radiation

Various rocks in the Earth, including granite, contain small percentages of radioactive uranium, thorium and potassium compounds. Our bodies also contain traces of radioactive materials as do the bricks and other building materials which are used to build our homes, schools and workplaces. In addition to these sources, we are also exposed to gamma radiation from the Sun.

These sources of natural radiation are referred to as background radioactivity or **background radiation**.

As this background radioactivity comes from a variety of sources, it is higher in some places than others. We are exposed to it all our lives. Normally it is very low and causes no risk to our health.

Although background radiation is usually very small, it must be taken into account when performing accurate experiments and measurements of radioactivity. Usually this involves subtracting the average count rate for background radioactivity from the measured count rate of the sample being tested.

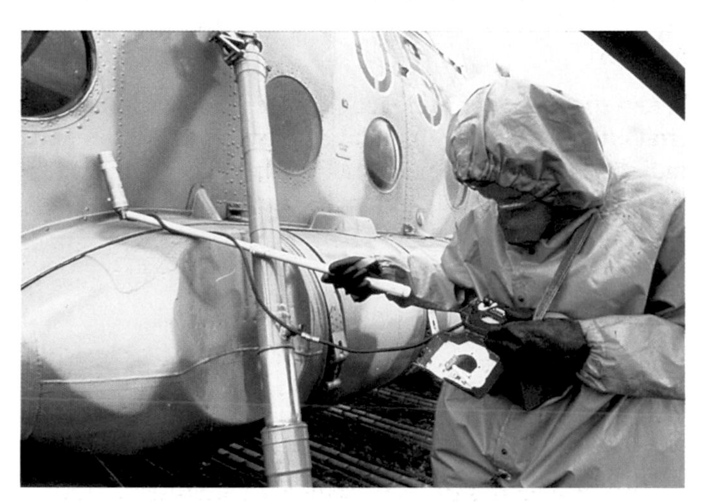

A technician using a Geiger–Müller tube and counter to check the levels of radiation on a helicopter used during the Chernobyl disaster. The radiation detector tube is located at the end of the metal rod in the technician's right hand

29.5 Half-life

Radioactive decay is a spontaneous and random process. You cannot tell when an unstable nucleus will break down. However, if there are large numbers of unstable atoms, then an average rate of decay should be proportional to the number of undecayed atoms remaining. This suggests that a sample containing ten million unstable nuclei of an element should decay at twice the rate of a sample containing five million unstable nuclei of the same element. As the unstable nuclei of a radioactive element decay, the rate of decay should fall.

Figure 29.2, over the page, shows the decay curve for a sample of iodine-131. This isotope is used by doctors to study the uptake of iodine by the thyroid gland in humans. The shape of the decay curve is similar to those for all other radioactive materials but the time scale can vary enormously.

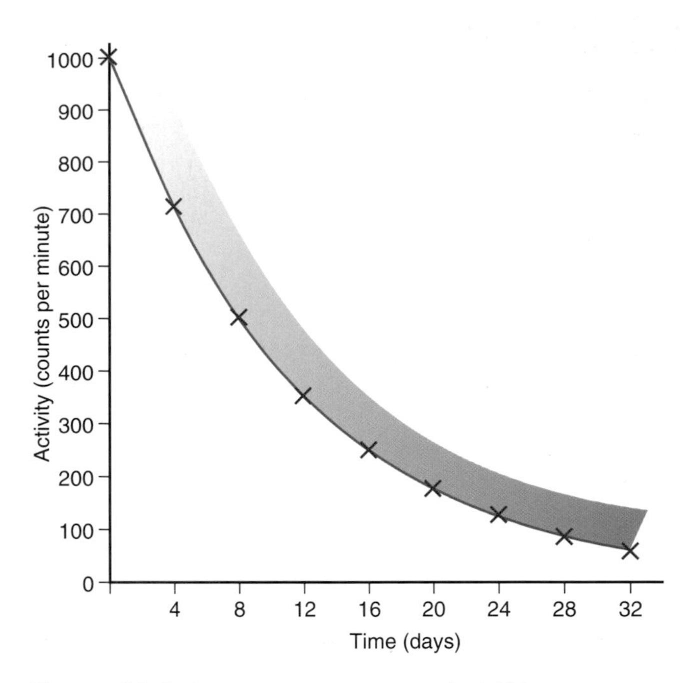

Figure 29.2 A radioactive decay curve for I-131

Notice from Figure 29.2 that:

- time taken for activity to fall from 1000 counts/min to 500 counts/min = 8 days
- time taken for activity to fall from 500 counts/min to 250 counts/min = 8 days
- time taken for activity to fall from 800 counts/min to 400 counts/min = 8 days
- time taken for activity to fall from 400 counts/min to 200 counts/min = 8 days

These results and Figure 29.2 show that the activity of the iodine-131 decreases with time and that the time taken for its activity to halve stays constant at 8 days.

> The time taken for the activity of a radioactive sample to halve is called the **half-life**.

Half-lives can be determined by plotting activity/time graphs similar to that in Figure 29.2 and then finding an average time for the activity to fall to half.

Half-lives can vary from a few milliseconds to several million years. The shorter the half-life, the faster the isotope decays and the more unstable it is. The longer the half-life, the slower the decay process and the more stable the isotope.

Uranium-238 with a half life of 4500 million years is 'almost stable', but polonium-234 with a half-life of only 0.15 milliseconds is quite the reverse.

Example
A hospital gamma ray unit contains 10 grams of cobalt-60 which has a half-life of 5 years. How much cobalt-60 is left after;

(i) 5 years,
(ii) 15 years?

Solution
(i) 5 years is one half-life, ∴ half of the 10 g will decay leaving only 5 g after 5 years.
(ii) 15 years is three half-lives, ∴ 10 g decays to 5 g after 5 years. 5 g decays to 2.5 g after 5 more years. 2.5 g decays to 1.25 g after another 5 years. ∴ after 15 years, 1.25 g remains.

29.6 The harmful effects of radiation

Radiation affects materials by causing the ionization of atoms and molecules. When an atom is ionized, one or more electrons are removed from it or added to it. If ionization occurs in our bodies, the ions produced may attack and destroy cells. Because of this, exposure to ionizing radiation can be harmful.

People who work with radioactive materials must be aware of the dangers from radiation. With penetrating radiation, these dangers include nausea, sickness, skin burns and loss of hair. Exposure to high doses can cause sterility, cancers and even death.

This site worker is wearing a special radiation-sensitive badge on his belt to indicate the level of radiation to which he has been exposed

Scientists and technicians who work with radioactive materials must wear special radiation badges containing film sensitive to radiation. The film is developed at regular intervals to indicate the level of exposure.

The effects of radiation depend on the energy and penetration of the emission as well as the amount of exposure. This means that the more penetrating gamma ray sources are generally more harmful than alpha and beta ray sources. So, people who work with dangerous isotopes emitting gamma rays must take extra safety precautions. These include:

- protection from the radiation by lead, concrete or very thick glass,
- remote control handling of isotopes and equipment from a safe distance,
- reducing exposure to any radiation to the shortest possible time.

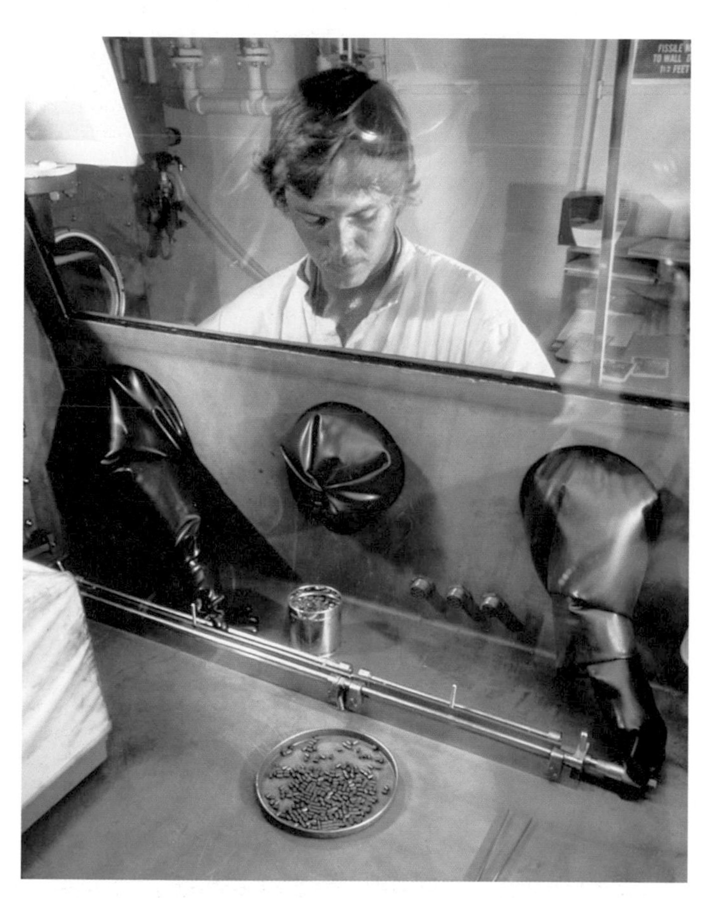

A scientist manipulating radioactive materials by remote control

29.7 Using radioactive materials

Radioactive isotopes are widely used in industry and medicine. Different isotopes are chosen for particular uses depending on their penetrating power and half-life.

Medical uses

Although radiation can be harmful to our cells and other living organisms, it can also be used in medicine to cure cancers. Radiotherapy may slow down the rate of growth of a tumour and in many cases, cause a cure. Penetrating γ rays from cobalt-60 ($^{60}_{27}Co$) are used to kill cancer cells and treat tumours within the body. The cobalt-60 has a relatively long half-life of 5.3 years and the dose can be adjusted to the same level over a few weeks of treatment.

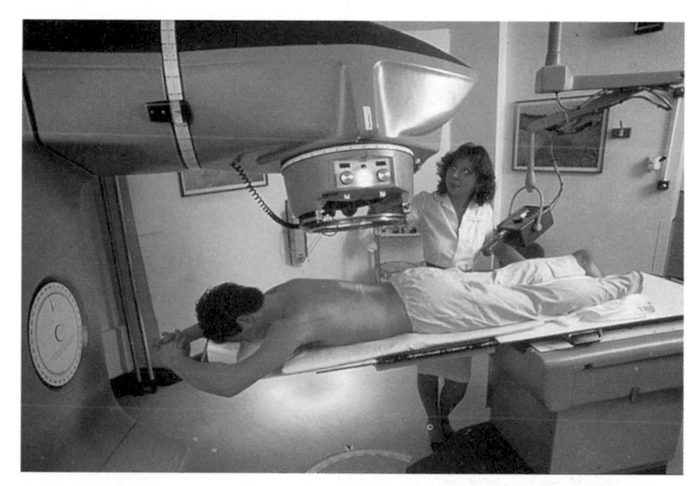

A patient being treated by radiation from cobalt-60. Gamma rays from the cobalt-60 can penetrate the body and kill cancer cells

On the other hand, skin cancer can be treated with less penetrating β rays. This is done by strapping a plastic sheet containing phosphorus-32 ($^{32}_{15}P$) or strontium-90 ($^{90}_{38}Sr$) on the affected area.

Nowadays, most medical equipment such as dressings and syringes are sealed in plastic bags and sterilized by intense gamma radiation from cobalt-60. The radiation sterilizes the articles by killing any bacteria on them.

Tracer studies
e.g. detecting leaks in pipes prior to digging

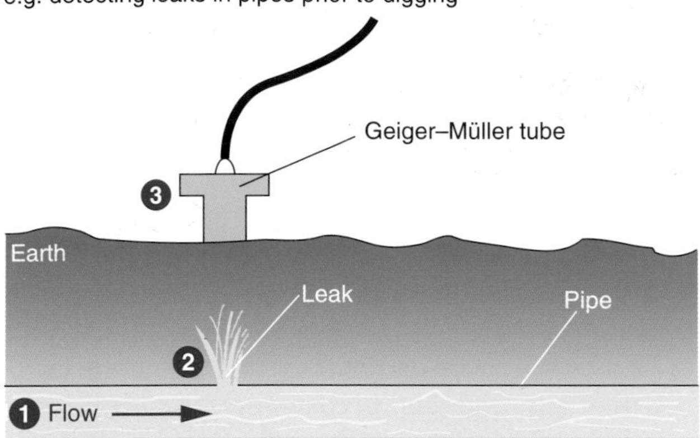

1 Small amount of radioactive tracer (emitting penetrating γ rays with a short half-life) is fed into pipe. γ ray emitter needed to penetrate the soil. Its short half-life allows radiation in the soil to disappear quickly.
2 Radioactive tracer leaks into soil.
3 Geiger–Müller tube is used to detect radiation and position of leak.

Thickness gauges
e.g. during the production of paper, plastic sheeting or aluminium foil

3
Modified Geiger–Müller tube
2
Sheet of material
Geiger–Müller counter
1

1 Radioactive source (emitting β rays and with a long half-life). β ray emitter, with less penetrating radiation used. The long half-life ensures the radiation emitted remains constant.
2 Long, modified Geiger–Müller tube detects radiation penetrating the sheet of material.
3 Geiger–Müller counter measures radiation level (the thicker the sheet, the lower the reading) and feeds back information to adjust the production process, if necessary.

Figure 29.3 Some industrial uses of radioactive isotopes

Industrial uses

Radioactive isotopes have a large number of industrial uses. These include their use as tracers (e.g. detecting leaks) and in thickness gauges (Figure 29.3).

Part of the Turin Shroud. Carbon dating showed that the shroud was about 700 years old. It could not have been used to cover the body of Jesus, as had been thought

Archaeological and geological uses – carbon dating

Almost all the carbon in living things is made up of the carbon-12 isotope ($^{12}_{6}C$) with a tiny constant percentage of carbon-14 ($^{14}_{6}C$). This gets into living things from the carbon-14 which is part of the carbon dioxide in the air. When an animal or a plant dies, the carbon-14 in it continues to decay. However, the replacement of decayed carbon-14 from food and carbon dioxide ceases. Scientists can measure the percentage of carbon-14 left in the remains of an animal or a plant using a mass spectrometer. Knowing the percentage of carbon-14 in living organisms and its half-life, it is possible to work out how long the animal or plant has been dead. The procedure is called carbon dating. Carbon-14 has a half-life of 5700 years. So, after 5700 years, the percentage of carbon-14 in the remains of a dead organism will be half that in a similar living organism.

Using a technique similar to carbon dating, geologists can work out the age of rocks and hence the age of the Earth. Most rocks contain traces of radioactive uranium-238 ($^{238}_{92}U$). This decays in a series of reactions to form lead-206 ($^{206}_{82}Pb$) with a half-life of 4 500 million years. If we assume that a rock contains no lead-206 when it is formed, then the present ratio of U-238 to Pb-206 in the rock can be used to calculate the age of the rock. For example, if the proportion of U-238 to Pb-206 in a rock sample is 1:1, then half of the U-238 has decayed to Pb-206. This means that one half-life has elapsed and the rock is about 4500 million years old.

1 a) Information on the average amount of radiation received by a person in a year is shown in the table.

Source	Amount (arbitrary units)
natural background	125.0
medical sources	55.2
occupational hazards	1.7
other radiation	2.0

i) Give **one** cause of natural background radiation.
ii) Explain why the dose of X-rays you receive should be measured.

b) i) When radon gas is kept in a sealed container for a long time small amounts of helium appear. Explain what is happening to the radon gas.
ii) Sketch a graph to show how the count rate of a radioactive substance will change with time (with time on the horizontal axes and count rate on the vertical axes).
iii) At 9 am a radioactive substance has a count rate of 400 counts per minute (cpm). At 10 am the count rate is 200 cpm. What will it be at 12 noon?

c) i) The diagram represents the structure of a helium **nucleus**.

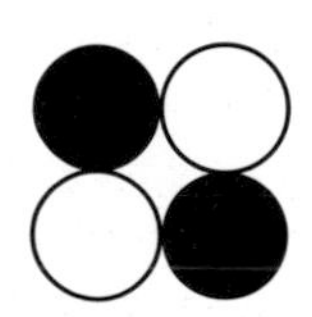

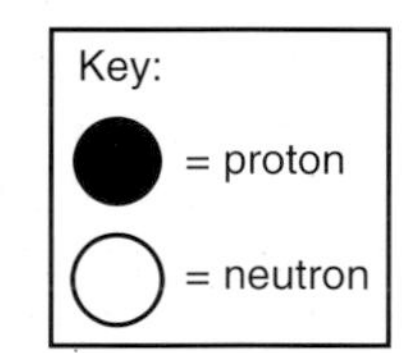

Copy and add information to the diagram so that it represents a helium **atom**.
ii) Use the information in the diagram to help you explain the difference between **mass number** and **atomic number**. **MEG**

2 A teacher used the apparatus shown in the diagram to demonstrate radioactivity to a class.

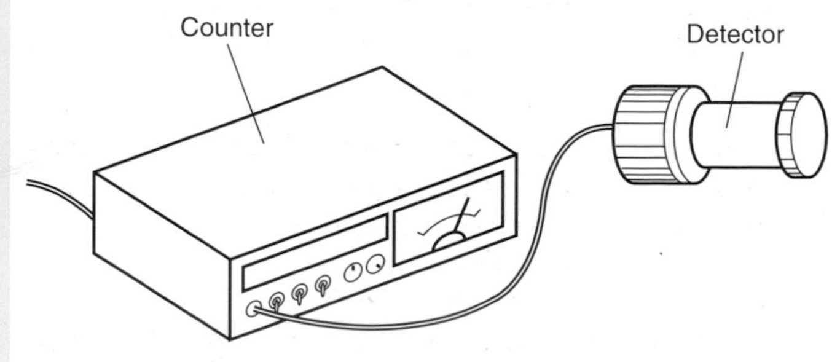

a) The teacher started the counter then stopped it after one minute. A count of **22** was shown. Why did the counter show a reading even though no radioactive source was being used?

b) The teacher then used the detector to investigate the types of radiation given off by a radioactive source.

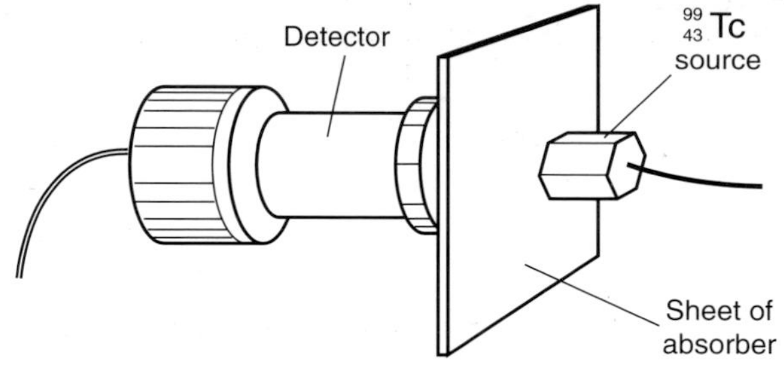

The teacher found the average number of counts per minute with and without absorbers present. The results are shown in the table.

Absorber used	Average counts per minute
No absorber	1973
Card 1 mm thick	1216
Lead 1 mm thick	22

State whether or not each of the following radiations is given out by the source. Give your reason for each answer.
i) Alpha,
ii) Beta,
iii) Gamma.

c) The graph shows how the radioactivity of an isotope of technetium changes with time.

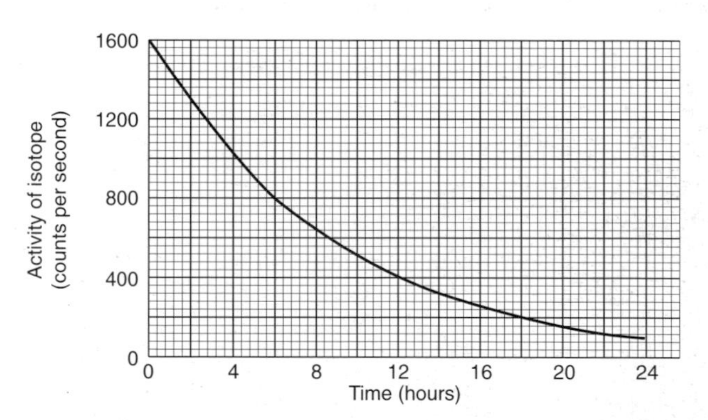

i) Explain what is meant by the term half life.
ii) Use the graph to calculate the half life of this isotope.

d) This isotope is used in medicine as a tracer.

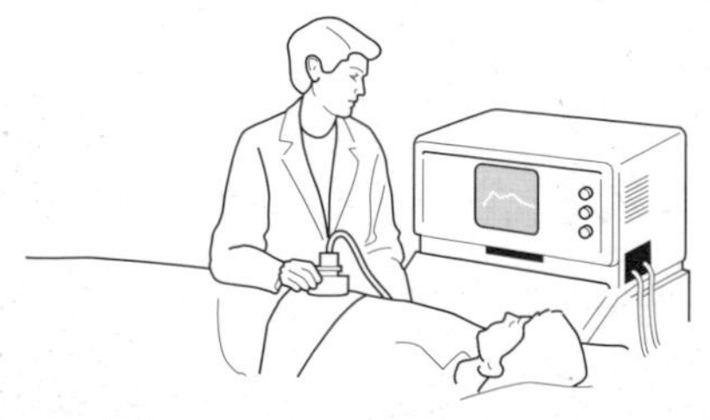

Give **two** properties of the isotope which make it suitable for this use. Explain your answer in each case. **NEAB**

CHAPTER

30

The Earth and beyond

30.1 Our Sun in the universe

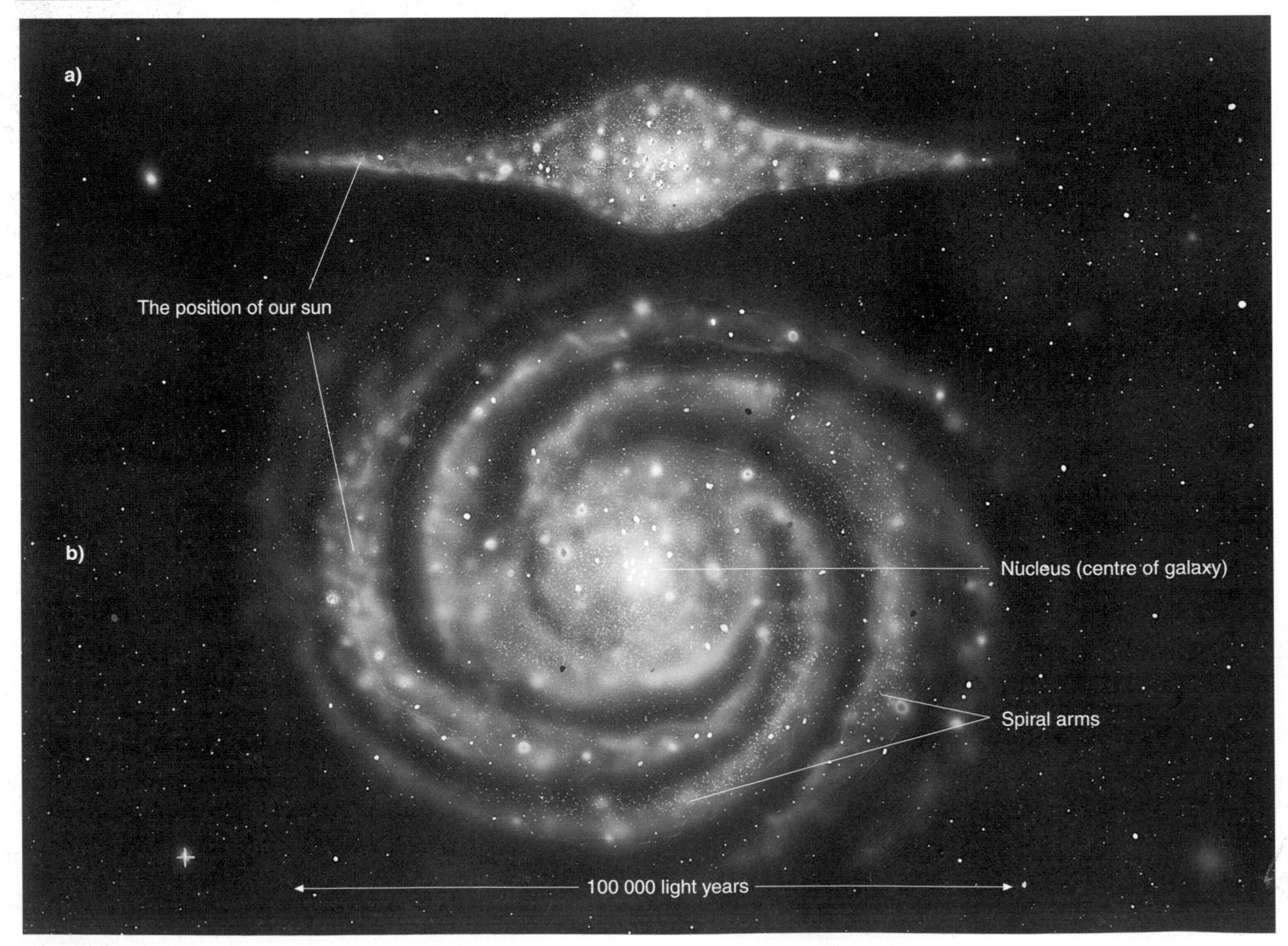

Figure 30.1 The Milky Way galaxy, a) viewed from the side, b) viewed from above

The four inner planets with the Sun (centre). The planets are (clockwise from upper left): Mercury (grey), Venus (yellow), Earth (blue) and Mars (red)

Looking up at the sky on a clear night you will see thousands of stars. Stars have formed and continue to be formed by the compression of gas (mainly hydrogen) and dust. As the gas and dust particles are compressed together, their temperature rises. Eventually, the material gets so hot that it starts to react. Reactions like those in a nuclear reactor take place, giving heat and light.

Our **Sun** is a star. It is the nearest star to Earth and therefore the brightest object that we see in the sky. The Sun and other stars are sources of heat and light. We see the planets and other bodies in the sky as a result of light which they reflect from the Sun and other stars.

Clusters of stars group together to form **galaxies** and billions of galaxies make up the whole **universe**. The Sun is part of the **Milky Way** galaxy (Figure 30.1).

There are about 100 000 million stars in the Milky Way. Viewed from the side, the Milky Way looks like two fried eggs back to back (Figure 30.1a) – long and thin, tapering at the ends with a bulge in the middle. Viewed from above (Figure 30.1b), the Milky Way looks like a huge whirlpool with giant spiral arms.

Our Sun is just one star in the Milky Way. It would take 100 000 years travelling at the speed of light (10^{16} metres per year) to cross the Milky Way. Just imagine how small the Earth is compared to the Milky Way, never mind the whole universe!

30.2 The Sun and our solar system

Let's now turn to a part of the universe where sizes and distances are easier to understand.

The Earth is only a very tiny object relative to the whole universe. It is a small planet, orbiting a small star on one of the spiral arms of the Milky Way galaxy.

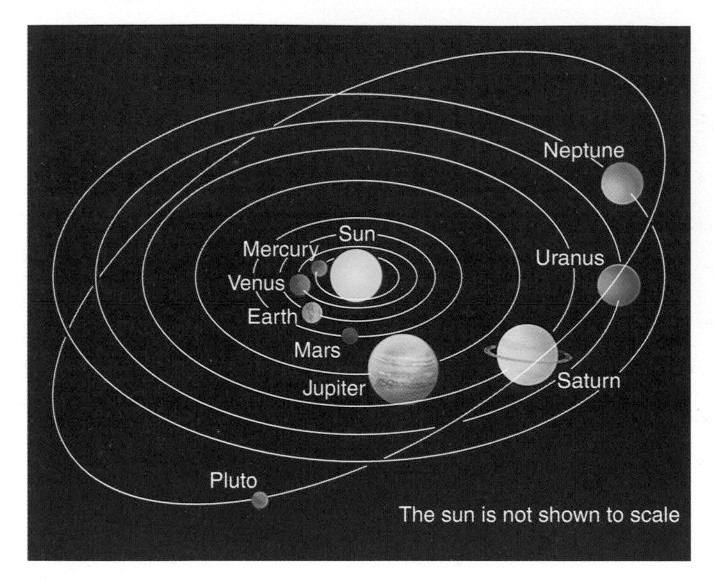

Figure 30.2 The planets in our Solar System

The Earth is one of nine planets orbiting the Sun and these make up our **Solar System** (Figure 30.2).

Notice that there are four planets (Mercury, Venus, Earth and Mars) relatively close to the Sun. These are sometimes called the **Inner Planets**. The other five planets which are further away (Jupiter, Saturn, Uranus, Neptune and Pluto) are sometimes called the **Outer Planets**.

All of the planets move in elliptical orbits in the same direction around the Sun. All of the planets lie in the same plane with the exception of Pluto whose orbit is at an angle to this plane (Figure 30.2). Table 30.1 shows important data for each of the nine planets.

The time that a planet takes to orbit the Sun depends on its distance from the Sun. As the distance from the Sun increases, so also does the orbit time (Table 30.1). For example, Mercury takes only 88 days to orbit the Sun compared to the Earth's 365 days. The four inner planets nearest to the Sun have hard, solid and rocky surfaces with thin, gaseous atmospheres. **Mercury**, closest to the Sun, has a cratered surface and baking temperatures reaching 350°C during the day. **Venus**, between Mercury and the Earth, has even higher

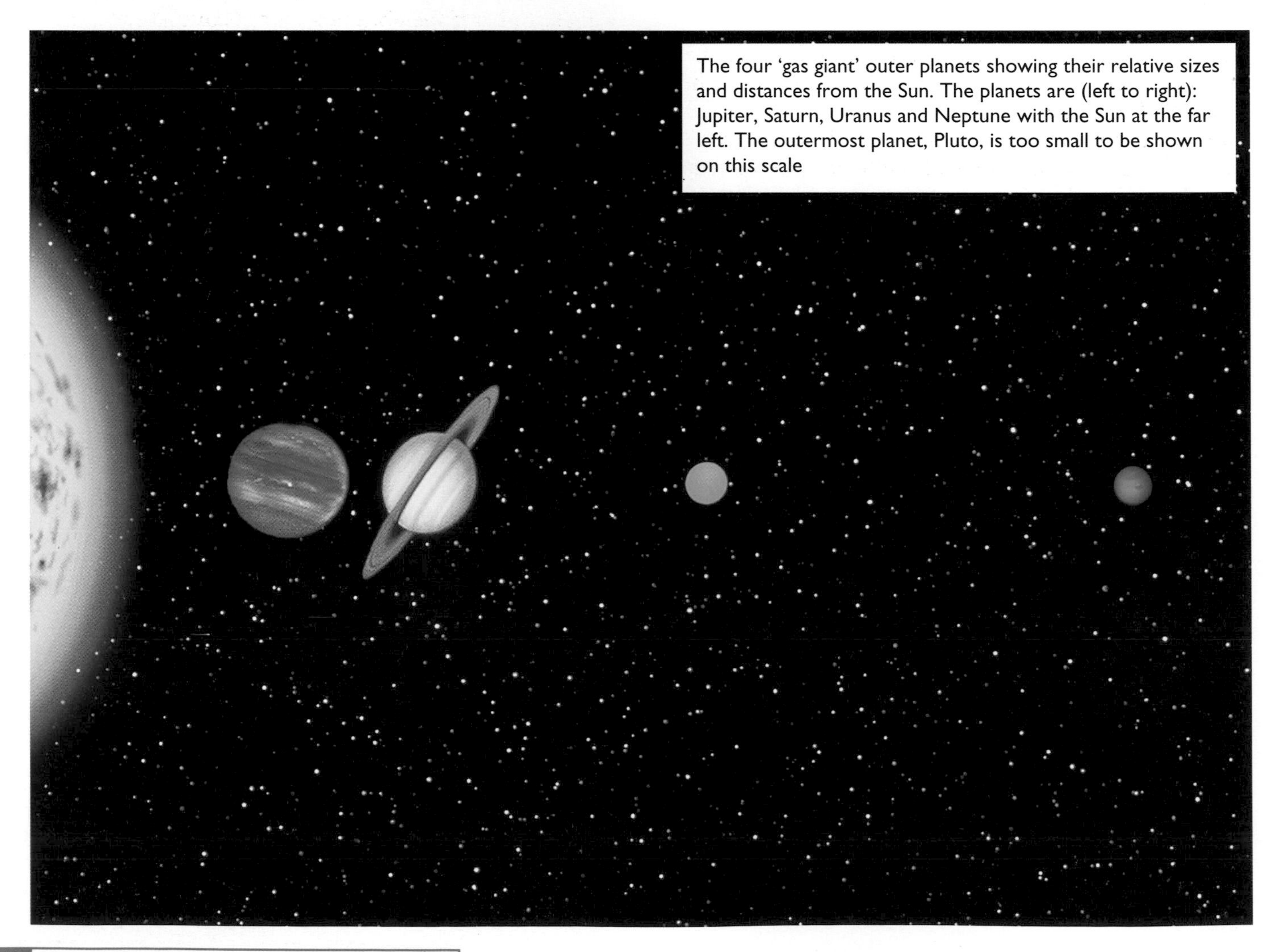

The four 'gas giant' outer planets showing their relative sizes and distances from the Sun. The planets are (left to right): Jupiter, Saturn, Uranus and Neptune with the Sun at the far left. The outermost planet, Pluto, is too small to be shown on this scale

Table 30.1 Important data for the planets in our solar system

Planet	Diameter of planet relative to diameter of Earth	Average distance from the Sun (millions of km)	Time taken to orbit the Sun	Average surface temperature (°C)	Number of moons
Mercury	0.4	58	88 days	350	0
Venus	0.9	108	225 days	450	0
Earth	1.0	150	365 days	20	1
Mars	0.5	228	687 days	−20	2
Jupiter	11.0	780	12 years	−150	14
Saturn	9.4	1430	29 years	−160	24
Uranus	4.0	2870	84 years	−220	15
Neptune	3.9	4500	165 years	−230	3
Pluto	0.3	5900	248 years	−230	1

temperatures reaching 450°C. These higher temperatures on Venus are the result of a thick atmosphere containing high concentrations of carbon dioxide which create a greenhouse effect (section 12.5).

In general, the temperatures on the planets fall as their distances from the Sun increase (Table 30.1). Further from the Sun, where it is colder even volatile substances have not evaporated fully. This leaves crystals and liquid mixtures of methane, ammonia and ice with dense, thick clouds of these substances mixed with helium and hydrogen. Because of this, **Jupiter**, **Saturn**, **Uranus** and **Neptune** are much larger planets with very small rocky cores surrounded by vast quantities of dense thick gas. Being mainly gas these planets will not have a hard surface like the Earth, which can be difficult to appreciate.

Pluto, furthest from the Sun, is thought to be a satellite of Neptune which has escaped and moved into its own orbit around the Sun. It is very small and rocky. However, our knowledge of Pluto is not great because it has not been visited yet by a space probe so there are no close-up photographs.

Notice that the conditions on a planet are affected by two major factors.

- **distance from the Sun** which determines the surface temperature and the evaporation of volatile substances,
- **relative size of the planet** which determines the gravitational pull on its atmosphere.

30.3 Gravitational forces

> Gravitational forces act between all masses and they get stronger if the objects involved have larger masses or if they get closer.

The largest object close to you is the Earth and the gravitational force between you and the Earth may be 600 N or more. This is your **weight**.

Gravitational forces from the Earth also act on the **Moon**. These gravitational forces hold the Moon in orbit and determine its movement around the Earth. The Moon orbits the Earth once a month. It is therefore a **satellite** or **planet** of the Earth. All the other planets, except Mercury and Venus also have moons (Table 30.1).

Gravitational forces of attraction hold planets in orbit around the Sun in the same way that they hold the Moon in orbit around the Earth. The orbit paths of most moons and planets are very close to circular. However, Mercury, the innermost planet, and Pluto, the outermost planet, have orbits more like squashed circles (they are *elliptical*).

The size of the Sun's gravitational pull on a planet depends on the distance of the planet from the Sun and the mass of the planet. Planets with almost circular orbits have an almost constant speed as they move around the Sun.

The orbit speed of a planet around the Sun or of a satellite around the Earth decreases with distance. Therefore, the Earth with an orbit diameter of 150 million km moves at a speed around the Sun equal to

$$\frac{\pi \times 150 \times 1\,000\,000}{365 \times 24 \times 60 \times 60} \text{ km/sec}$$
$$= 15 \text{ km/sec}$$

Neptune, with an orbit diameter of 4500 million km and an orbit period of 165 years moves at a speed around the Sun equal to

$$\frac{\pi \times 4500 \times 1\,000\,000}{165 \times 365 \times 24 \times 60 \times 60} \text{ km/sec}$$
$$= 2.7 \text{ km/sec}$$

Mercury, being nearest the Sun, has the fastest average speed of 48 km/sec but its speed varies due to the elliptical shape of its orbit.

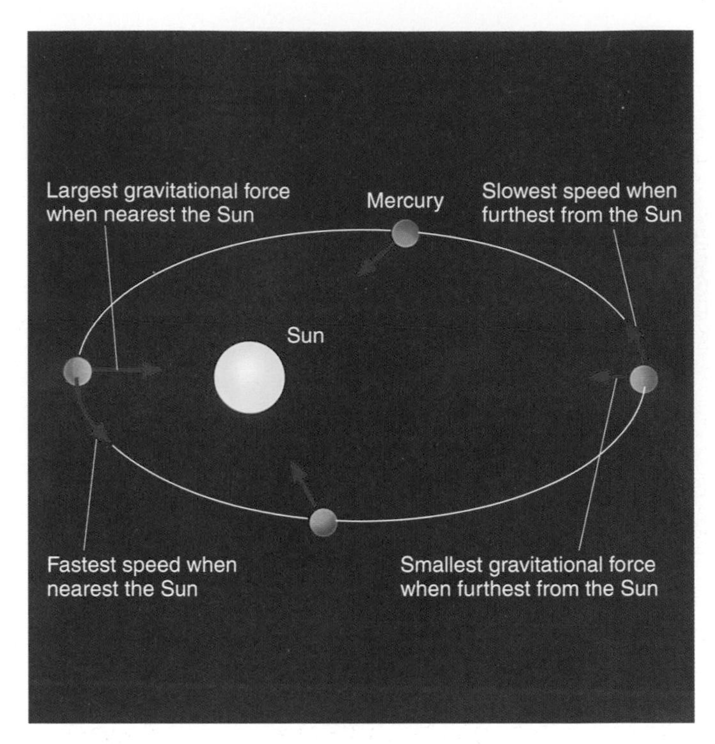

Figure 30.3 The relative size and direction of the gravitational forces and the speed of Mercury during its orbit around the Sun

The comet Hyakutake showing its glowing head and tail. Hyakutake passed within 15 million km of the Earth in 1996

Figure 30.3 shows the elliptical orbit of Mercury around the Sun, the relative size and direction of the gravitational forces and the changing speed.

The changing gravitational force on Mercury causes it to speed up as it approaches the Sun and to slow down as it moves away.

Like Mercury and Pluto, **comets** have elliptic orbits around the Sun. Their paths are, however, much more elongated than the planets. The nucleus of a comet is approximately 1 km in size. It is thought to consist of rocky material covered with vast amounts of frozen water and other gases. Comets become visible with a long tail when they get close to the Sun. The heat of the Sun causes the frozen water and gases to vaporize, forming the tail.

The orbit periods for comets can vary from a few years to centuries or more. Many comets do, in fact, spend most of their orbit time travelling very slowly at great distances from the Sun. As they approach the Sun, the increasing gravitational attraction causes them to speed up and pass around the Sun very rapidly. Once the comet has passed the Sun, the gravitational pull from the Sun slows it down again.

The artificial Earth satellite, ERS–1, launched by the European Space Agency. ERS–1 monitors weather measurements, changes in shore lines and ocean currents

In addition to the Moon, there are hundreds of other satellites orbiting the Earth. These artificial satellites can be used;

- to communicate between places which are hundreds of miles apart on the Earth and
- to monitor and predict the weather.

Communications satellites are usually put into orbit high above the equator so that they move around the Earth at exactly the same rate as the Earth spins. This means that they are always above the same point on the Earth. Because of this, they are called **geostationary satellites**.

In contrast, **monitoring satellites** are usually put into orbits circling the poles so that they scan the whole Earth as it spins beneath them each day.

30.4 The evolution of stars

Stars are not permanent. They are born, live for billions of years and then die. Our Sun is a middle-aged star, about ten billion (10^{10}) years old. In another ten billion years, it will stop emitting light and die.

A piece of coal is normally black. As it starts to burn, it becomes dull red and then bright yellow when it is burning furiously. Like the burning coal, stars indicate their age and temperature by their brightness and colour. The stages in the life cycle of a star are shown in Figure 30.4.

Figure 30.4 The life cycle of a star

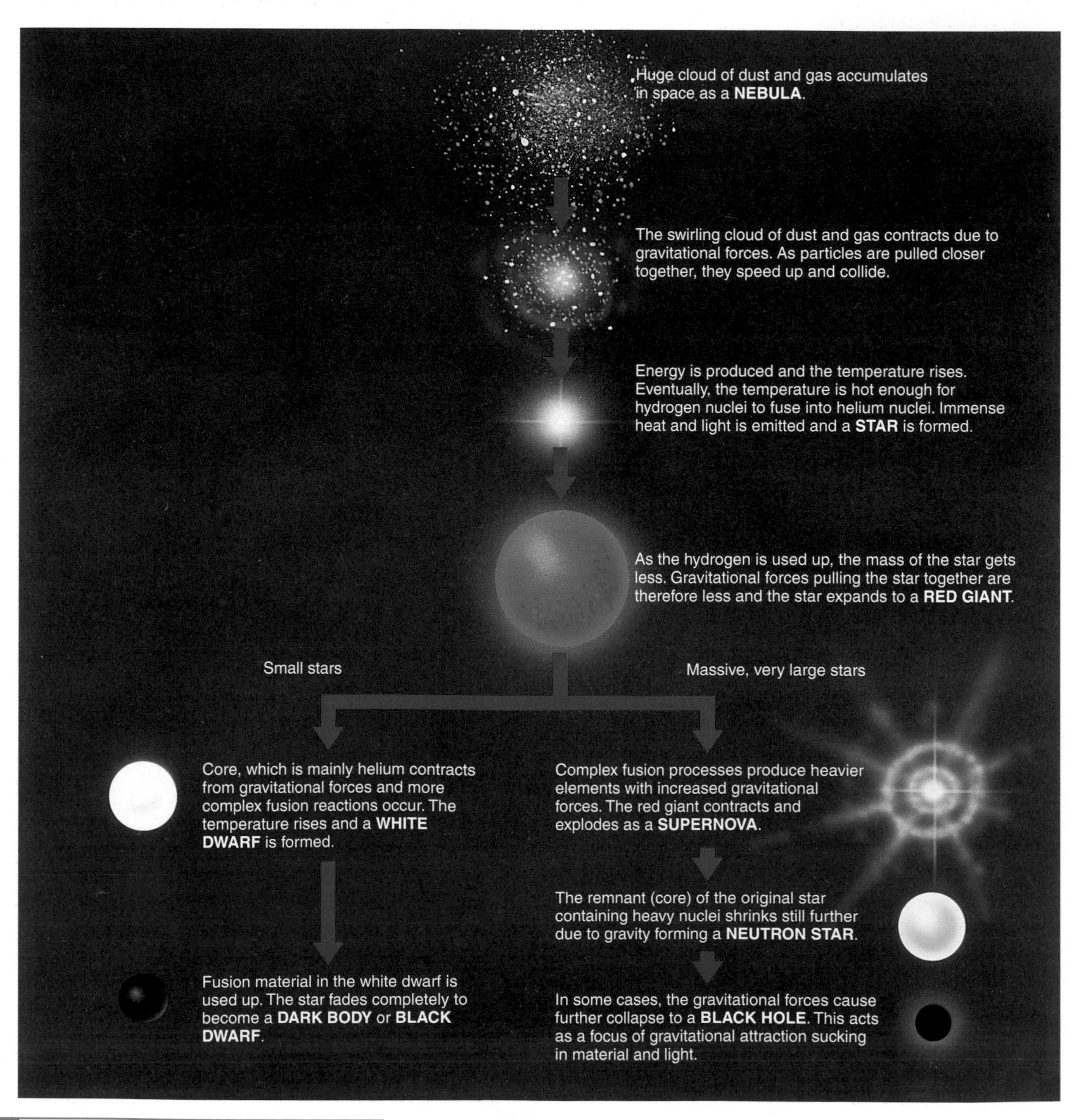

- **Stars are born** when a huge cloud of dust and gas accumulates in space. This is called a **nebula**. The amount of dust and gas is enormous and this is pulled closer together by gravitational forces. Smaller masses may also form and become attracted by the star to form planets.
- The swirling cloud of dust and gas slowly contracts due to the inward pull of its own gravity.
- As the particles of dust and gas are drawn even closer due to the gravitational forces, they speed up and collide emitting energy and the temperature rises.
- **A star is formed** when the temperature is hot enough for hydrogen nuclei to fuse into helium nuclei, emitting immense amounts of heat and light.

During these nuclear processes, two ordinary hydrogen nuclei (^1_1H) first fuse to form one heavy hydrogen (deuterium, ^2_1H) nucleus and a positron. A positron is like a positively charged electron.

$$\oplus + \oplus \longrightarrow \begin{matrix}\circ\\ \oplus\end{matrix} + \text{positron} + \text{heat} + \text{light}$$

$$^1_1H + {}^1_1H \longrightarrow {}^2_1H + {}^0_1e + \text{heat} + \text{light}$$

Two heavy hydrogen (^2_1H) nuclei then fuse to form a helium nucleus (^4_2He).

$$\begin{matrix}\circ\\ \oplus\end{matrix} + \begin{matrix}\circ\\ \oplus\end{matrix} \longrightarrow \begin{matrix}\circ\oplus\\ \oplus\circ\end{matrix} + \text{heat} + \text{light}$$

$$^2_1H + {}^2_1H \longrightarrow {}^4_2He + \text{heat} + \text{light}$$

- During the main stable period of a star, which may last for billions of years, the gravitational forces pulling the star together are balanced by forces from the high temperatures causing it to expand. Our Sun is at this stage of its life.
- As the hydrogen is used up, and the fusion processes convert matter into energy, the mass of the white hot star gets less. The gravitational forces holding the star together are smaller and so the star expands to form a **red giant**. The temperature of the red giant is less than that of the white star but its size is larger due to the reduced gravitational forces.

Two things can now happen depending on the size of the star.

- **For small stars**, like our Sun, the core which is mainly helium contracts under gravitational forces. Eventually, it becomes hot enough for helium nuclei to fuse together forming carbon. The red giant becomes a **white dwarf**. Finally, light from the white dwarf fades as the fusion reactions decrease. The star fades completely and becomes a **dark body** or black dwarf.
- **For very large, massive stars**, the second stage fusion processes within the core produce even heavier elements than carbon. This may allow the star to shine brightly again as the gravitational forces from the heavier elements cause the star to collapse. The intense heat produced as the star contracts causes it to explode spectacularly as a **supernova**. Outer layers of dust and gas are thrown out into space to form a new nebula during the supernova.
- The remnant of the original star, containing the nuclei of heavier elements, shrinks still further to form a very dense **neutron star**. This may be millions of times denser than any matter on the Earth. Some neutron stars emit radiation as radiowaves and are known as pulsars.
- In some cases, the neutron star shrinks still further due to intense gravitational forces to become a **black hole**. This is so dense that it acts as a centre of gravitational attraction, sucking in any material and light near it.

Our own Sun is thought to have formed as a **second generation star** from the nebula of dust and gases of an exploding supernova. Indeed, the nuclei of the heaviest elements in the Sun are also present in the inner planets which further suggest that our solar system was formed from the material of a supernova.

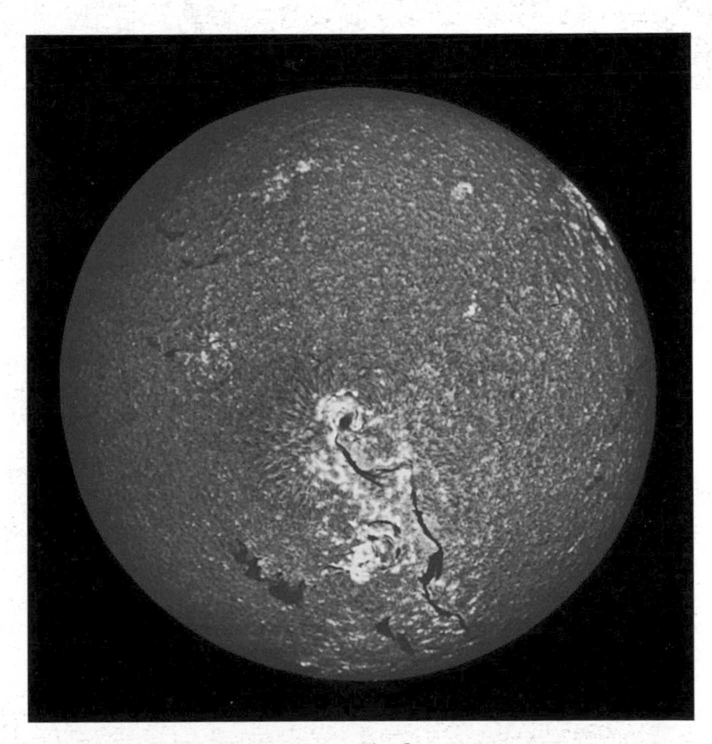

When nuclear reactions occur on the Sun, enormous amounts of heat and light are emitted and temperatures in the core of the Sun reach fifteen million degrees Celsius. The darker spots are 'Sun spots'. These tend to appear when the Sun is hotter

30.5 The evolution of the universe

Astronomers are interested in both the past and the future – what happened at the beginning of the universe and what will happen as it evolves.

In the early twentieth century, astronomers realised that our solar system was just a very small part of an enormous galaxy of stars which they called the Milky Way. They also realised that the universe contained millions upon millions of similar galaxies.

The first astronomer to observe and measure other galaxies was the American, Edwin Hubble. Hubble discovered two very important rules from his observations.

1 *When he measured the light from other galaxies, he found that the wavelength was* **longer** *than expected.* Hubble called this the **red shift** because red light has longer wavelengths than other colours of the spectrum. The change in wavelength is not enough to make the light red in colour, but it moves towards the red end of the spectrum.

2 *The further a galaxy is from the Earth, the greater is its red shift.* This is usually called **Hubble's law**.

Any theories about the origin of the universe must obviously account for Hubble's observations.

Hubble explained his first observation by suggesting that the red shift occurred because other galaxies are moving away from us. As these galaxies move away at great speed, the light emitted from them will appear to have a longer wavelength (i.e. the waves are being pulled out).

Hubble explained his second observation by suggesting that galaxies further away must be moving away faster.

Hubble's suggestions supported the **'big bang' theory** for the origin of the universe. According to this theory, the universe began with an immense explosion millions of years ago from one place. This created the dust and gases that have since formed the stars and planets that exist today. Since the 'big bang', the universe has been expanding and cooling as galaxies move away from each other.

By measuring the acceleration of galaxies and their actual speed away from us, it is possible to estimate how long they have been moving and hence when the 'big bang' occurred. Astronomers estimate that the 'big bang' happened between 10 and 20 million years ago and this is their best guess for the age of the universe.

There are various possibilities for the future of the universe depending on the total mass in the universe and the speed at which galaxies are moving apart.

- The total mass of the universe may not be sufficient for gravitational forces to stop its expansion. In this case, it will continue to expand and cool.
- A second possibility is that the mass of the universe is just sufficient to prevent further expansion. This would allow the universe to attain a constant, steady size.
- If, however, the mass of the universe is greater than this, then gravitational forces will eventually make the universe contract again. This would ultimately pull all the matter together, cause tremendous heating and may lead to another 'big bang'.

QUESTIONS

1 a) Rewrite the following in order of size **starting with the smallest**.

galaxy, the Moon, planet, solar system, universe.

b) i) Name the attractive force that acts between the Earth and Moon.
ii) State **two** factors which affect the size of the attractive force between any two objects. **WJEC**

2 a) The figure below represents the Sun, Earth and Moon.

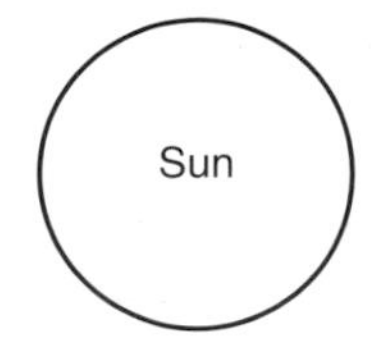

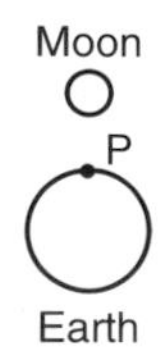

A person standing at point **P** on the Earth can see the Moon.
i) Copy the figure and draw lines to show how light from the Sun enables the person to see the Moon.
ii) The following figure shows some of the ways the Moon can appear against the background of a dark sky.

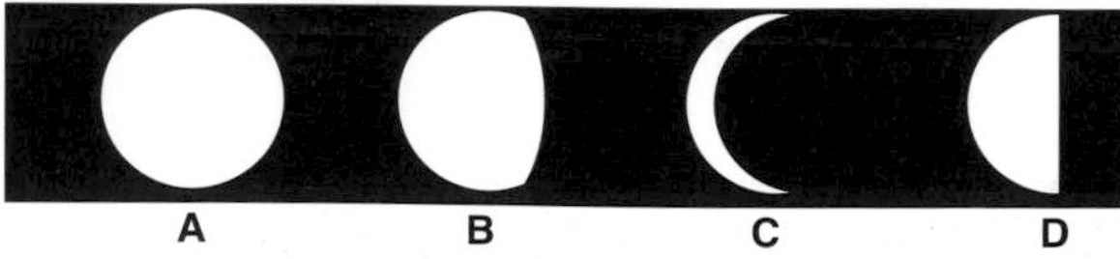

Write down the letter from this figure which shows what the person at P is likely to see.
iii) A lunar eclipse occurs when the Moon moves into the Earth's shadow. Mark an L on your diagram from part i) to show a possible position of the Moon during a lunar eclipse.

b) Approximately two thirds of the Earth's surface is covered in water. The gravitational force of the Moon causes the water to bulge.

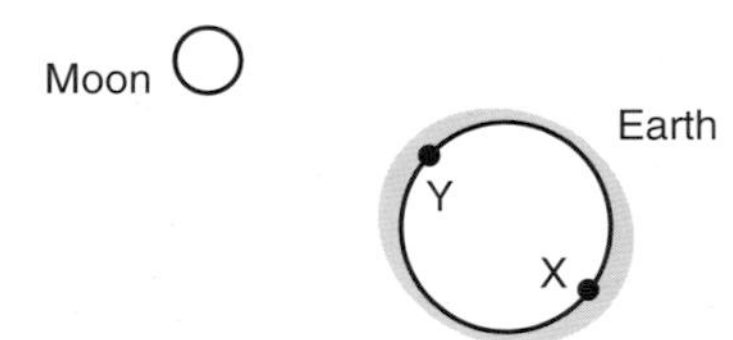

i) Explain why the Moon's gravitational force on the water at **X** is weaker than that on the water at **Y**.
ii) Describe what will happen to the water level at the place **X** as the Earth rotates once on its own axis.

c) Tides are caused by the combined effect of the Sun and of the Moon. The figure below shows one position of the Moon where the combined effect is to 'flatten' the bulge of water.

i) Copy this figure and mark with an **X** another position of the Moon where the effect would be similar.
ii) Mark a **Y** in each of **two** positions of the Moon where there would be the highest tides.

d) Jupiter has several moons. One of them, Io, is similar in size and mass to the Earth's moon. Early astronomers made three predictions about Io:
1 Io would be colder than our Moon
2 Io would have no atmosphere
3 Io would be covered in impact craters.
i) Suggest **one** reason for each prediction.
ii) Recent discoveries show that Io has more volcanic activity than any other object in the Solar System.
Suggest **two** ways in which this volcanic activity is likely to make the conditions on Io very different from those predicted by the early astronomers. **MEG**

3 a) The figure shows a television receiving aerial. The television signal is detected by the horizontal rods.

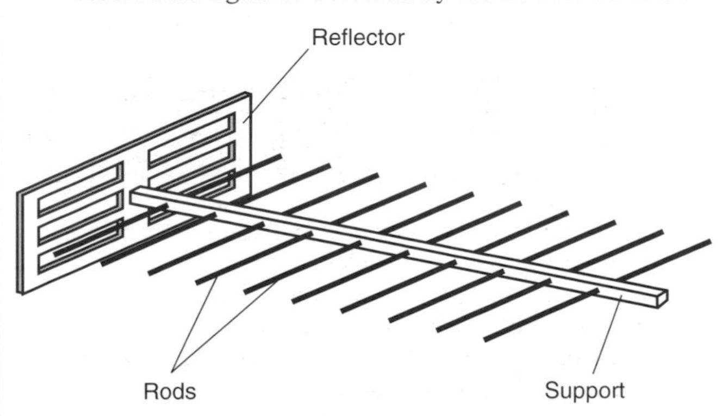

When the aerial is turned through 90° so that the rods are vertical no signal is detected.
What does this tell you about the waves that carry the television signal?

b) Dish aerials are used to transmit television signals to satellites. The figure below shows how the dish focuses the radio waves into a narrow beam which is aimed at the receiving aerial of the satellite.

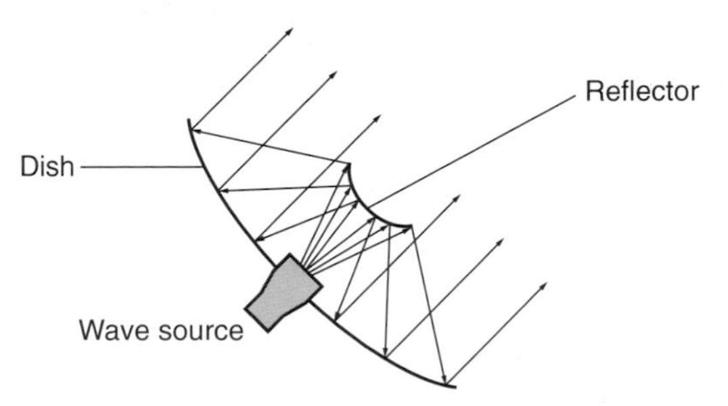

The dish aerials are designed to minimise the effects of diffraction.
i) Explain why this is necessary.
ii) Radio waves are diffracted at the dish in the same way that light and sound are diffracted when they pass through an opening.
What **two** factors determine the amount of diffraction that occurs?

c) The time taken by a satellite to orbit the Earth depends on its height above the Earth's surface. A satellite used to transmit television signals is placed in orbit directly above the equator and has an orbit time of 24 hours. The figure below shows such an orbit.

Explain why 24 hours is a suitable orbit time for a satellite that transmits television signals.

d) Some communications satellites are in elliptical orbits. The gravitational force on a satellite in an elliptical orbit changes as it goes round the Earth. The figure below shows a satellite in an elliptical orbit.

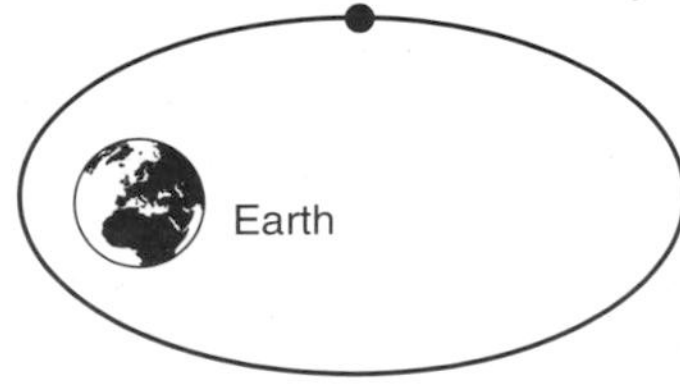

i) Copy the figure and draw an arrow on the satellite to show the direction of the force acting on it.
ii) Describe what happens to the size and direction of this force as the satellite gets nearer to the Earth.
iii) Explain how the speed of the satellite changes as it orbits the Earth. **MEG**

Index